P9-CBG-283

Affordable Automation

Affordable Automation

Sabrie Soloman

McGraw-Hill

New York San Francisco Washington, D.C. Auckland Bogotá
Caracas Lisbon London Madrid Mexico City Milan
Montreal New Delhi San Juan Singapore
Sydney Tokyo Toronto

Library of Congress Cataloging-in-Publication Data

Soloman, Sabrie.
Affordable automation / Sabrie Soloman.
p. cm.
Includes index.
ISBN 0-07-059633-6
1. Automation. 2. Engineering design. 3. Manufacturing process. I. Title.
T59.5.S645 1996
670.42'7—dc20 95-51480
CIP

McGraw-Hill

A Division of The ***McGraw-Hill*** *Companies*

1 2 3 4 5 6 7 8 9 0 DOC/DOC 9 0 1 0 9 8 7 6

ISBN 0-07-059633-6

The sponsoring editor for this book was Harold B. Crawford, the editing supervisor was Caroline R. Levine, and the production supervisor was Suzanne W. B. Rapcavage. It was set in Century Schoolbook by Ron Painter of McGraw-Hill's Professional Book Group composition unit.

Printed and bound by R. R. Donnelley & Sons Company.

This book is printed on acid-free paper.

This book is dedicated to the one who stood by me in times of distress and in times of joy, in times of plenty and in times of scarcity, in times of sowing and in times of harvest, in times of tears and in times of laughter, to the one who has elated my heart—my beloved Rochelle Good.

Contents

Foreword

How do you get from pure science to technology? Many inventors can't bridge that gap because they aren't practical enough to apply their scientific skills to develop technology and equipment. So when I met Sabrie Soloman, I had my skeptic's hat on. What went through my mind was how could anyone without any background in my area of expertise grasp the fundamentals and be able to design equipment that could revolutionize an industry.

I started working in meat plants in the summer of 1951 and have a clear recollection of what the factories looked like. So when I compare what existed then with what we have now, its like viewing a split screen. Major changes have occurred in sanitation, product quality, and general plant condition, but in the pork business, much of the basic technology of the 1950s is intact in factories of the 1990s. So this year when we were exploring building a new plant, we were determined to do something different. We did not want to build the last great piston aircraft.

Enter Sabrie Soloman. It's not often that people surprise me, but Sabrie fell into that category. We met, and his logic and intensity convinced me to let him visit our factories to see what we do, how we do it, and how it could be changed.

His ability to use inductive reasoning enabled him to understand the end goals so that he could determine the means to the end. He was not inhibited by the stagnant thought processes that tend to exist when you've been part of an industry for a long time. He could readily bypass decades of bad practices and old techniques. So Sabrie was able to travel the route from pure science through applied science and then to technology.

But is it affordable? Breakthrough technology is always expensive. Each project has its share of unknowns. Feasibility studies raise the odds of success; however, before you start any R&D program, you need to decide where you want to be in 5 or 10 years and recognize that you're not going to get there without taking some risks. The vision has to be broad enough to know the project cost is low if the new equipment works; and if the equipment doesn't work, you have to be able to recognize the expense as a research and development cost. Affordable automation is not an oxymoron like military intelligence or jumbo shrimp. It's a practical approach to twenty-first century savings and modernization.

Donald Slotkin
President,
Swift & Company
Greely, Colorado

Preface

One of the deepest traditions in science is that of according respectability to what is quantitative, precise, rigorous, and categorically true. It is a fact, however, that we live in a world that is pervasively imprecise, uncertain, and about which it is hard to be categorical. It is also a fact that precision and certainty carry a cost. Driven by our quest for respectability, we tend to close our eyes to these facts and therefore lose sight of the steep price we must pay for high precision and low uncertainty. Another visible concomitant of the quest for respectability is that in much of the scientific literature, elegance takes precedence over relevance.

A case in point is the traveling salesman problem, which is frequently used as a test bed for assessing the effectiveness of various methods of solution. What is striking about this problem is the steep rise in assessing time as a function of precision of solution. Barely lowering the accuracy reduces the cost of assessing time ten-fold.

A more familiar example that illustrates the point is the problem of parking a car. We find it relatively easy to park a car because the final position of the car is not specified precisely. If it were, the difficulty of parking would increase geometrically with the increase in precision, and eventually parking would become impossible.

These and many similar examples lead to the basic premise and guiding principle of affordable automation. However, the ideology of affordable automation does not necessarily collide with the quest for high precision. It opposes the rising cost of *high* precision. Affordable automation propagates the use of *innovation* to achieve precision at low cost. Therefore, the concept of affordable automation can be regarded as a collection of methodologies that aim to exploit tolerance for imprecision to achieve tractability, robustness, and ultimately, low solution cost. Its principal constituents are off-the-shelf components, ready-made standard mechanisms, and innovative technology. Affordable automation is likely to play an increasingly important role in many application areas, including low volume and intermittent production.

In many industries, production cycle times have become shorter, batch sizes have shrunk, and product mix is rising. Also add the reality of shorter product life cycles, and one begins to understand the need for more intelligent, faster, and adaptable assembly and fabrication processes with reduced set-up and development times. With the wide range of component form factors and feeding methods found in most product assembly today, there is a need for reliable, affordable, and adaptable tools. Machine vision, properly applied to a variety of assembly processes, provides the required capabilities.

Competitive trends have prompted manufacturers to look into faster and more adaptable assembly processes. Choices had been limited to either high-speed, moderately priced, near-dedicated assembly equipment for high-volume production with limited flexibility, or high-priced, robot-based "islands of automation" that attempted to deliver great flexibility while maintaining some justifiable production rate. Affordable automation deals with both choices.

In order to survive in brutally competitive world markets, it is essential that manufacturers have the capability to deploy rapidly affordable automation with minimum downtime. This ability to adapt automatically to a changing workcell environment results in cost savings and increased production which, in turn, sets the stage not only for survival but for success. The role model for affordable automation is the provision of the human mind.

Sabrie Soloman

Acknowledgments

I am immensely grateful to my brother Dr. Nasier Soloman for his valuable insight. This book also was made possible by the efforts of my colleagues and friends in various universities and industries, particularly, Dr. Tamer Wasfy at the Center for Computational Structures Technology at NASA Langley Research Center, and by the encouragement and wisdom of the staff of mechanical and industrial engineering of Columbia University.

Chapter 1

Introduction

Engineering design is a crucial component of affordable automation. It is estimated that at least 70 percent of the life cycle of most products is determined during design. Effective engineering design, as some foreign firms especially have demonstrated, can improve quality, reduce cost, and speed time to market (thereby better matching products to customer needs).

Unfortunately, the overall quality of engineering design in the United States is poor. The best engineering design practices are not widely used in U.S. industry, and the key role of engineering designers in the product realization process is often not well understood by management. Partnership and interaction among the three players involved in this endeavor—industries, universities, and government—have diminished considerably. Engineering curricula focus on a few conventional design procedures rather than on the entire product delivery process, and industry's efforts to teach engineering design tend to be fragmented. A revitalization of university research in and teaching of engineering design has begun, but is not well-correlated with the reality of design practice, and research results are not effectively disseminated to industrial firms. Finally, the U.S. government has not recognized the importance of revitalization.

1.1 United States Competitiveness

This state of affairs virtually guarantees the continued decline of U.S. competitiveness. To reverse this trend will require a complete rejuvenation of engineering design, manufacturing practices, education, and research, involving intense cooperation among industrial firms, universities, and government.

1.2 Product Realization Process

To use design effectively as a tool for affordable automation to turn business strategy into effective product, a firm must follow these steps:

1. Commit to continuous improvement both of products and of design and production processes.
2. Establish a corporate product realization process (PRP) supported by top management.
3. Develop and/or adopt and integrate advance design practices into the PRP.
4. Create a supportive design environment.

Converting to operation under the discipline of a PRP is not easy. Often, complete reorganization from top to bottom and a dramatic change in the way of doing business are required. An effective PRP generally incorporates the following steps:

1. Define customer needs and product performance requirements.
2. Plan for product evolution beyond the current design.
3. Plan concurrently for design and manufacturing.
4. Design the product and its manufacturing processes with full consideration of the entire product life cycle, including distribution, support, maintenance, recycling, and disposal.
5. Produce the product and monitor product and processes.

The PRP is a firm's strategy for product excellence and continuous improvement; design practices are its tactics. Because not all practices are applicable to or useful in the design of a given product, each company must carefully identify a set appropriate to its goals and incorporate them into its PRP. Practices (such as Taguchi's methods) and tools [such as computer-aided design (CAD) and computer-aided engineering (CAE)] must be fully integrated into the PRP if they are to have more than minimal effect. Companies must also develop means of assimilating new practices as they are developed by researchers and others because currently effective practices are being improved and even superseded.

Design is a creative activity that depends on human capabilities that are difficult to measure, predict, and direct. An understanding of the design task and characteristics and needs of people who design effectively is essential to the creation of a stimulating and nurturing design environment for affordable automation.

1.3 Prerequisite for Affordable Automation

Effective design and manufacturing, both necessary to produce high-quality products, are directly related to affordability. However, effective design is a prerequisite for effective manufacturing and automation; quality cannot be manufactured or tested into a product, it must be designed into it.

This book presents recommendations for engineering design for affordable automation of assembly and fabrication. The recommendations are based on studies of the manufacturing practices and in various industrialized countries.

The studies on which this book is based had the following goals:

1. Determine the importance of engineering design to U.S. industry's competitiveness in world markets.
2. Articulate how the practice of engineering design in the U.S. can be improved to achieve affordable automation.
3. Propose actions to improve undergraduate and graduate education in engineering design and manufacturing.
4. Propose a national effort to improve the practice of engineering design through research and development.
5. Recommend to government, industry, and academe mechanisms for improving engineering design practices, education, and research for affordable automation.

1.4 U.S. Decline in World Market Share

It is now widely believed that U.S. industry's extended period of world dominance in product design, manufacturing innovation, process engineering, productivity, and market share has ended.[1] The once globally dominant U.S. automobile and steel industries have lost market share at home and abroad, and U.S. products have all but disappeared from the consumer electronics market. There is consensus that U.S. industry as a whole is not as productive as it might be, and that its rate of productivity increase is lower than that of industries in many other countries.[2] This loss of competitiveness with foreign firms has been keenly felt in some areas in job losses and

[1]National Research Council, *Toward a New Era in U.S. Manufacturing,* National Academy Press, Washington, D.C., 1986, p.5.

[2]S. Berger et al., "Toward a New Industrial America," *Scientific American,* vol. 20, no. 9, June 1989, pp. 39–47.

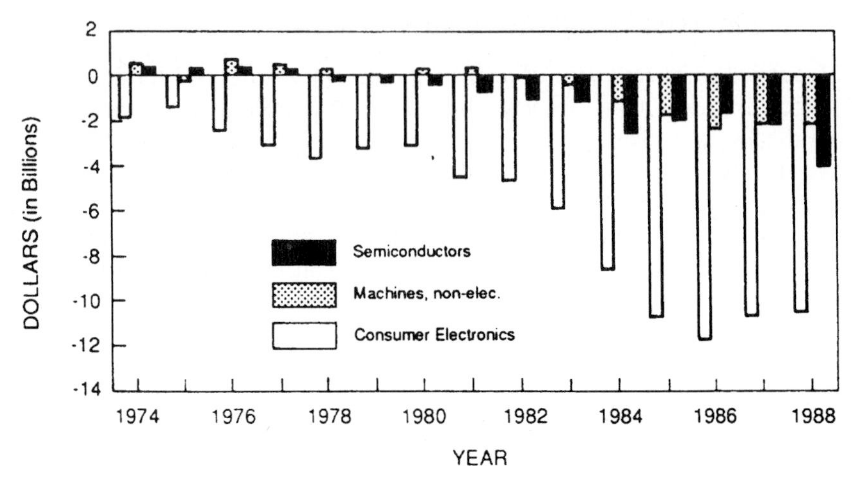

Figure 1.1 U.S. trade deficit in three key industries.

plant closings. Profitability continues to decline in many key industries, threatening further loss of market share and jobs. U.S. citizens, from the individual consumer to the senior corporate executive, daily observe evidence of the decline of the nation's "industrial might."[3] Figures 1.1 and 1.2 illustrate the declining performance of some important U.S. industries.

The decline of U.S. international competitiveness has been ascribed to many factors, among them national fiscal and trade policies, exchange rates, national culture, deficiencies in manufacturing, industrial management and accounting practices, unfair foreign trade practices, and methods of providing capital. A crucial factor that is not often recognized is the quality of engineering design and automation in U.S. industry. Engineering design and manufacturing automation are the key technical ingredients in the product realization process,[4] the means by which new products are conceived, designed, developed, produced, brought to market and supported. (Various other names, in-

---

[3]President's Commission on Industrial Competitiveness, Global Competition: *The New Reality,* U.S. Government Printing Office, Washington, D.C., 1985; U.S. Congress, Office of Technology Assessment, *Making Things Better: Competing in Manufacturing,* OTA-ITTE-443, U.S. Government Printing Office, Washington, D.C., February 1990; and MIT Commission on Industrial Productivity, *Made in America: Regaining the Productive Edge,* MIT Press, Cambridge, Mass., 1989.

[4]The process also includes determining customers' needs and translating those into engineering specifications.

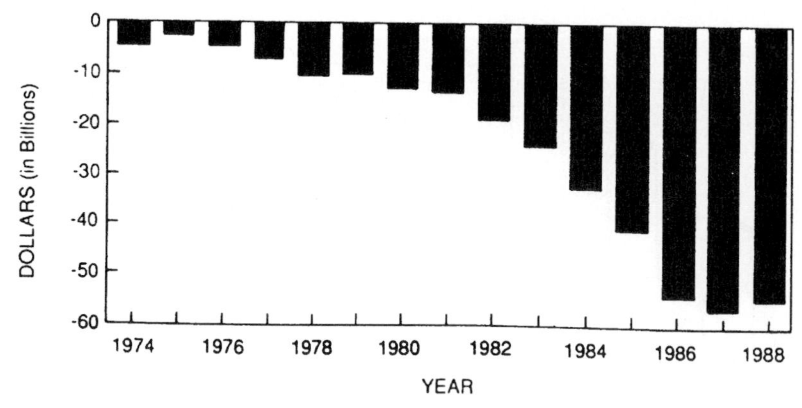

Figure 1.2 U.S. trade deficit in the auto industry.

cluding *concurrent engineering,* are in use for the product realization process or major parts of it.)

The ability to develop new products of high quality and low cost that meet customer needs is essential to increasing profitability and national competitiveness. The link between quality and profitability has been convincingly demonstrated by studies using the PIMS[5] database. Figure 1.3 summarizes the results of a study, using the PIMS database, of the effect of quality and market share on profitability for a large group of U.S. industries, predominantly manufacturers.[6]

1.5 The Central Role of Engineering Design for Affordable Automation

High-quality products satisfy customer needs for reliability, serviceability, and acceptable life-cycle cost, as well as for functionality and aesthetics. Competitiveness demands high-quality products, which require high-quality in their components and the systems and processes used in their production. Effective design and manufacturing, both necessary to produce high quality products, are closely interrelated, but effective design is a prerequisite for effective and economical manufacturing; quality cannot be manufactured or tested

[5]The Profit Impact of Market Strategy database, compiled by The Strategic Planning Institute of Cambridge, Mass., includes operating and quality data on approximately 3000 business units in 450 companies for periods ranging from 2 to 10 years.

[6]*The PIMS Principles,* The Free Press, New York, 1987.

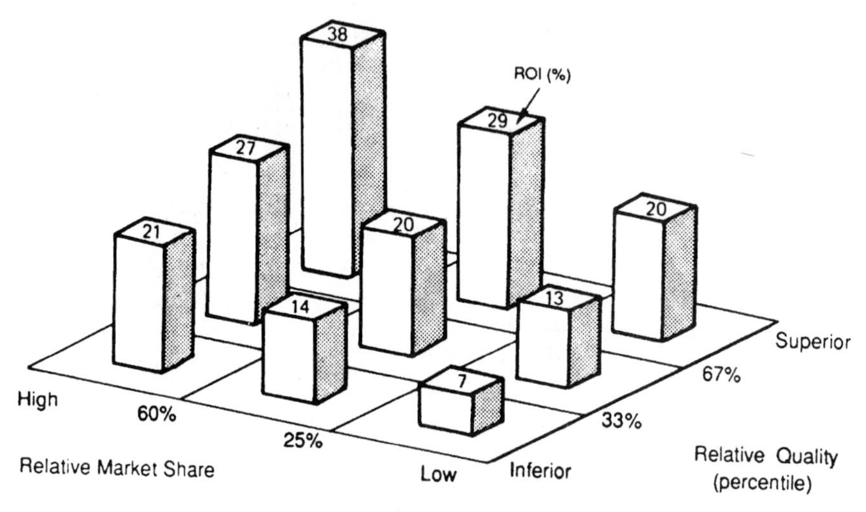

Figure 1.3 Return on investment as a function of quality and market share.

into a product, it must be designed into it.[7] Figure 1.4, derived from studies done at Westinghouse and General Motors, suggests that a major fraction of the total life-cycle cost for a product is committed in the early stages of design.[8]

As products become complex, manufacturing yield falls dramatically without parts and manufacturing operations of extremely high quality. The sensitivity of final product quality to component quality as complexity increases may be demonstrated readily. Assume that a final product requires n components and operations, each with a probability of being acceptable P_j. Then the probability P of the final product being acceptable is

$$P = \prod_{j=1}^{n} P_j \tag{1.1}$$

If $P_j = p$ is the same for all n components and operations, then Eq. (1.1) simplifies to

[7]J. R. Dixon and M. R. Duffey, "Quality Is Not Accidental—It Is Designed," *New York Times,* June 26, 1988.

[8]Adapted from Chap. 1 of J. L. Nevins and D. E. Whitney, eds., *Concurrent Design of Products and Processes,* McGraw-Hill, New York, 1989. An earlier study giving similar results is reported in W. G. Downey, "Development Cost Estimating," Report of the Steering Group for the Ministry of Aviation, (Her Majesty's Stationery Office, London, 1969). Reference from E. J. Leech and B. T. Turner, *Engineering Design for Profit,* John Wiley, New York, 1985.

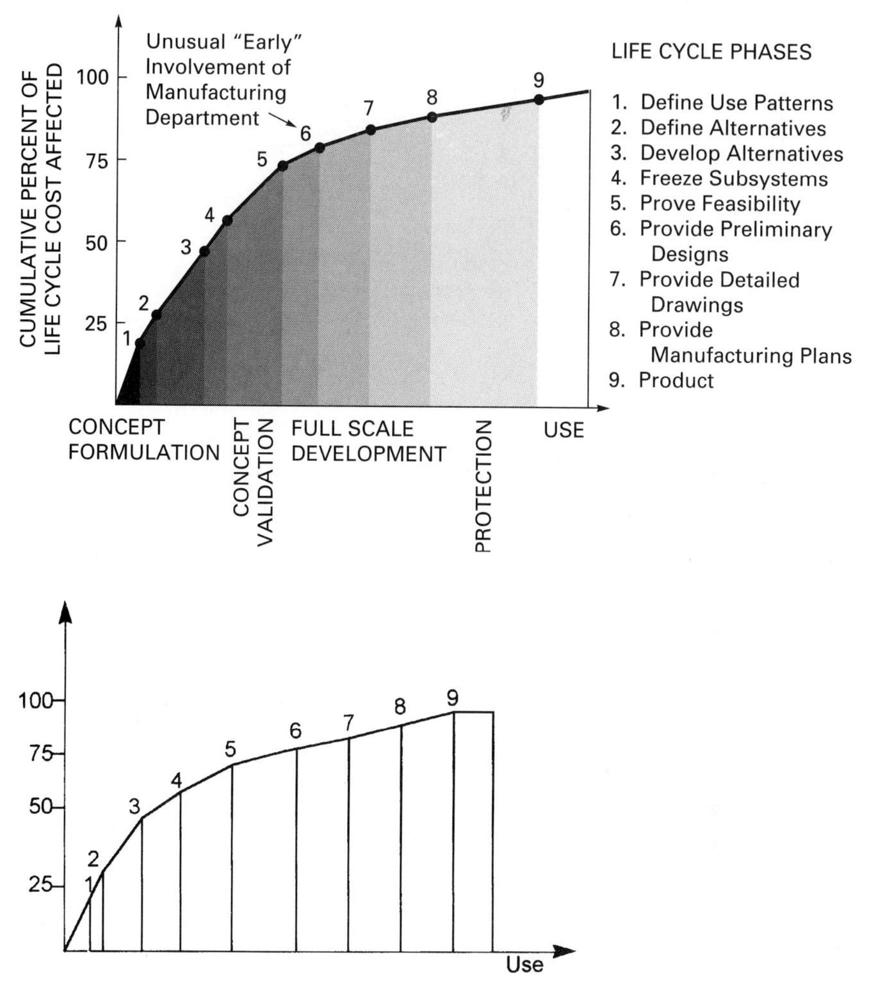

Figure 1.4 Life-cycle cost commitment.

$$P = p^n \tag{1.2}$$

Figure 1.5 is a parametric plot of Eq. (1.2) which shows very high quality in all components and assembly operations is required to obtain acceptable yields even for products with only a few hundred parts or assembly operations. Note that a component quality of 0.99999 (10 defective parts per million parts) is required to obtain yield in the 99 percent range for a system composed of 400 parts.

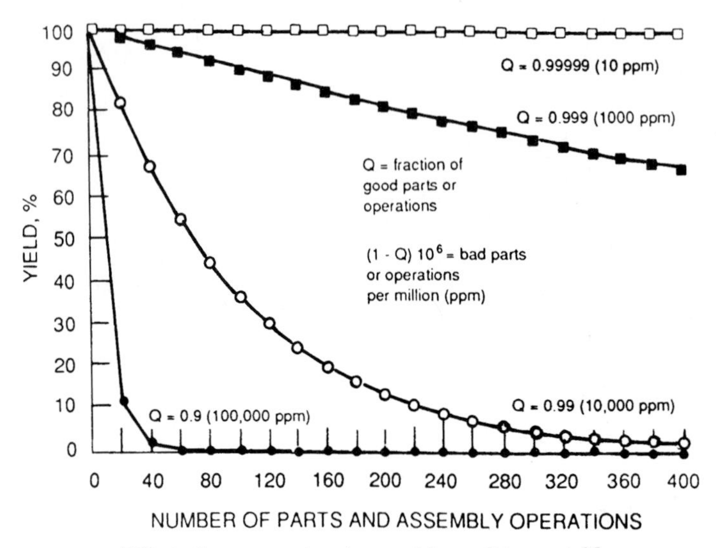

Figure 1.5 Effect of component and assembly quality on yield.

U.S. performance in engineering design can be compared to that of other nations on the basis of the speed and cost with which new products are brought to market. The greater time from concept to delivery for U.S. products compared to Japanese products is illustrated by Fig. 1.6.[9]

Manufacturing performance, including adherence to design specifications, flexibility, and efficiency, is also involved, but effective design is at the heart of the concept of continuous accumulated improvement to make a product better year after year.

When measurements are made, it becomes clear that U.S. industry's loss of market share results from poor performance in the very areas in which successful foreign companies, particularly some Japanese companies, usually excel.[10] Loss of market share resulting from poor

---

[9]K. B. Clark and T. Fujimoto, "Overlapping Problem Solving in Product Development," K. Ferdows, ed., *Managing International Manufacturing,* North-Holland, Amsterdam, 1989.

[10]In "Turning Ideas Into Products," *The Bridge,* vol. 18, no. 1, Spring 1988, pp. 11–14, R. E. Gomory, a senior vice-president of IBM, states that IBM's "most effective foreign competition has been characterized by tight ties between manufacturing and development, an emphasis on quality, the rapid introduction of incremental improvements...of preexisting product, and a tremendous effort by those actually in the product cycle to be educated on the relevant technologies, on the competition's products and on what is going on the world."

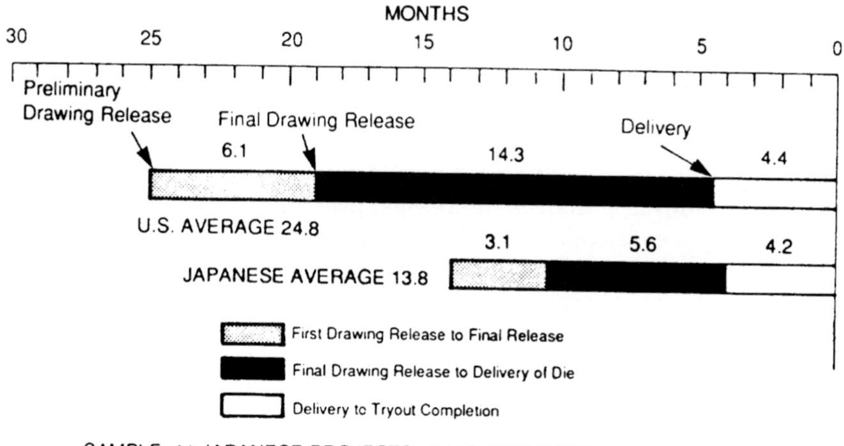

Figure 1.6 Lead time for a major body die (months).

design is likely to spread as foreign competition expands into other industries. Aerospace, large appliances, and cosmetics industries are likely near-term targets.

1.6 The Nature of Engineering Design for Affordable Automation

As the key technical ingredient in the product realization process, engineering design bears responsibility for determining in detail how products will be made to meet performance and quality objectives for the customer at a cost that permits a competitive price. It thus plays a key role in the ability of business to excel.

Engineering design has both technological and social components. The technological component includes engineering science, design models, engineering models, materials, engineering, and computers. The social component includes corporate organization and culture, team design methods, the nature of the design task and of the designer, customer attributes, and employees' involvement.

An ever-evolving problem-solving activity, engineering design encompasses many different and increasingly advanced practices, including methods for converting performance requirements into product features, computer-integrated manufacturing, cross-functional teams, statistical methods, comparative benchmarking of products, computerized design techniques, and use of new materials and manufacturing processes. These and other methods used by the most competitive companies worldwide do not exist or operate independently, but rather are integrated into a unified process.

A recent study by the Committee on Engineering Design Theory and Methodology considered a broad range of engineering design activities, including practices, processes, principles, methodologies, and techniques in companies large and small. Although the study did not focus on very large scale integration design or software design because these are narrower domains, significant successes in these areas are ascribed to the close coupling of product and process design, and they thus provide lessons for all areas of design.

1.7 The Current State of Engineering Design for Affordable Automation in the United States

The recent study mentioned above benchmarked the firms under investigation against leading competitors. Significant benchmarks included cycle time, the number of iterations of the design cycle, and the number of design changes. From this background, a set of questions was developed and posed to a number of leading-edge companies. The study concluded that these U.S. firms believe significant efforts will be needed to attain the advantages already enjoyed by their most effective foreign competitors. In addition, attaining a similar level of competence involves a moving target. The companies investigated were large and well-supported. The investigation did not explicitly study small and medium-sized firms, whose situation is starkly documented in various reports that show that their level of adoption of even computer-aided design (CAD) substantially lags that of foreign firms.

The status of research in design theory was assessed by a different approach. A panel of experts in the field drafted, refined, and ranked a set of topics covering various areas of investigation. They estimated the minimum support necessary to obtain "above-threshold" progress in these areas and compared these desired levels to existing levels of support. They concluded that current levels of support are far less than needed to advance the field.

In regard to design education, the following general statements by the Committee are offered with no intent to cover all cases; subsequent sections of the study identify wide variations in industrial and educational practice throughout the nation. Nevertheless, the investigation's findings support the following statements.

1.7.1 Engineering design practices in U.S. industry

The best engineering design practices are not widely used in U.S. industry.[11] Many U.S. companies limited by existing practices are unwilling to try new ones, often because of management rather than technical barriers. Those U.S. companies that do try to identify and absorb current best practices are still often outstripped by their best foreign competitors, which continue to evolve new and still better practices. A higher rate of new product introduction in these foreign firms results in more rapid learning, which translates into more rapid improvement of design and manufacturing processes.[12] Improvement migrates slowly in the U.S. because the process of sharing and dissimilating design knowledge among companies remains dependent on informal networking of individuals.

1.7.2 The role of designers in the product realization process

The key role of designers in the PRP is often not well-understood by management. Most designers take on, often by default and without portfolio, an enormous range of new activities in support of the PRP, and management often does not recognize the importance of these nontraditional design activities. Motivation and support of designers is complicated because there is no way to use data from traditional cost accounting systems to evaluate the contribution of design to profit or to compare the effectiveness of different designs. In recognition of this problem, proposals for different cost accounting systems have recently been published.[13]

1.7.3 Changing the goals and culture of U.S. industry

Some U.S. firms use design effectively, but they have had to change their goals and culture to do so. To improve from stable, high-volume, slowly changing production to continuous improvement requires profound cultural change; firms that have made this shift have adopted an all-enterprise approach, employing dedicated agents to catalyze and support change. These firms use the product realization process

---

[11]The phrase "best engineering design practices" means that set of practices that is best for a particular company. Best practices will vary from company to company.

[12]R. E. Gomory and R. W. Schmitt, *Science,* vol. 240, May 27, 1988, pp. 1131–1204.

[13]For example, R. S. Kaplan, "Management accounting for advance technological environment," *Science,* vol. 25, August 25, 1989, pp. 819–923.

as the vehicle for involving people at all levels and in all functions in defining, designing, and producing the product and moving it to market. They choose design practices to support the PRP and design the product, and they set metrics to guide the process.

1.7.4 Partnership interactions among U.S. industry, research, and education

Partnerships and interactions among industry, research, and education are so limited that each is poorly served by the others. With few exceptions, education and research on engineering design is divorced from industry needs. For its part, industry does not articulate its requirements, support changes in the design component of curricula, or view education as an incubator of design talent. University design research efforts are often isolated from industry, and industry rarely uses the results of university research. Although some companies have served well despite this environment, most (particularly medium-sized and small companies) suffer the consequences of outdated methods and poorly prepared new engineers in product quality, market share, competitiveness, and international trade.

1.7.5 Engineering curricula and the product realization process

Current engineering curricula do not focus on the entire product realization process. Most curricula emphasize a few steps of conventional, essentially technical, design procedures. Curricula as a whole lack the essential interdisciplinary character of modern design practice and do not teach the best practices currently in use in the most competitive companies. The result is engineering graduates who are poorly equipped to utilize their scientific, mathematical, and analytical knowledge in the design of high-quality components, processes, and systems. Few have experienced design as part of a team, even fewer understand the multiple goals that motivate design, and most lack understanding of statistics, materials, manufacturing processes, cost accounting, and product life-cycle considerations. Industrial training courses try to fill these gaps at considerable cost and with varying degrees of success.

1.7.6 Industry efforts in engineering design for affordable automation

Industry's internal efforts to teach engineering design, intended to compensate to some degree for these shortcomings in education, are too fragmented and not institutionalized as a natural component of doing business. These efforts, affordable only by the largest companies, are not based on the fundamental understanding of design processes that could be provided by design research. Yet most engineers, including new employees, must currently learn modern design techniques for affordable automation from industrial training courses.

1.7.7 Responsibility of universities

Although universities nominally bear responsibility for producing practices and practitioners, they do not fill this role in engineering design in the United States. The breakdown extends beyond curricula. Universities do not, in general, value engineering design as an intellectual activity, either in research or in teaching. Lack of instructional materials and experienced faculty and the need for time-consuming interaction with students make courses in design difficult to teach. Many who do teach design have little experience and are unaware of the most recent design techniques. The few efforts to revitalize university research and teaching in engineering design are fragmented, insufficiently funded, and not well enough coupled to the needs of industry to produce either well-prepared new engineers or useful research results.

1.7.8 University research in engineering design

A revitalization of university research in engineering design has begun. Unfortunately, it is not well-correlated with the realities of the full scope of design for comparative products, and results are not well-disseminated to industry. The National Science Foundation (NSF) program in engineering design theory and methodology is funded at too low a level and is not yet recognized by the research community as a stable source of research leadership and support. NSF's engineering research centers, some of which have design-oriented research thrusts, are a step in the right direction, but, again, funding for design efforts is inadequate.

1.7.9 Superior engineering design as national priority

The U.S. government has not recognized the development of superior engineering design as a national priority. Though engineering design is a primary determinant of competitiveness over the entire spectrum of manufacturing industries, it has not received the level of support that has been accorded specific product areas such as semiconductors and superconductors.

This state of affairs virtually guarantees the continued decline of U.S. competitiveness over the long term. A complete rejuvenation of engineering design practice, education, and research—aimed at future needs rather than just catching up to competitors' current standards—is fundamental to gaining and maintaining U.S. industrial competitiveness. An objective of this magnitude requires intense cooperation among industries, universities, and the government.

In the United States, federal and state government policies have not traditionally been directed toward helping private enterprises enhance their competitiveness through adoption of advanced technologies, in part because technology-based industries have in the past faced little serious competition from foreign firms. However, nearly all foreign competition in high-value-added products is strengthened to some extent by various foreign government measures to increase the technological strength of key industries. Consequently, traditional government policies warrant intense restudy and, in all likelihood, revision.[14]

1.8 Achieving Affordable Automation

The consequences of better design practice, education, and research for affordable automation may lead to major gains in U.S. national economy. Improving engineering design practice to achieve affordable automation in U.S. industry will result in shorter development time, lower cost, and better match of products to customer wants. The fastest way to realize these benefits is for the vast majority of U.S. companies to learn to use the advanced design practices that have already been implemented by leading-edge companies in the United States and abroad. It has taken these pioneering companies 5 to 8 years to change their practices, yet many are willing to share their lessons, enabling other companies to learn and implement advanced design practices in a much shorter time.

On a slightly longer time scale, better engineering design education will improve the practice of engineering in the United States. If the recommendations of the investigation are followed, in a few years universities will begin to graduate students whose knowledge of engineering design, contact with industry during their schooling, and awareness of good design practices will better attune them to the needs of industry and the realities of engineering design and dispose them to continuing education throughout their careers. These gradu-

[14]U.S. Congress, Office of Technology Assessment, *Making Things Better: Competing in Manufacturing,* OTA-ITEE-443 U.S. Government Printing Office, Washington, D.C., February 1990, Chap. 7.

ates will augment and eventually replace a generation of designers who received limited coherent engineering design education. Students who emerge from graduate engineering design programs familiar with current advances in theoretical foundations of design and forefront methodologies will not only contribute to engineering practice, but also will be prepared to create new design tools, teach design to next-generation students, and conduct research in design.

The benefits of extended design research will take longer to accrue—even with improved dissemination of research results to U.S industry and greater eagerness on the part of industrial firms to use the results—but many have the greatest impact on productivity. Indeed, given the best result, it could provide the means for leaping ahead of the competition. Research will provide new design methods and principals to support more rapid development of further improved design practices. It will provide tools for faster and more complete learning of design methods by both practicing engineers and students, multiplying both the quantity and quality of design engineers. Research results will be further developed into computer programs, databases, visualization devices and techniques, methods of predicting behavior and cost early in the design process, and other valuable, but today unforeseeable, mechanisms.

It is crucial that improvements be made in each of the three areas of design—practice, education, and research. Halfway measures will not suffice. Simply adopting the design practices of foreign companies will doom U.S. industry to perpetual follower status. Educating new designers and performing research relevant to the needs of industry will require both the development of new faculty and intellectual and financial support from the companies at the forefront of engineering design practice. New research is needed to enable U.S. industry, when it is ready and able to accept new design methods and tools, to leap ahead of competitors to practice effective affordable automation.

1.9 Affordable Automation for Competitive Advantage

Affordable automation is the fundamental determinant of both the speed and the cost with which new and improved products are brought to market and the quality and performance of those products. Design excellence is thus the primary means by which a firm can improve its profitability and competitiveness.

Yet few U.S. firms have adopted either contemporary design practices or product realization processes, and there seems to be inadequate understanding of how to go about improving current design practices. It is important to outline here the necessary steps to improve design for affordable automation practices and cite sources of information that should assist in this process.

Designing for affordable automation for competitive advantage requires much more than the adoption and use of new design practices. Firms that utilize design most effectively were found to:

1. Commit to continuous improvement
2. Follow a product realization process tailored to their products[15]
3. Use a set of design practices to implement their PRP that fosters a supportive design environment

1.10 Corporate Commitment and Action for Affordable Automation

Though many U.S. companies doubt their ability to win the competitive battles they are waging, a few have recognized the challenge that faces them and have successfully fended off foreign assaults on profitability and market share. What these companies have in common is a shared corporate resolve to change the internal corporate culture in response to that challenge. Companies such as Xerox, Hewlett-Packard, and Ford, among others, have changed their internal cultures and reshaped the way they do business. From what it learned from these and other companies, the investigation of the Committee on Engineering and Design Theory and Methodology suggested that the first and most important step in introducing improved design practice for affordable automation is to generate corporate awareness of the leverage design can provide and the need for change to utilize that leverage. Change must begin with recognition of the importance and impact of design deficiencies and knowledge of possible routes to improvement. The analysis and the collective experience of the members conducting the analysis suggest that denial that a problem exists is the major obstacle to the introduction of new design processes and methods for affordable automation.

Denial is particularly prevalent in industries not yet besieged by significant foreign competitors. Until they have faced competitors that use superior engineering design practices, companies rarely recognize the advantages to be gained by improving their own design practices.

[15]J. Hauser and D. Clausing, "The House of Quality," *Harvard Business Review,* May-June 1988, pp. 63–73; R. B. Chase and D. A. Garvin, "The Service Factory," *Harvard Business Review,* July-August 1988, pp. 61–69; and G. Stalk, "Time—The Next Source of Competitive Advantage," *Harvard Business Review,* July-August 1988, pp. 41–53.

Thus, many companies begin to improve their design practices for affordable automation only after they have lost significant market share to competitors that made such improvements years ago. Years of playing catch-up could be avoided, and competitive advantage gained, if enlightened management committed to continuous improvement under a PRP[16] in anticipation of—rather than as a result of—competition.

Businesses that have successfully incorporated state-of-the-art design practices for affordable automation have done so in an all-enterprise way. They have recognized engineering design for affordable automation as a vital part of the product delivery capability rather than as just another department in the company. This view ultimately required them to change many parts of the company beyond the design department; indeed, it usually spawned a totally new way of doing business.

Once a company recognizes the need to improve design, it must begin to identify solutions. Since deficiencies are rooted in organizations, technique, and infrastructure, the main avenue of response are recognition, adoption of formalized product realization processes, and involvement in research and education. In companies that successfully design for competitive advantage, the extent of external and internal change is often striking, reflecting a degree of self-examination rarely seen outside crisis situations. Successful programs of change typically feature strong top-management leadership in setting corporate goals for improved design, development of metrics to measure progress toward these goals, creation of corporate centers of design excellence, extensive training programs for new hires and experienced designers, and effective relationships with universities for research and technology transfer.

Knowledgeable observers point out that real change cannot be accomplished in a large organization without the impetus of a change agent, a group or department whose sole responsibility is to initiate change. Change agents are necessary because people whose main responsibilities lie else where usually have neither the dedication nor the time to initiate significant change themselves. Xerox has assigned approximately 300 people (out of a corporate total of 113,000) to change-agent roles; Hewlett-Packard, about 1000 (out of 89,000). Education and training programs supported by senior corporate leadership and applied at the enterprise level are effective and necessary supports for the change agents.

Support for change must include the following parameters:

1. Programs to determine which practices worldwide would be most useful to the firm.

---

[16]As noted earlier, various other names are used for *product realization.*

2. Methods for securing support for the introduction of new practices
3. Coordination of the change throughout the firm.

Designers must be made a part of the change team, and the engineering design methods introduced must be explained as part of an evolving whole rather than as a series of unrelated fads. Unless engineers are educated in the value, goals, and necessity of a change plan, they will continue to use demonstrably inferior design practices. Because changing the product realization process affects the entire company, all employees, not just engineers, must be made part of the change process.

Though discussion to this point has targeted practice in large companies, much of the design and manufacturing in the United States is conducted in small and medium-sized companies (i.e., 500 or fewer employees) that often cannot afford extensive training programs or even separate design departments. Nevertheless, all of the principles stated here are applied in and are crucial to the success and competitive position of smaller companies as well. Indeed, the integration and cross-communication implied in the product realization process may be more readily accomplished in smaller operations. Firms that cannot afford to conduct actions such as extensive training courses in-house can avail themselves of external courses and workshops. Large companies' training programs, for example, are often open to their suppliers.

1.11 Plan for Continuous Improvement

Companies that design successfully have carefully crafted product realization processes that extend over all phases of product development from initial planning to customer follow-up. The PRP is their plan for continuous improvement. The decision to develop and operate under a PRP is a corporate one. Successful operation of a PRP requires extensive cooperation among a firm's marketing and sales, financial, design, and manufacturing organizations.

PRPs are not static, but evolve continuously. They change in response to feedback from production and incorporate new methods and tools. Design is an essential element of the PRP, and designers play a broad role in formulating and carrying the steps of the PRP. The description that follows is an idealized composite of the various elements found in current processes, which vary from company to company.[17]

---

[17]Product realization processes developed by Polaroid and Hewlett-Packard.

1.11.1 Definition of customer needs and product performance requirements

A good product realization process begins with an exploration of business, marketing, and technical opportunities, followed by a few definitions of customers needs and product performance requirements, including quality, reliability, durability, and other important factors such as aesthetics.[18]

The new product's essential technologies are reviewed to ensure that inventions will not be required to produce it, and comparable products are analyzed to establish benchmarks for it.

1.11.2 Planning for product evolution

The technology review in the design phase indicates where technological advances or inventions can improve performance or reduce cost. In some industries, an entire range of products in the same line that requires further invention, research, or development is mapped out, with planned evolution of future capabilities, during this review. Core technologies for the future products are identified, and product performance specifications are defined with inputs from manufacturing, marketing, engineering, and finance.[19]

1.11.3 Planning for design and manufacturing

Cross-functional teams with representatives from marketing, design, manufacturing, finance, sales, and service are established. The design and implementation of manufacturing processes and production systems are projected. Necessary training programs are begun.

1.11.4 Product design

The product is designed by the members of the cross-functional teams, including suppliers of purchased components, whose different objectives are expected to balance one another.[20] The engineering ef-

---

[18]The interplay of the various factors that enter into this phase of definition is particularly well-described in papers by D. Garvin. See, for example, D. A. Garvin, "What Does Product Quality Really Mean?," *Sloan Management Review,* vol. 26, Fall 1984, p. 25.

[19]The process of arriving at appropriate specifications is well-described in J. Hauser and D. Clausing, "The House of Quality," *Harvard Business Review,* May-June 1988, pp. 63–73.

[20]D. E. Whitney, "Manufacturing by Design," *Harvard Business Review,* July-August 1988, pp. 83–91.

fort aims at achieving a design that will exhibit little performance variation despite wide variation in the operating environment and product parameters and even customer errors.[21,22] Simplification and standardization are applied to reduce the number and variety of parts and to make the product easily manufacturable. Conscious attention is paid to interfaces within the product and its manufacturing process and to the designer's planned evolution to the next model.

1.11.5 Manufacturing process design

The cross-functional teams establish requirements for product fabrication, assembly, and testing. They analyze tolerances, estimate costs, identify the best processing methods, plan assembly steps and sequences, lay out the factory, and determine training requirements for factory personnel. All processes, manual and automated, are studied to determine whether they can consistently deliver products that meet specifications for quality, reliability, durability, and other attributes. Specifications are set for acquisition of in-process data needed early in the design process to help define, as accurately as possible, the capabilities of any new machine or process that is to be used. Layout, production plan, and logistics for the factory and its suppliers are designed for minimum inventory and high flexibility.

1.11.6 Production

Statistical Process control and in-process checks are used continuously. Inputs from these measurements and observations from manufacturing personnel are continuously fed back to improve both manufacturing and design processes and to aid in planning and follow-up products.

1.11.7 Difficulties in the design of complex products

In the foregoing idealized account of the product realization process, everyone cooperates, desired quality is achieved, and the product succeeds in the marketplace. In practice, the process is difficult and full of conflict and risk. Converting a concept into a complex, multitechnology product involves many steps of refinement. The design process

---

[21]Design practices used in this phase, which include designed experiments, Taguchi's robust design protocols, and specific programs such as Motorola's 6σ program, are utilized.

[22]H. B. Bebb, "Quality Design Engineering: The Missing Link to U.S. Competitiveness," keynote address, National Science Foundation Engineering Design Conference, Amherst, Mass., June 1989.

requires a great deal of analysis, investigation of basic physical processes, experimental verification, complex tradeoffs between conflicting elements, and difficult decisions. For example, there may be insufficient space for a desired function unless a costly development is undertaken, or space is taken from another function, affecting quality, fabrication yields, or ease of assembly. The original concept may not function as planned, and additional work may be required, affecting the schedule or requiring a change in specifications.[23] Satisfying the different and conflicting needs of function, manufacturing, use, and support requires a great deal of knowledge and skill.

1.12 Contemporary Design Practices for Affordable Automation

If the product realization process is a firm's strategy for continuous improvement, design practices for affordable automation are its tactics. Most advanced engineering design practices are not particularly complex or difficult to understand and use. Indeed, many are becoming accessible in computer software packages, short courses, and books. Confusion exists because there are too many practices, with different, and sometimes overlapping, functions. Some (e.g., Taguchi methods) cover more than on practice. Because not all practices are applicable or useful in the design of a given product, each company must carefully identify those appropriate to its uses and incorporate them into its PRP. Companies must also establish means of assimilating new practices as they are developed. In order to leap ahead of competitors, companies must continually develop (or work with others who are developing) new practices to meet changing needs.

The following sections describe traditional design practices, modern practices for setting strategy and specifications, and modern practices for executing designs.

1.12.1 Traditional practices

The following traditional practices remain important and continue to evolve.

Searching and studying patents and literature. Patents and the literature, an extremely fruitful source of information generated by inventors, researchers, and other practitioners, can help designers avoid

---

[23]J. L. Nevins and D. E. Whitney, eds., *Concurrent Design of Products and Processes,* McGraw-Hill, New York, 1989, Chap. 8.

wasting time and money on approaches that would not work. Return per dollar of engineering effort invested is probably as great for patent and literature search and study as for any engineering activity. But because it is not recognized as a mainstream design activity and management fails to adequately motivate it, many designers shun this work. Consequently, the practice is underutilized in the United States. Efforts to review foreign literature are especially meager. In contrast, Japanese firms assign engineers to this specific task; purchases of rights under U.S. patents are among Japanese firms' most effective investments.

Using standards. Use of standards of all types for components, procedures, computer-aided engineering/design (CAE/D), etc. can save design time, reduce uncertainty in performance, and improve product quality and reliability. It can also lead to economies of scale. Companies often define standard components and procedures with the goal of obtaining these advantages and then fail to enforce their use. Many new designers, failing to recognize the advantages of standards, tend to choose parts from their own knowledge or from the most familiar or convenient catalog. Unless a firm establishes standards and makes their importance known, any benefits that might result from their use will almost certainly be foregone.

Setting tolerances and the methods for checking them. Greater understanding of physical factors that contribute to variations in parameters and excellent metrology tools are powerful aids to setting and checking tolerances. There is, nevertheless, a pressing need to better understand relationships between design tolerances and product quality and cost. Designers must have information and supporting tools to choose appropriate, cost-effective, and robust methods.

Prototyping. Prototyping is an important tool for reducing time to market and creating models for evaluation of quality and producibility. In the past, prototyping was done by trial-and-error methods that were slow and cumbersome. At present, prototypes that are faithful representations of the final product are frequently required for use in experiments to optimize the product and work out assembly procedures. It is highly desirable to make these models with the same labor force and on the same line on which the product will be produced. However, this is not always possible, so better means of providing models are needed.

Analytical models. Both conventional analytical models and correlational models from design histories are powerful aids to engineering design, and they continue to evolve. Correlational models, which relate design variables to performance measures using empirical data, are valuable tools in complex and incompletely modeled situations.

Utilizing design reviews. Although they are time-consuming and expensive and take reviewers away from their own projects, peer design reviews are immensely helpful in finding and avoiding faults and suggesting alternative approaches. For design reviews to be effective, management must motivate designers to participate and reward them for doing so.

1.12.2 Modern practices for setting strategy and specifications for affordable automation

New practices that have emerged to support the PRP are used to provide estimates of the cost and quality of new or redesigned products, in strategic evaluation of a firm's position relative to its competitors, in recognition of the various contributors to a design, and even in negotiating with vendors and customers.

Product quality-cost model tools. Models that give the designers the means of evaluating product quality and cost in the design phase are essential, inasmuch as 70 percent or more of product cost is committed early in the design phase. New accounting methods, such as activity-based costing, provide accurate data on previous designs that can be used to generate quality-cost models, which are rapidly finding applications in the design of both products and processes.[24]

Comparative benchmarking and quality function deployment. The most successful firms benchmark continuously on not only their own product performance and features, design tools and techniques, technology, production approach, and facilities, but also those of their most successful competitors. Reverse engineering is often a part of the activity. Quality function deployment (QFD)[25] is a process that seeks to ensure that products not only are technically correct and manufacturable but also reflect customer needs. In QFD, an interfunctional team identifies product attributes consistent with customer needs and ranks them in an order determined by the customer. An adequate

---

[24] AT&T Bell Laboratories conduct research on product quality-cost models for semiconductor and printed wiring board design and fabrication processes. Research at Bell Labs yielded the Carter-Dishman theory that provides a guide to the economical application of very large scale integration (VLSI), taking into account the many factors that enter into integrated circuit development and design.

[25] J. Hauser and D. Clausing, "The House of Quality," *Harvard Business Review,* May-June 1988, pp. 63–73.

weight is assigned to each attribute, and the attributes are converted into measurable parameters. The team then benchmarks these characteristics against the competition, chooses and incorporates in its own design the best that others have done, and develops only those features that provide comparative advantage. QFD is used by AT&T, Digital Equipment Corp., Ford, Hewlett-Packard, IBM, and Xerox, among other companies.

Metrics for evaluating design practice. Generating metrics to judge a design can produce useful feedback, both during a design and when earlier designs are viewed. Metrics are extremely difficult to craft, and the search for better ones, such as the number of engineering change orders or warranty costs, continues. Hewlett-Packard uses metrics based on break-even time (BET) to guide and evaluate product realization projects. BET is defined as the time at which net operating profit (sales less cost of sales) equals total cost (TC) of design and development (Fig. 1.7).

The S curve. Almost all products follow an S-shaped life-cycle curve. A product progresses from a stage in which its contribution is much greater than the cost of keeping it viable to a state in which an ever-increasing investment of engineering effort and capital is required to keep it in the market. It is important to know where each of the firm's products and each competing product is in its life cycles in order to gauge when to move to a new technology or approach with further growth potential. Some companies test the viability of its product by

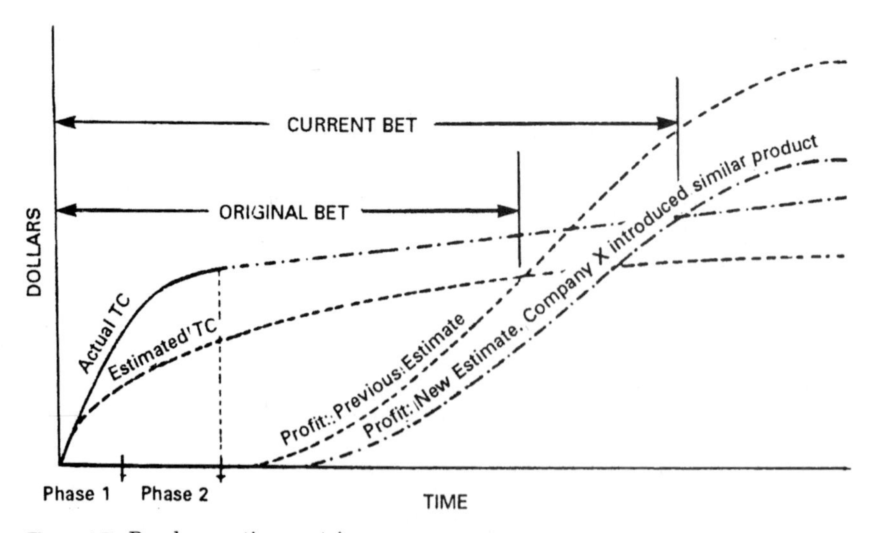

Figure 1.7 Break-even time metric.

establishing teams that play the part of competitors with a product on or approaching an S curve with a growth rate that surpasses that of the firm's own product.[26]

New management accounting systems. A design's leverage derives from the extent to which it determines product quality, cost, and time to market. For complex products especially, design is often a substantial fraction of total product cost. Because most companies cannot determine the contribution of design to profitability, track design improvement, or effectively compare different products and process designs, the research and development budget for design and processes is usually determined by applying a research and development to sales ratio "about right for this industry" or "about equal to what we think our best competitor spends."

This situation results from the use of a cost accounting system, originally designed for other purposes, that provides only delayed and aggregated data, perhaps based on labor or labor-and-material costs.[27] The situation also results from the fact that research and development costs, being charged when incurred, are not associated with any product or process.

New methods use detailed real-time information, obtained product by product, sometimes through computer-integrated manufacturing (CIM) systems, to provide current information, at an affordable cost, that applies to design as well as to manufacturing. They can operate as overlays on existing processes and so need not supplant traditional cost accounting systems initially. It seems clear that these methods will eventually be widely employed in manufacturing to provide the data used to control production and track product, and also in design, to:

1. Determine the contribution of design to profitability
2. Identify avenues to design improvement
3. Establish product life-cycle cost
4. Provide accurate information for budgeting and planning new products
5. Document savings that result from reducing transactions

---

[26]R. N. Foster, *Innovation,* Summit Books, New York, 1986.

[27]Manufacturing Studies, "Towards a New Era in Manufacturing" National Academy Press, Washington, D.C., 1986; R. S. Kaplan, *Measures for Manufacturing Excellence,* Harvard Business School Press, Boston, 1990, and "Management Accounting for Advanced Technological Environments," *Science,* August 25, 1989, p. 819

These methods can also be used to generate, and associate with process design efforts, important nonfinancial measures such as quality, number of transactions, and manufacturing cycle efficiency.

These new methods are of two generic types:

- Operational control and performance measuring systems that use broad-based real-time data from production
- Activity-based costing methods that associate engineering and marketing cost, as well as labor, materials, overhead, energy, and machine and process time, with individual product.

In summary, new accounting methods make it possible to determine the contribution of product and process design to quality and profitability so that allocations for research and development can be made intelligently, and to determine explicitly the contribution of individual designers for purposes of recognition and compensation.

The quality-loss function. Taguchi defines quality in terms of quality loss: quality is measured by total loss to society due to fundamental variations and harmful side effects resulting from the manufacture and use of a product. Working from this definition, he introduces a quality-loss function[28] (QLF) to replace conventional go, no-go specifications. Because it varies smoothly and continuously as a product parameter varies from specification, the QLF carries more information and hence is more useful than go, no-go specifications. By providing a common cost measure, it facilitates interactions between divisions in vertically integrated companies, between vendors and suppliers, and in resolving conflicts that arise from varying definitions of quality within marketing, manufacturing, and design.[29] In a typical application, the customer quantifies and supplies to the vendor the cost of departures from nominal specifications. The designer can then optimize these parameters and know what the customer is willing to pay for them. Both win. AT&T uses this concept to obtain agreement on transfer costs between divisions, and Texas Instruments' cost-of-ownership approach, used in working out integrated circuit supply contracts, is based on it.

---

[28]Sometimes called *quadratic-loss function,* a somewhat inappropriate name since not all QLFs are quadratic and the utility is vastly broader than that for the quadratic case.

[29]The conflicts that can arise because of differing quality definitions among the various functional organizations in a firm are discussed in D. A. Garvin, "What Does Product Quality Really Mean?," *Sloan Management Review,* vol. 26, Fall 1984, p. 25.

Motorola's 6σ approach,[30] a derivative of the QLF and of Taguchi's robust design methodology, mandates designs that yield components that operate satisfactorily within $\pm 6\sigma$ from the mean (where σ is the standard deviation of a product) specified by the customer. This means that the product will exhibit only about 3.4 defects per million if the process mean shifts by 1.5σ in either direction. For example, 3.4 ppm defective means that the throughput from a process that uses 300 such parts and has 500 such assembly operations is about 99.73 percent as illustrated in Fig. 1.5. Products produced thus are considerably more tolerant in the customer's application.

1.12.3 Modern practices for executing designs for affordable automation

Modern design practices for affordable automation can also be classified into the following categories:

1. CAD and CAE
2. Design for *X* [DF(*X*)]
3. Design rating systems
4. Concurrent design
5. Simplification
6. Incremental improvement
7. Robust design
8. Designed experiments

CAD and CAE. Computer-aided design and computer-aided engineering have evolved over a period of 20 years into powerful tools that provide the ability to design mechanical, electronic, and architectural objects on a computer screen and transfer the design to manufacturing processes. In some cases, particularly for electronic objects, this transfer is seamless and entirely computerized. However, apart from electronics design, the use of CAD and CAE in U.S. industry, is surprisingly limited, and in even fewer cases is the output of the CAD system directly linked to computer-aided manufacturing systems or numerically controlled tools. A manager at one large automotive manufacturer estimated that only one-third of the company's designers

[30]M. J. Harry, "The Nature of 6σ Quality," Government Electronics Group, Motorola, Inc., Sunnyvale, California.

used conventional CAD, and only a tiny fraction of those used three-dimensional solid modeling.

The capabilities of CAD and CAE systems do not meet the needs of many designers. Most often, the systems are used as little more than electric pencils that enable superior graphic presentation of designs. Only a few engineering systems permit any mathematical analysis to be performed on designed objects, other than in the well-established areas of finite-element analysis and electronic circuit simulation. Methods are needed to link designs of interactive individual parts for purposes such as establishing tolerances or performing assembly analysis. Similarly, methods are needed to link product designs to other kinds of business data, such as inventory control, cost predictions, and factory modeling.

DF(*X*). *Design for X*, where *X* stands for almost any operation (e.g., manufacture, assembly, test), is a ubiquitous technique.[31] General Electric, for example has an excellent program for designing with plastics that helps designers decide which type of plastic material to use in a given application. DF(*X*) techniques capture, in a standard procedure, all of the factors known to be important in a particular design activity. In the usual instances, costs are evaluated at each stage and at each interaction. These programs often provide examples and incorporate guidelines that help keep costs in the forefront, encourage the use of experience and standards, and prevent oversights. Though these programs are often specialized within a firm, progress is being made on generic methods of design for assembly and design for manufacture. DF(*X*) techniques are continuing to evolve, and new ones continually appear.

Design rating systems. Design rating systems such as those devised by Boothroyd and General Electric–Hitachi provide an impetus for design simplification and a method for tracking improvement.[32] These methods count parts of various types, promote the use of standard parts and the reuse of parts and subassemblies, and classify the motions required in assembly to provide estimates of quality and manufacturability.

---

[31]Recent references to current DFM and DFA techniques are K. A. Swift, *Knowledge-Based Design for Manufacture,* Kogan Page, London, 1987, and M. N. Andraesen, S. Kahler, and T. Lund, with K. Swift, *Design for Assembly,* 2d ed., IFS Publications, London, 1988.

[32]S. Kiyawaka and T. Ohashi, "The Hitachi Assemblability Evaluation Method," *Proceedings,* First International Conference on Product Design for Assembly, Newport, R.I., April 1986; G. Boothroyd and P. Dewhurst, *Product Design for Assembly Handbook,* Boothroyd Dewhurst Inc., Wakefield, R.I., 1987.

A number of companies use design cost evaluation systems to compute the costs of capital expenses required by competing designs in order to obtain more realistic comparisons. Though neither perfect nor foolproof, such systems intelligently applied can reduce risks in cost, schedule, and design time. AT&T uses a computer-based system that evaluates designs transmitted electronically to a manufacturing facility by designers at 14 remote locations and flags designs that cannot be manufactured without manual intervention. Within one year of operation, more than 99 percent of the designs received by the system did not require manual intervention.

Concurrent design. Concurrent design involves product designers, manufacturing engineers, and representatives of purchasing, marketing, and field service in the early stages of design in order to reduce cycle time and improve manufacturability.[33] This practice helps resolve what is sometimes called *designer's dilemma*—coping with the fact that most of a product's cost, quality, and manufacturability are committed very early in design before more detailed information has been developed. Assembling a multidisciplinary design team can permit pertinent knowledge to be brought to bear before individuals become wedded to their approach and much of the design cost has been invested. Differences are more easily reconciled early in design, and reductions in design cycle time that result from use of this method invariably reduce total product cost. Though the use of concurrent design concepts has met with success, little is known about how to organize and manage concurrent processes and cost-functional teams effectively.

Simplification. Simplifying a product by reducing the number and variety of parts and interfaces is often extraordinarily effective in reducing cost and improving quality and manufacturability. IBM's Proprinter development, GE's redesign of its electrical distribution and control product line for CIM production,[34] and Cincinnati Milacron's redesign of its plastic injection-molding machines[35] are well-known examples of projects that applied simplification effectively. Reduction in the number of interfaces between parts and processes, a facet of simplification that is

[33]D. E. Whitney, "Manufacturing by Design," *Harvard Business Review,* July-August 1988, pp. 83–91; J. L. Nevins and D. E. Whitney, eds., *Concurrent Design of Product and Processes,* McGraw-Hill, New York, 1989.

[34]W. J. Sheehan et al., "The Application of State-of-the-Market CIM to GE's Electrical Distribution and Control Business," *Electro 88 Conference Record,* 1988.

[35]G. T. Rehfeldt, "The Return of Competitiveness in American Manufacturing Companies—Lessons Learned," SRI Meeting on Strategic Management of Technology, San Francisco, January 26, 1988.

often overlooked, has proved to be particularly fruitful for AT&T. Simplification, though not difficult, is another nontraditional activity that must be made a specific design goal to be used to advantage.

Incremental improvement. This technique builds on accumulated experience and developing technology to reduce product cost and improve quality. Warranty costs and experience from field returns are continuously monitored for opportunities for improvement. Technology is monitored to find particular parts or subassemblies that can be replaced with lower-cost, more reliable ones. Often, simplification is applied. An incrementally improved product can usually be introduced to the market more quickly and with less risk than a new design. The successive stages of incremental improvement are readily discernible in the development of video cassette recorders, compact disc players, and cameras by Japanese firms.

Robust design. Robust design is a systematic three-stage process, pioneered by Taguchi, to optimize a product or process. It calls for designers to examine all possible ways of eliminating quality loss in order to find the most economical one.[36] Following this protocol,[37] design commences with a *system design* phase in which required features and functions, including materials, parts, and tentative product parameter values, are selected. In the next phase, called *parameter design,* the designer systematically studies all parts to determine which do not significantly affect reliability or manufacturability. For these, the designer seeks low-cost, commercial-grade parts. For example, a punched part might be specified rather than a machined one or a ±20 percent resistor rather than one of higher precision. In the third phase, called *tolerance design,* the designer determines the tolerances required for the remaining parts to provide the broadest possible margins in manufacturing and operation. Because the number of parts is now fewer, more detailed analysis of the sensitivities of the design to parameter variation due to aging, environment, etc., can be performed. Often, cost-performance tradeoffs can be made specific.[38] A variety of tools

[36]There are several statements of the principles of robust design. M. S. Phadke states it is to "minimize the effect of the cause of variations without controlling the cause itself." J. G. Elliott says, "Americans remove the cause of the effect. Japanese remove the effect of the cause."

[37]As described in D. M. Byrne and S. Taguchi, "The Taguchi Approach to Parameter Design," *Proceedings of the ASQC Quality Congress,* 1986.

[38]A good exposition and examples of the technique are provided in M. S. Phadke, *Quality Engineering Using Robust Design,* Prentice-Hall, Englewood Cliffs, N.J., 1989.

can be used to facilitate this analysis. For electronic circuit the group of programs generally referred to as *SPICE*[39] permits designers to optimize circuit operating margins given real or assumed statistical descriptions of component values and operating conditions. AT&T has an equivalent mechanical and electronic design program.

SPICE was developed in 1970 at the University of California at Berkeley by a team under Professor Donald Peterson. It has been enormously successful, and many companies now offer customized versions of it. Much of the software is in the public domain, and copies of the current version, SPICE 3D-2, are available from Professor Peterson's group at the University of California at Berkeley for a nominal fee.

Designed experiments. Appropriately designed experiments are used for determining the relative importance of many different factors to reliability or process yield. Experiments can be constructed to use all of the experimental data in several ways, often reducing by an order of magnitude the amount of experimental data that must be selected compared to the traditional approach of varying one parameter at a time while holding others fixed. The experimental approach is most useful when the number of variables is large, the effects are relatively substantial, and interactions among the various parameters are unknown. Use of this approach is rapidly being made easier by the availability of good personal computer–based software tools.[40]

The methods used in this approach were developed by R. A. Fisher, who applied them in agricultural experiments in England during the 1920s.[41] G. E. P. Box and others extended Fisher's method and applied them in many industrial applications. Professor Box's approach is straightforward and satisfactory for most problems.[42] Taguchi has

---

[39]There are many books and references on SPICE in its various versions, such as P. Tuinenga, *SPICE: A Guide to Circuits Simulation and Analysis,* Prentice-Hall, Englewood Cliffs, N.J., 1988. A good survey and historical account of the development of SPICE in its several forms is contained in a paper by A. Vladimierestu, *Proceedings of the Bipolar Circuits and Technology Meeting,* September 1990.

[40]G. Hahn and C. Morgan, "Design Experiments with Your Computer," *Chemtech,* November 1988. The American Suppliers Institute, Dearborn, Mich., provides PC-based software that helps in the design and guides the execution and analysis of design experiments using Taguchi's techniques. Texas Instruments has announced a PC-based expert system that will make it possible for the user to conduct experiments of this type with no additional training.

[41]R. A. Fisher, *Design of Experiments,* Hafner Publishing Co., New York, 1951.

[42]G. E. D. Box, J. S. Hunter, and W. D. Hunter, *Statistics for Experimenters,* John Wiley & Sons, New York, 1978; D. C. Montgomery, *Design and Analysis of Experiments,* 2d ed., John Wiley & Sons, New York, 1984.

promoted and applied these methods in design and troubleshooting.[43] Japanese and U.S. automobile industries use these techniques extensively, sometimes performing tens or even hundreds of experiments during various stages, particularly early stages, of product design.

1.13 Understanding, Motivating, and Supporting the Designer for Affordable Automation

The design of products and processes is a creative activity that depends on human capabilities not easily measured or predicted. The most effective designs are acts of creativity that rank with those in the fine arts. We are not within sight of the time when machines can perform the design function, though tools can certainly aid the designer. Dependency on designs makes it vitally important that companies understand the nature of the design task and the nature, characteristics, and needs of people who design effectively in order to be able to create an environment that stimulates and nurtures them.

The design environment is set in large part by the organization of and strategy for design. A formal, well-supported product realization process can make an important contribution to a productive design environment. In the following sections, we discuss briefly the nature of the design task and the designer, and some steps that can be taken to provide a supportive design environment.

1.13.1 The design task for affordable automation

The design task for affordable automation, which once could have been adequately defined as achieving a function at a specified cost, has broadened under competitive pressures to include at least three areas of endeavor:

1. Designing products and processes to meet many constraints
2. Developing and improving design tools and processes, including the PRP

---

[43]M. S. Phadke, *Quality Engineering Using Robust Design,* Prentice-Hall, Englewood Cliffs, N.J., describes these techniques and provides examples; J. G. Elliott, *Statistical Methods and Applications,* is a Taguchi "cookbook" that describes how to apply this method using examples from the automobile industry; Taguchi's contributions and the relationship between Taguchi's methods and traditional design of experiments are described clearly in G. E. P. Box and S. Bisgaard, *The Scientific Context of Quality Improvement,* Center for Quality and Productivity Improvement, University of Wisconsin, Madison, 1987.

TABLE 1.1 Touchstones for Design

Customer—Who is the customer? What does he or she really need?

Stakeholders—Understand the positions of those who have stakes in the product's success or the status quo.

Ease-of-use—Human factors design needs to be addressed early in the process.

Documentation—Essential; match to user's needs; start early.

Cultural change—If development or production of this product or process requires cultural change, its introduction will not be easy or swift.

Patent/copyright—Plan for this early to avoid pitfalls and to get high quality coverage.

Legal/regulatory—Consider early. Such obstacles have delayed or damaged many projects.

Environmental impact—Determine if the manufacture or use of any product may adversely affect the environment.

Manufacturability—Has the manufacturing engineer been on the team?

Aesthetics—These hard-to-define characteristics are also critical.

Dynamics—How does the product or process behave in non-steady-state conditions?

Testability—How will the product be tested? Where, by whom, at what cost?

Prototypes—Consider how the final product may differ from the prototype if prototype and production processes are not identical.

Universality—Universal solutions almost never work.

Simplicity—Strive for beautiful, simple designs. They often work well.

Appearance—If the design does not look right, watch out!

Interfaces—Many otherwise sound designs fail because of unanticipated problems at interfaces.

Maturity—Where is the product on its S curve? Is it time to jump to a new approach?

Partitioning—Consider partitioning to provide additional degrees of freedom.

Models—Do the mathematical models used in design apply over the anticipated range of use?

Scale-up—Do not undertake this lightly. Proceed by small increments.

Transportation—What happens to the product in transportation.

3. Standardizing parts and generating specifications

The various practices described previously present only a part of the designer's task in designing products and processes for affordable automation. Table 1.1 lists some of the factors besides quality, cost, and time to market that can make or break a design.

In addition to designing products and processes to meet many constraints, designers often have the task of integrating numerous separate procedures into complete processes. This function, which may include some tool development, controls the flexibility of the resulting process and the time required to execute design. For example, a designer may develop a computer program to link the output of a CAD system to an automatic part insertion machine, eliminating manual data transfer, and thus saving time and reducing errors.

Designers also work with other parts of the firm, with customers, and with vendors as they use many of the tools and techniques described earlier to obtain the information needed to set product specifications. As the principal agents for PRP, they must have strong interpersonal skills as well as sound technical skills and creative ability.

The breadth of knowledge required by the practicing engineer today is immense, encompassing many topics not emphasized or included in standard engineering curricula.

1.13.2 The designer

Who does design? In most firms today, design is not limited to those who are educated as designers or who spend most of their time designing products or processes. Many more engineers and scientists participate in design than those whose job assignments are design. An increasing number of people are involved in activities such as comparative benchmarking and reverse engineering that are more analytic than synthetic in nature. To derive information useful to the designer, these people must understand design. Those who do process design and systems integration must also have knowledge of the design process, as must the many engineers and scientists who work on CIM or the PRP. Manufacturing engineers who work on teams with designers and marketing people must understand design as thoroughly as manufacturing in order to arrive at manufacturable products.

Chapter

2

Part Design and Handling for Affordable Automation

2.1 Introduction

It is hard to envisage a technical field that is more exciting than affordable automation in advanced manufacturing technology. This field is extremely challenging to engineers and managers of many disciplines, including electrical, industrial, mechanical, and manufacturing engineers. It is to the application of innovative cost reduction techniques to manufacturing operations that this text is primarily addressed—the manufacture of such products as automobiles, refrigerators, electronic circuit boards, and computers. This is the manufacturing category that is the concern of the four categories of engineers just cited.

In this chapter, we consider design of parts so that they can be fabricated and assembled, with high-quality results, by affordable automation. We first present a case study—flexible printed circuit manufacture—that demonstrates the interrelationship of part and process design. We then look at ways of designing parts and processes with the aid of statistics and probability to ensure a given level of performance.

2.2 The Role of Affordable Automation in Manufacturing Enterprises

Today's businesses face an ever-increasing number of challenges. The manufacturers that meet these challenges will be the ones that develop more effective and efficient forms of production, development, and marketing.

Affordable automation can make a fundamental contribution to manufacturing solutions based on simple and affordable integration.

Utilizing affordable automation, one can integrate manufacturing processes, react to rapidly changing production conditions, help people to react more effectively to complex qualitative decisions, lower product cost, and improve product quality.

The first step in achieving such flexibility is establishing an affordable manufacturing system in a synchronous or an asynchronous form that can be reshaped whenever necessary. This will enable the manufacturing enterprise to respond to changing requirements. This reshaping must be accomplished with minimal cost and disruption to the operation.

Undoubtedly, affordable automation will play a key role in the manufacturing information system. Affordable automation will shorten lead time, reduce inventories, and minimize excess capacity. Integrating various sensors with appropriate control throughout the manufacturing operation will play a key role to successful implementation of affordable automation.[1] The result is that individual manufacturing processes will be able to flow, communicate, and respond together as a unified cell.

In order to develop an affordable automated manufacturing and control system that will achieve these objectives, the enterprise must start with a long-range architectural strategy, one that provides a foundation that accommodates today's needs and takes tomorrow's into account. These needs include supporting new manufacturing processes, incorporating new data functions, and establishing new databases and distribution channels. The tools for this control and integration are available today.

Affordable automation is of paramount importance to the economic well-being of any nation in these times of great competition in the manufacturing arena. It can ensure reduced design-to-prototype lead times, fewer problems with engineering design change implementations, flexible manufacturing capabilities, maximum production rates commensurate with reduced costs, and many other production benefits.

2.3 Case Study: Flexible Printed Circuit Manufacture

Manufacture of flexible printed circuits is a complex process that offers an excellent example of how affordable automation techniques can improve product yield, reduce costs, and allow rapid response to market demands.

---

[1]Sabrie Soloman, *Sensors and Control Systems in Manufacturing,* McGraw-Hill, New York, 1994.

How often does one find that an ill-fated flexible circuit began with parts ordered as individual pieces, without serious considerations of the assembly operation as a whole? Automatic fabrication and automatic assembly of flexible circuits should go hand in hand in order to achieve the manufacturer's economic goals.

Depending upon the specific flexible circuit, automatic assembly may be defined as either (1) the mechanized orientation and selections of bulk discrete parts or (2) the integral fabrication of component parts, with their controlled placement, joining, and test, to produce a useful assembly.

2.3.1 Part handling and orientation techniques

It is nearly impossible to handle flexible circuit parts (Fig. 2.1) through a standard means of orientation such as a vibratory feeder. It is also extremely costly to handle flexible circuits automatically as patterned sheets in a magazine (Fig. 2.2). Moreover, flexible circuits often suffer abrasion and physical damage when handled as large sheets. Reel-to-reel automatic etching currently is the only system known to combine the production of flexible circuits with automatic handling, assembly, and testing.

The very flexibility of the flexible circuits makes them difficult to handle. Indexing a blanking die to separate each flexible circuit from the sheet cannot be done successfully at an affordable cost. Each cen-

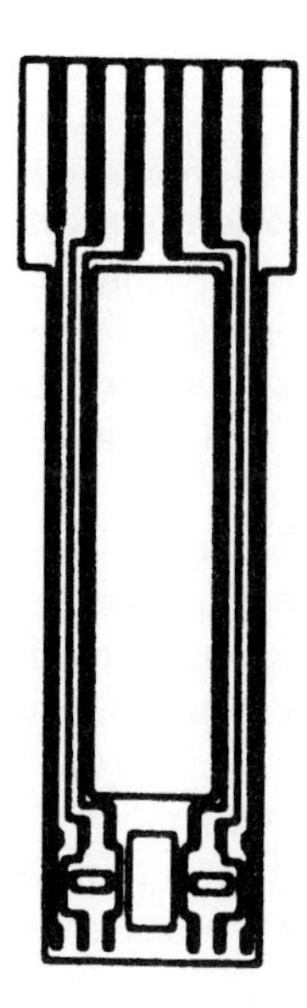

Figure 2.1 Flexible circuit piece part.

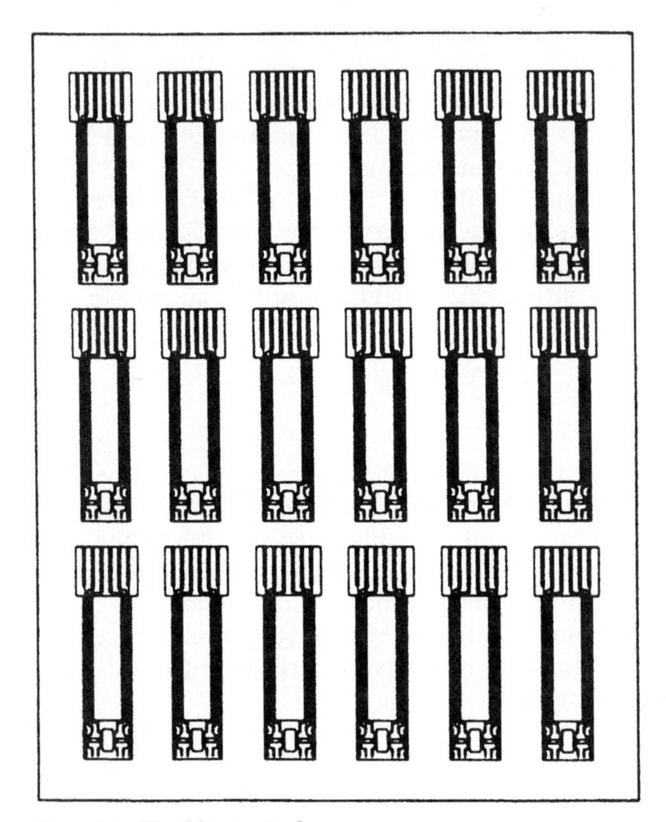

Figure 2.2 Flexible circuit sheet.

ter line of a flexible circuit has a ±0.0005-in tolerance. Insufficient rigidity (which creates skew) makes it nearly impossible to maintain dimensional integrity between each flexible circuit and the adjacent one. And once the flexible circuit has departed its original etched sheet, it automatically loses its flatness. It is difficult to restore this flatness, which is critical for automatic handling, during further automatic assembly.

In assembly, every component must be oriented and placed in specific relationships with all other component parts. Substantial portions of the cost of assembly are incurred in reestablishing proper orientation of parts initially presented to the assembly machine in random orientation.

It is a fact, however, that every fabricated part has, at the time of its creation, some specific (and known) orientation. Whether a part is molded, stamped, turned, or extruded, the manufacturing process gives it a specific orientation at its discharge from the fabrication equipment.

A systems approach to automatic fabrication and assembly of flexible circuits can retain this orientation. The author has developed such a system in designing affordable automation for etching and assembling flexible circuits.

2.3.2 Criteria for automatic etching and assembly

Various methods of coupling the fabricating processes to the assembly machine were considered in designing the affordable automation system. The concept of automatically indexing the blanking die over a table carrying a sheet of etched flexible circuits could not achieve the desired accuracy, as noted above, for the following reasons:

1. *Indexing.* The imbedded wires were frequently destroyed because of inconsistency of the dimensions between flexible circuit center lines resulting from skew (see Fig. 2.3).

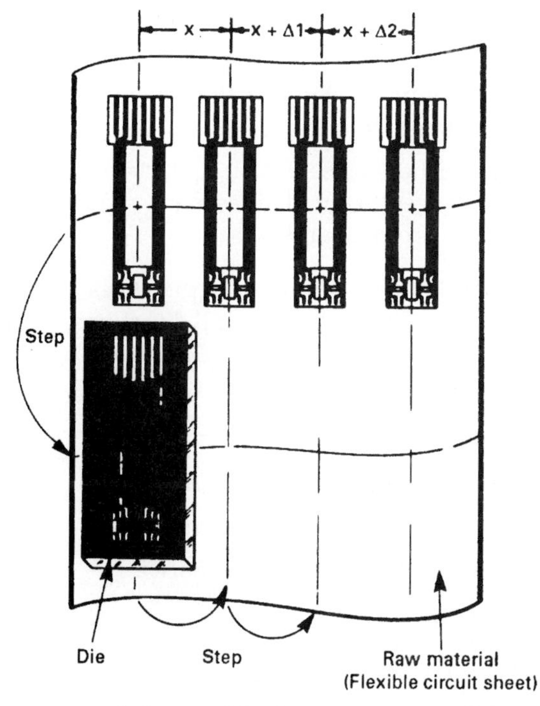

Figure 2.3 Indexing of flexible circuit for blanking operation.

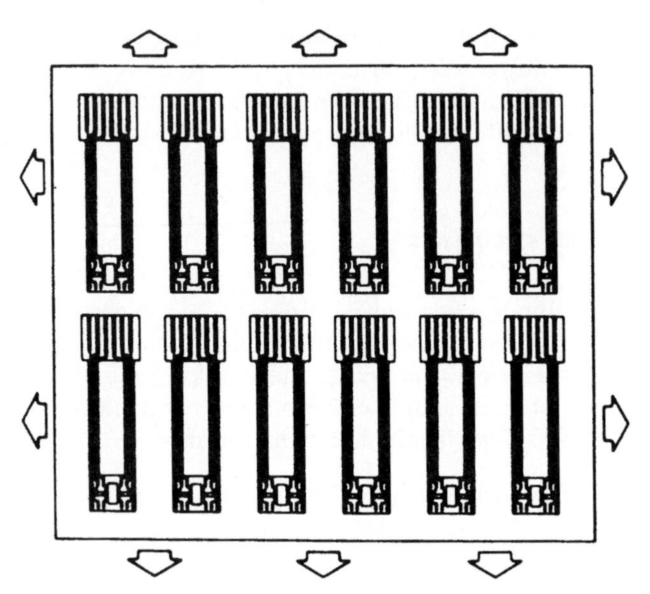

Figure 2.4 Maintaining equal tension on flexible circuit sheet.

2. *Tension.* It is obvious that a shorter index produces better tolerances in a sheet of flexible circuits, as long as the tension in the sheet is controlled in all directions, as in Fig. 2.4.
3. *Flatness.* A circuit loses its flatness as soon as it departs from the sheet (Fig. 2.5). To restore flatness to the defined form requires a complicated pickup and placement mechanism.

It is evident that a rigid clamp should be attached to the edges of the flexible circuit sheet to provide the tension needed in all directions. But this would complicate and add to the cost of producing flexible circuits.

The affordable system was expected to:

- Create a narrow strip of flexible circuits.
- Index each flexible circuit within ±0.0005-in accuracy.
- Maintain constant tension between each flexible circuit.
- Produce acceptable parts that can be handled automatically.
- Produce parts at reasonable cost.

The above parameters were investigated extensively with various manufacturers of flexible circuits, nationally and internationally. It was apparent that special-purpose etching and chemical systems would be required.

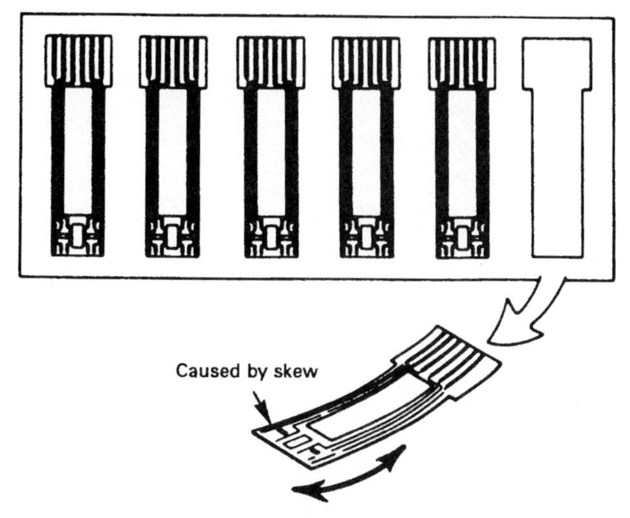

Figure 2.5 Loss of flatness.

2.3.3 Manufacturing system concept

The investigation led to the following plan:

Use a reel-to-reel strip to carry a single row of flexible circuits.

Develop a mechanical indexing system to index between each flexible circuit with a ±0.0005-in tolerance.

Use a mechanical sprocket punching system to produce holes at the edge of the strip carrying the flexible circuits.

Use the mechanical indexing system to maintain a specified tension force.

Check each flexible circuit for continuity, bridging, and shortening.

Hand over each flexible circuit to the automatic assembly machine in its original flexible form.

The reel-to-reel transport system for flexible circuits is much like that for conventional motion picture film.

A reel of copper-Kapton material is needed. The nominal width of the flexible circuit reel is 2.25 in. The width of a standard roll is 9.0 in. Therefore, a slitting machine was developed to cut 2.25-in-wide strips, which were wound on reels (Fig. 2.6). The length of each cut strip varies between 100 and 500 ft.

A reel of resist material, to be laminated on the copper-Kapton strip, is also required. The resist is a light-sensitive material that is exposed to an image of flexible circuits. Dimensions of reels are simi-

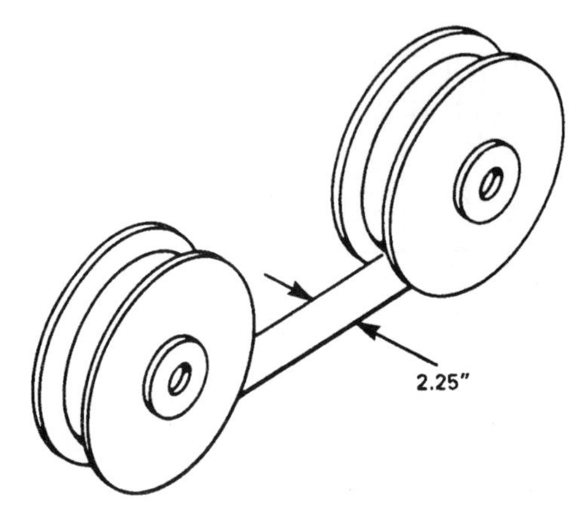

Figure 2.6 Reels of flexible circuit material.

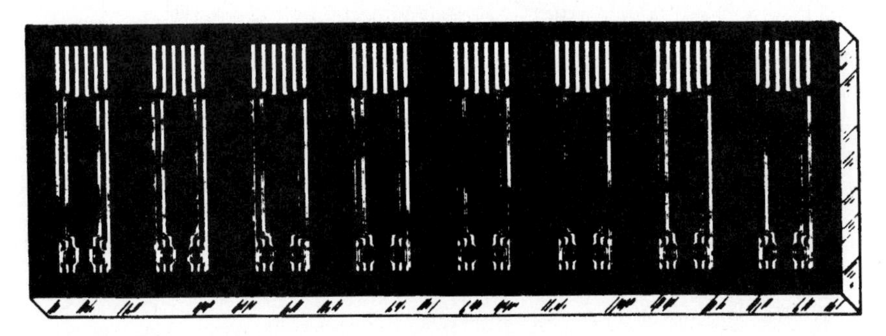

Figure 2.7 Etched glass artwork.

lar to those of the copper-Kapton. The image of the flexible circuit is imprinted on a glass panel (Fig. 2.7).

2.3.4 An innovative affordable automatic production system

The standard existing equipment for producing flexible circuits is, in general, very costly. It was primarily developed to accommodate much larger widths than 2.25 in.

A simple laminator machine (Fig. 2.8) with a tension control system was developed to accommodate a narrow strip of flexible circuits. The strips of copper-Kapton and resist are laminated together and wound on a reel. It is imperative to create registration holes whose accumulative center-to-center variation is no greater than ± 0.0005 in for a span of 21 in of flexible circuit. Therefore, a sprocket punching machine was also developed (Fig. 2.9). The laminate of copper-Kapton and photosensitive material is punched by the die

Figure 2.8 Laminator machine.

Figure 2.9 Sprocket punching machine.

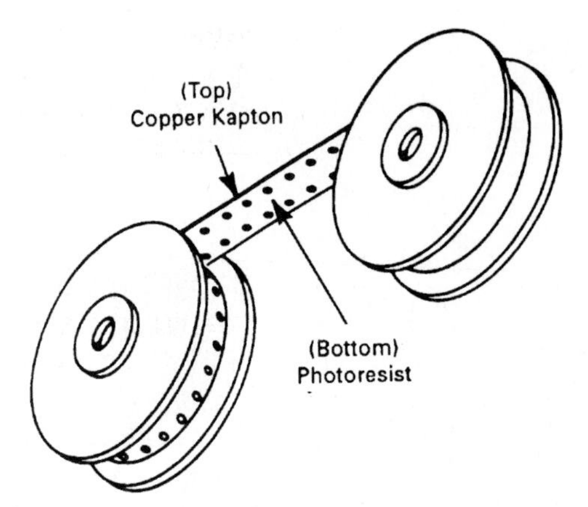

Figure 2.10 Punched flexible circuit material.

pins and advanced to another winding reel (Fig. 2.10). The sprocket-punched material can then be passed through an exposure machine (Fig. 2.11).

The exposure machine is essentially a camera with automatic wind and rewind reels. The light source is located under the machine to create image of the flexible circuit on the photosensitive material (Fig. 2.12). Key components of the exposure machine are a step motor, a device for locating fixtures with artwork, a control system, and etching and plating systems, which are described in the following sections.

Step motor. The step motor advances the reel of flexible circuits and winds it on a take-up spool. The linear accuracy of the step motor advancement must be better than ± 0.0005 in.

Locating fixtures with artwork. The artwork is permanently fixed on the exposure machine beneath two locating fixtures (Fig. 2.11). The locating fixtures contain equally spaced locating pins to maintain proper alignment between the artwork and the strip laminate. A pressure pad on the locating fixtures keeps the contact between the sensitive laminate and the artwork free from foreign particles. (Particles such as dust and air bubbles will cause wire discontinuities, which lead to rejection of flexible circuits.)

Control system. The control system advances the flexible strip in precise increments of length. It also controls the sequence of operations of the two locating fixtures (Fig. 2.13). It coordinates the advancement of the strip of flexible circuits with the up-and-down motion of the locating fixtures.

Figure 2.11 Exposure machine.

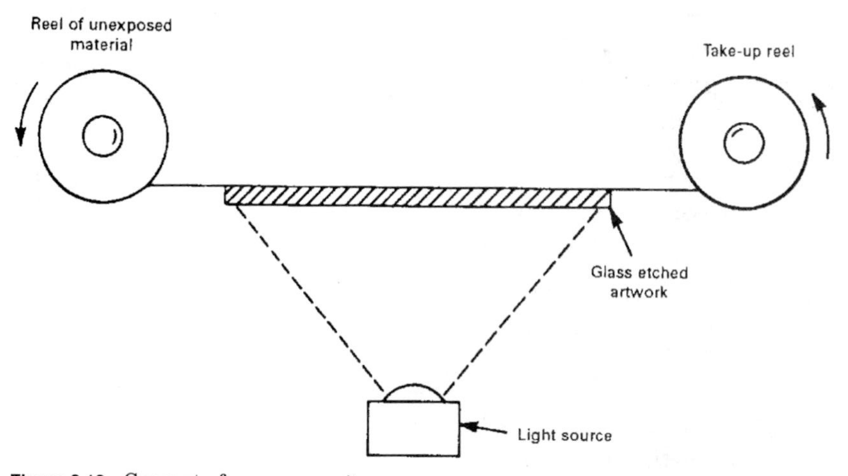

Figure 2.12 Concept of exposure unit.

Etching and plating. The exposed reel can then be taken to an automatic reel-to-reel etching system. This system (Fig. 2.14) continuously etches the relatively narrow strip of flexible circuits.

To prevent corrosion of the copper wire of the flexible circuit, it must be plated with solder. In a reel-to-reel solder plating system (Fig. 2.15),

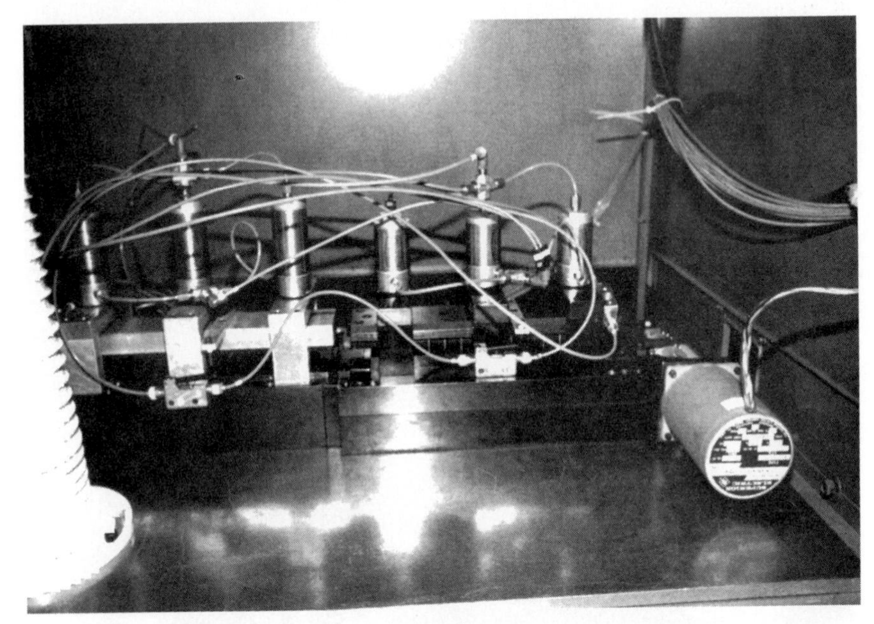

Figure 2.13 Locating fixtures with control systems.

Figure 2.14 Reel-to-reel etching system.

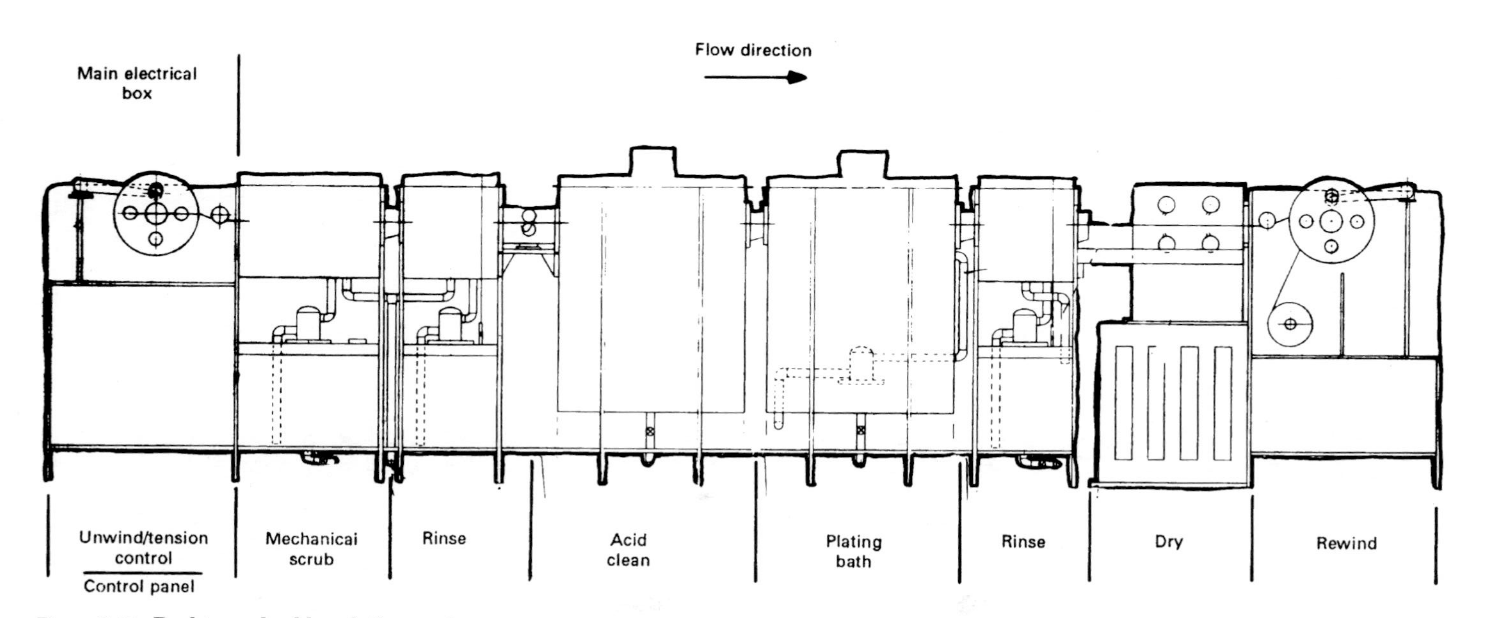

Figure 2.15 Reel-to-reel solder plating system.

the reel of material is unwound, passed through a tension control mechanism and into a tank where it is mechanically scrubbed to accept the solder, and driven into a series of tanks where it is rinsed and chemically cleaned. It then enters a solder electroplating tank, where rollers maintain electrical contact to the strip as it continuously passes through the electroplating bath on a festoon-type roller system. The material is then rinsed, dried, and rewound on a reel. Again, tension is controlled throughout the system. The end result is a continuous strip of flexible circuits wound on a reel. The reel is then taken to assembly machine where the strip is automatically inspected and parts are assembled to it.

2.3.5 Economic analysis

Table 2.1 lists the costs of developing the automatic fabrication equipment for flexible circuits in house and buying it from outside vendors as standard products.

TABLE 2.1 Cost of Equipment for Automatic Fabrication of Flexible Circuits

		Cost, $		
Machine	Function	Vendor	In House	Decision
Slitting machine	Slit material to proper width	4,300	4,300	Purchase
Chemical cleaning machine	Clean copper-Kapton for lamination	32,000	32,000	Purchase
Lamination	Laminate photosensitive material on copper-Kapton for exposure	60,000	12,000	Develop in house
Sprocket punching machine	Create mechanical references between flexible circuits	80,000	15,000	Develop in house
Exposure machine	Create photographic image on sensitive material	125,000	20,000	Develop in house
Etching machine	Develop film, etch copper, and strip remaining film	50,000	50,000	Purchase
Solder plating machine	Place protective cover on copper to prevent corrosion and enhance solderability	98,000	23,000	Develop in house
Total		$449,300	$156,300	

TABLE 2.2 Cost per Circuit for Reel-to-Reel Fabrication

Cost of copper-Kapton	5¢
Cost of resist	1¢
Cost of burdened labor	1¢
Depreciation of capital equipment	0.33¢
Total:	**7.33¢**

TABLE 2.3 Cost per Circuit for Piece Fabrication

Volume	Cost/circuit, $
1000–10,000	3.10
10,000–50,000	2.85
50,000–150,000	2.25
150,000–500,000	2.13
500,000–1000,000	1.98

Quantitative analysis. The vendors' costs, as shown in Table 2.1, were significantly higher than the in-house costs for most equipment. Exceptions were machines 1, 2, and 6 which were purchased; therefore in these cases standard equipment was adapted to narrow, continuous-reel processing. A breakdown of the cost per circuit for reel-to-reel fabrication on the mostly in-house equipment is given in Table 2.2.

The total cost per circuit for reel-to-reel processing is substantially less than piece processing. Although the piece processing cost decreases with increasing circuit quantity, as Table 2.3 indicates, the minimum savings with reel-to-reel processing is $1.90 per circuit.

Qualitative analysis. Not only is cost lower, but quality is higher for the reel-to-reel process. Table 2.4 compares dimensional accuracy of flexible circuits made by mostly in-house reel-to-reel equipment with that of flexible circuits made as piece parts by outside vendors. The greater accuracy of the reel-to-reel circuits facilitates automatic handling and assembly.

2.4 Applicability of Design for Affordable Automation to Manual and Semiautomated Processes

The product itself must be designed for producibility and, in particularly, *assemblability.* The advent of affordable automation has focused attention on product design to facilitate orientation, positioning, and mating of components. The unexpected benefit is that such attention to automated producibility in the product design phase has resulted

TABLE 2.4 In-House versus Outside Vendor Precision

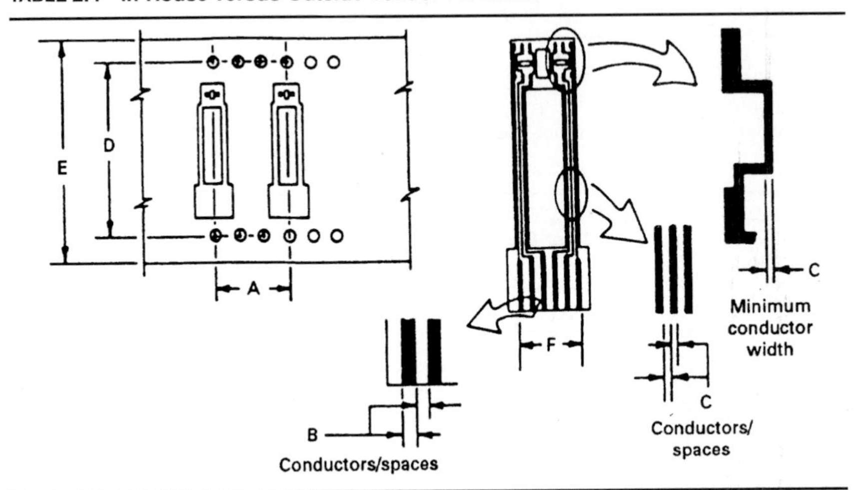

Drawing location	Nominal dimensions, in	Actual dimensions, in	
		In house	Outside vendor
A	0.750 ± 0.005	0.74915	0.7508
		0.74920	0.7502
		0.7500	0.7512
		0.7502	0.7518
		0.7503	0.7509
B	0.030 ± 0.005	0.0313	0.0339
		0.0315	0.0340
		0.0317	0.0348
		0.0318	0.0351
C	0.010 ± 0.002	0.0114	0.0119
		0.0115	0.0118
		0.0117	0.0120
		0.0117	0.0123
		0.0118	0.0121
D	1.8000 ± 0.0002	1.79955	1.8002
		1.79980	1.8003
		1.8000	1.8002
		1.80005	1.8002
		1.80010	1.8009
E	2.265 ± 0.005	2.2628	2.2690
		2.2636	2.2670
		2.2638	2.2695
		2.2639	2.2693
		2.2639	2.2648
F	0.330 ± 0.005	0.3316	0.334
		0.3321	0.337
		0.3321	0.336
G	0.006 min	0.0083	0.0079
		0.0084	0.0083
		0.0084	0.0084
		0.0088	0.0073
		0.0088	0.0069

TABLE 2.5 Cost Breakdown of Conventional Automation versus Affordable Automation

Convention assembly system component*	Cost, $	Affordable assembly cost, $ system component†	
Robot	175,000	3-D pickup and placement mechanism	17,000
Vision	65,000	High-precision guiding dowel pins	2,000
Cost/system	240,000		19,000
Total cost	240,000 (one system is required)		57,000 (three systems are required)‡

*4-second cycle time.
†10-second cycle time.
‡2.5 systems are required for a 10-second cycle time system. Two systems will be operated at 100% utilization, while the third system will be operated at 50% utilization.

in more efficient manual production as well. In fact, in a substantial percentage of cases, products are made by a highly efficient semiautomated process that may not even include an assembly machine.

Although some automated assembly systems are being equipped with sophisticated vision systems for orienting and handling parts, the majority of applications will not support the additional cost of such systems. Therefore, the automated handling or assembly system must be well designed to work without costly peripheral devices for part orientation.

Table 2.5 compares an affordable automatic assembly system for 3½-in disk drives to a conventional assembly system employing robotics and machine vision.

2.5 Probability, Statistics, and Reliability in Design for Affordable Automation

There is ample evidence of a growing interest in using probability and statistics in engineering design for affordable automation. Probabilistic design makes sense because engineering variables are modeled more completely by statistical methods; e.g., stress-strength interactions are probabilistic. Only recently has a performance-related measure, reliability (R), been used to express adequacy or level of safety of design for affordable automation.

The factors that must be considered in an assessment of design for affordable automation based on probability and statistics include:

1. *Important concepts from probability theory.* How are they applied in engineering?
2. *Distribution theory.* Which distributions describe engineering variables and which distributions constitute workable models?
3. *Variability in functions of random variables.* What are the statistics of distributions of the functions of random variables; that is, what are their mean values and standard deviations?
4. *Method of describing load-induced stress and strength.* How are their distributions and statistics defined?
5. *Synthesis plan.* Given the required performance measure R and answers to the preceding items, what plan can be developed to carry out the design synthesis for affordable automation?
6. *Optimization plan.* Given a probabilistically synthesized design, how can it be optimized and implemented at a specific reliability level?

The probabilistic design for affordable automation must include the unique relationships among design variables. A second tier of modeling is required, i.e., one which preserves the essential features of each variable in a function, in particular, its multivalued aspects and random nature.

2.5.1 Design for reliability

Careful and rigorous analyses can become worthless if results are diluted through the use of empirical multipliers selected arbitrarily. Conventional practices like this, adopted to circumvent variability problems, often produce overly conservative designs. To ensure a realistic analysis, fundamental questions must be answered:

- How safe is a design, expressed by a measure of performance?
- How safe should a design be, relative to the consequences of failure?
- How can design to specified levels of adequacy be ensured?

Answers to these questions involve probabilistic methods. The field of engineering is fraught with uncertainty. Presently available mathematical tools have not been well-suited to the task of dealing with uncertainty; i.e., they use real numbers almost exclusively. The algebra of real numbers produces unique single-valued answers in the evaluation of a function. However, one must take into account more than a single value (such as average) in representing the behavior of a variable and accounting for uncertainty. Statistical uncertainty has been ignored, for instance, in the search for guaranteed minimum values of load and for ultimate strength.

If load is expressed as

$$L = F(x_1, x_2, \dots, x_n)$$

it is a deterministic model, whereas if it is expressed as

$$(\bar{L}, \leq L) = f[(\bar{x}_1, \leq x_1), (\bar{x}_2, \leq x_2), \dots, (\bar{x}_n, \leq x_n)]$$

it is a statistical model.

The magnitude of almost all engineering variables changes in a random fashion; variables are characterized by a spectra of values, not by unique values. The result of a series of tests or measurements is a population of nonidentical observations: $x_1, x_2, \dots, x_n$. Plotted on a graph as magnitude versus frequency, such (hystographic) displays tend to approach a stable, predictable distribution, such as the normal, lognormal, gamma, or Weibull.

Many engineering variables x are well-represented by a mean value $\bar{x}$ and a measure of variations $\leq x$. These are estimators of the mean μ and the deviation σ_x.

For example, $\bar{S}_y$ represents the mean tensile yield strength of a sample and $\leq S_y$ represents its standard deviation. Consequently, a variable may be described by a couple such as $(\bar{S}_y, \leq S_y)$ and a statement of its distribution. These two numbers can uniquely describe any two-parameter distribution.

For classical strength-limited design, once the criterion of failure is identified, the rule for an adequate design is

$$\text{Strength} > \text{stress}$$

or, to cover imponderables,

$$\text{Strength} > \text{design factor} \times \text{stress}$$

The concept of design factor conveys the idea that the mean strength and the mean stress have been intentionally separated to provide the requisite level of safety. But design factors can be grossly inaccurate because they do not take into account the inherent variability of stress and strength and the existence of stress and strength distributions.

When the distributions of strength and stress are known, the adequacy of a component or system may be estimated from the interference depicted in Fig. 2.16. The curves in Fig. 2.16 predict a finite incidence of failure.

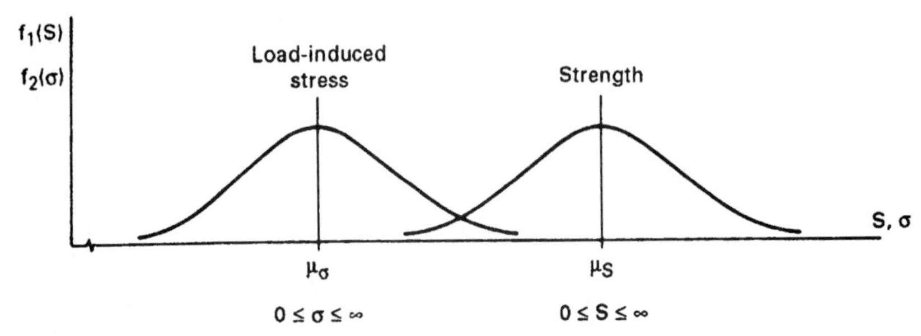

Figure 2.16 The interaction of distributions of stress and strength when the mean strength exceeds the mean stress.

The following steps are necessary when probabilistic models are used:

1. Invent a concept and give it a preliminary form.
2. Describe external forces to be imposed on the product, their range, and their frequency.
3. Analyze the preliminary system. Estimate force intensities by statistical models.
4. Select materials according to physical and mechanical properties, considering cost, availability, and feasibility.
5. Describe strength and failure characteristics of materials as probability density functions to account for variability.
6. Make quantitative estimates of strength and failure characteristics at the component level.

In a strength-limited design, the results will be a joint probability function

$$P[S > \sigma] = P[S - \sigma > 0] \geq R$$

$$R = \int_{-\infty}^{\infty} f_1(S) \, [\int^{\infty} s \, f_2(\sigma) \, d\sigma] \, dS$$

where f_1 = strength probability density function
f_2 = load-induced stress probability density function
S = significant strength
σ = significant load-induced stress

The design is to ensure that

$$S > \sigma$$

Failure may be defined as:

- Fracture
- Yield
- Fatigue cracking
- Deformation

The classical definition of probability requires a complete knowledge of all possible outcomes, as in coin flipping or dice rolling. It is applicable to simple, well-defined situations having known outcomes. The relative frequency definition of probability is applicable to more complicated situations where the outcomes cannot be easily specified and where the likelihood of outcomes may be in doubt. In such cases, one must make a large number of trials and count the various outcomes. This is the procedure followed in engineering and in practical statistical determinations. It is the only procedure available today in studying natural phenomena.

Probability statements such as

$$p(x > x_1)$$

$$p(x < x_2)$$

$$p(x_1 < x_2)$$

may be interpreted as areas on hystographic or continuous probability-density diagrams (Fig. 2.17).

2.5.2 Descriptive statistics

The basic goal of designers of affordable automation is to develop systems that balance cost and risk in response to human needs. Probabilistic design methods are an appropriate step in this direction. They will gradually replace classical deterministic design procedures. In the classical approach, all design variables are idealized and treated as deterministic (Fig. 2.18):

$$\sigma = F(X_1, X_2, \ldots, X_n)$$

$$S = G(Y_1, Y_2, \ldots, Y_n)$$

and an empirical design factor provides the margin against uninvestigated or unforeseen variability. In the probabilistic approach, statistical information is employed in the design algorithm and maximum al-

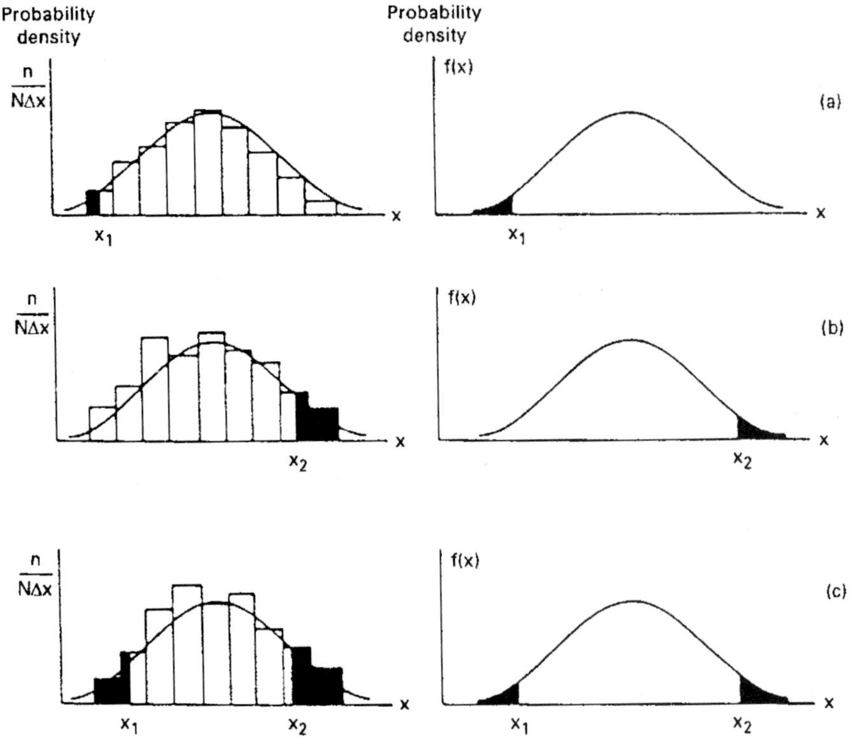

Figure 2.17 Probability statements. (*a*) $p(x > x_1)$; (*b*) $p(x < x_2)$; (*c*) $p(x_1 < x_2)$.

lowable probability of failure provides the proper margin against variability. Such a simple shift in the design viewpoint has far-reaching ramifications.

In the design of mechanical components for strength and/or deflection, attention centers on the characteristics of the variables that influence component behavior. Evidence now supports the contention that engineering variables tend to be random variables (see Table 2.6 and Fig. 2.19). Most design random variables are continuous, and can be modeled by known continuous distributions. Also, design random variables interact functionally. Consider the classical (deterministic) and probabilistic treatment of load functions. If the real conditions were such that each variable $X_1, X_2, \ldots, X_n$ was single-valued and the relationship of these variables to the variable x was described by the functions $F(X_1, X_2, \ldots, X_n)$, then the transformation from $X_1, X_2, \ldots, X_n$ to σ would be a single value (Fig 2.18*a*). If, however, the variables $X_1, X_2, \ldots, X_n$ were each multivalued, then the transformation for $X_1, X_2, \ldots, X_n$ to σ would be a distribution of probable values of σ (see Fig. 2.18*b*). This is similarly true for strength functions. In Table 2.7, statistical algebra for simple functions of random variables is summarized.

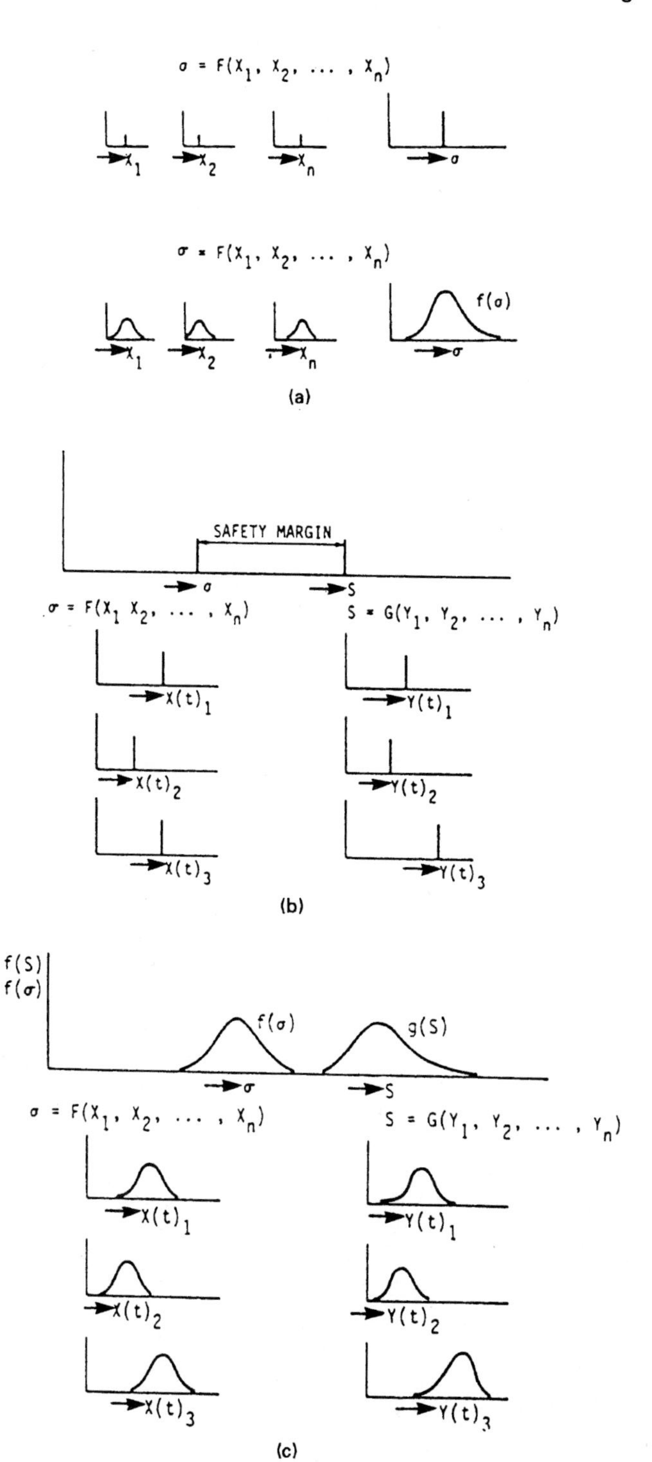

Figure 2.18 (*a*) Transformation to σ is a single value. (*b*) σ is a distribution. (*c*) Probabilistic distributions.

TABLE 2.6 Basic Definitions Utilized in Probability Theory

A **random variable** is a real number corresponding to the outcome of a nondeterministic experiment, e.g., the ultimate strength S_u of a material in a tensile test. S_u will take on different values when the experiment is repeated.

Probability (observational concept): Probability of occurrence of event D:

$$P(D) = \lim_{n \to \infty} \frac{n_o}{n}$$

where n = number of times that the experiment is repeated, and n_o = number of times that event D occurred during n trials.

Mathematical descriptions of a continuous random variable: X = random variable, x = specific value of X.

Cumulative distribution function (CDF):

$$F(x) = P(X \le x)$$

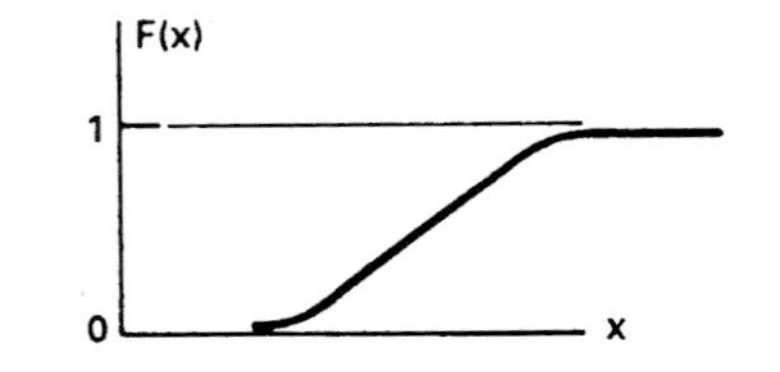

Probability density function (PDF):

$$\text{Area} = P(a \le X \le b)$$

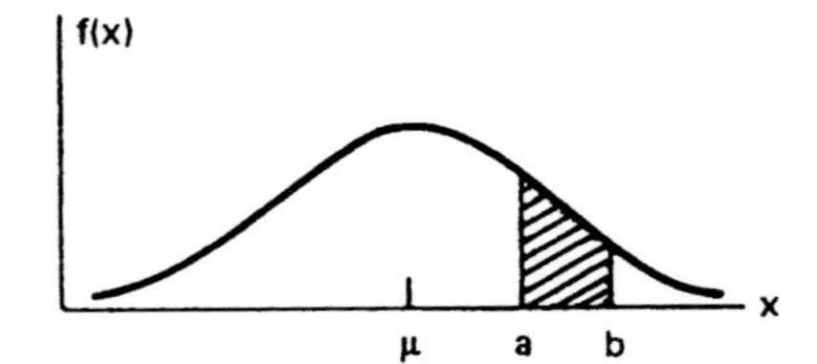

Mean (or average value): A measure of the central tendency of X:

$$\mu_1 = \int_{-\infty}^{\infty} x f(x)\, dx$$

Geometric interpretation: Centroid of area of PDF.

Variance: A measure of the dispersion or variability of X:

$$\sigma_x^2 = \int_{-\infty}^{\infty} (x - \mu_1)^2 f(x)\, dx$$

Geometric interpretation: Second moment of area about centroid of PDF.

Standard deviation:

$$\sigma_x = \sqrt{\sigma_x^2}$$

This is often used by engineers in lieu of variance to describe dispersion because it has the same units as X and μ_x

Coefficient of variation: A dimensionless alternative measure of dispersion:

$$C_x = \frac{\sigma_x}{\mu_x}$$

This is commonly used in design because it has been observed that many design parameters, e.g., strength, have nearly constant coefficients of variation.

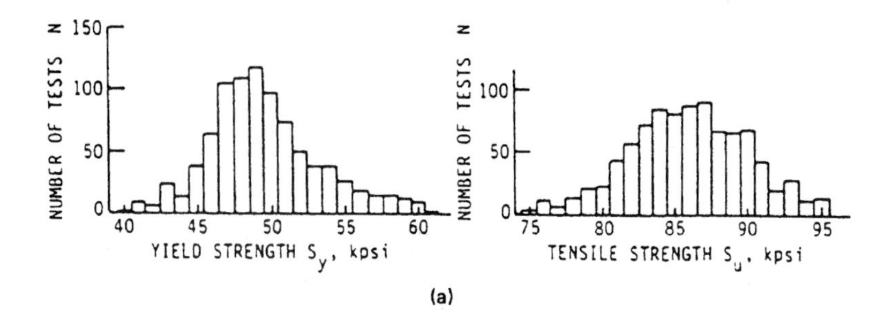

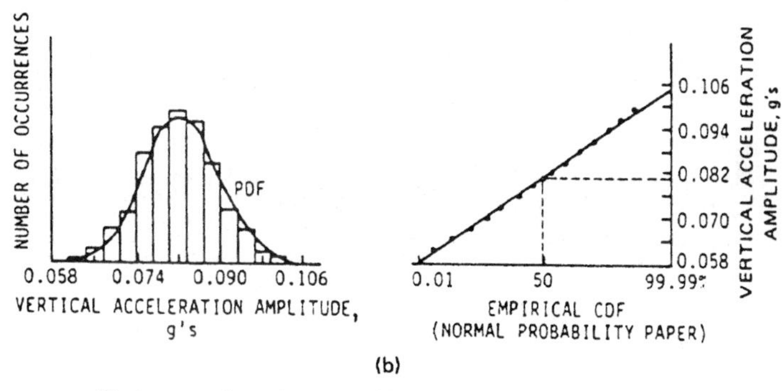

Figure 2.19 Various engineering variables.

TABLE 2.7 How to Compute Mean and Standard Deviations of Functions of Independent Random Variables *X* and *Y* (*a* Is a Constant)*

Function	Mean μ_Z	Standard deviation σ_Z
$Z = a$	a	0
$Z = aX$	$a\mu_X$	$a\sigma_X$
$Z = X \pm a$	$\mu_X \pm a$	σ_X
$Z = X \pm Y$	$\mu_X \pm \mu_Y$	$[\sigma_X^2 + \sigma_Y^2]^{1/2}$
$Z = XY$	$\mu_X\mu_Y$	$[\mu_X^2\sigma_Y^2 + \mu_Y^2\sigma_X^2]^{1/2}$
$Z = X/Y$	μ_X/μ_Y	$[\mu_X^2\sigma_Y^2 + \mu_Y^2\sigma_X^2]^{1/2}/\mu_Y^2$
$Z = X^2$	$\mu_X^2 + \sigma_X^2$	$[4\mu_X^2\sigma_X^2 + 2\sigma_X^2]^{1/2}$
$Z = 1/X$	$1/\mu_X$	σ_X/μ_X^2

*The algebra of random variables shares many elements in common with the algebra of real numbers.

It is more meaningful to state "This unit has a probability of 10^{-4} [a performance-related measure] of failing after 250,000 cycles of operation" than to state "This unit has a factor of safety of 2.3." Reliability is performance-related, a probability statement providing a valid and complete description of mechanical performance (Fig. 2.18*c*). Probability-based information can be used to develop rational policies in pricing, warranties, and spare-parts requirements, for example.

The algebra of random variables shares many elements in common with the algebra of total numbers:

Sum or product is unique.

Associative law applies in addition or multiplication.

Identity element exists for addition or multiplication.

Unitary minus operation leads to subtraction; multiplicative inverse leads to division.

Commutative law applies in addition or multiplication.

In combinations of addition and multiplication, the distributive law holds. In general, if Z is a function of k independent random variables,

$$Z = Z(X_1, X_2, \dots, X_k)$$

The following approximate formulas give reasonable results if the coefficients of variation of each X_i are less than 0.20:

$$\mu = Z(\mu_1, \mu_2, \dots, \mu_k)$$

$$\sigma_Z{}^2 = \sum_{i}^{k} = 1\,(\partial Z / \partial X_i)_\mu^2\,\sigma_{Xi}^2$$

where the partial derivatives are evaluated at the mean values. Note that when Z is a complicated function of the X_i's, a computer program may be used to evaluate X, as in finite-element analysis. This same program may be used to numerically estimate $\partial Z/\partial X_i$.

2.5.3 Concepts for mathematical component design

In the probabilistic approach, design variables are recognized as random variables. Definitions of random variables and other necessary basic concepts from probability theory are given in Table 2.6. A complete probabilistic description of a random variable S is contained in its cumulative distribution function (CDF) or probability density function (PDF), as also given in Table 2.6. The mean value μ_X and standard deviation σ_X are real numbers which together with the PDF of the distribution of X uniquely describe the random variable (for any two-parameter distribution). Often an engineer's first contact with uncertainty in engineering phenomena occurs in studies of experimental data. If the data yield an estimate of the PDF or CDF, a probability statement can be made of the sort depicted in Fig. 2.17. An aspect of statistics is that probability models can be constructed from observed data.

It is necessary in probabilistic design to develop μ_X and σ_X for each design variable X and further to identify the likely form of the PDF in order to compute probabilities. Table 2.8 illustrates the statistical analysis of data. The sample mean value X and sample standard deviation $\leq X^*$ approximate μ_X and σ_X, respectively, and the shape of the histogram approximates that of the PDF. In general, the larger the sample size n, the closer the approximation. The distribution of data can be approximated graphically. The number of occurrences in an interval is plotted in a bar chart (histogram) which indicates the form of the underlying PDF as the number of data n becomes large.

Many design variables suggest normal distributions, characterized by symmetrical bell-shaped PDFs (Table 2.6). Other statistical models have been used to describe design parameters,[1] including the lognormal, gamma, Weibull, and exponential models.

2.5.4 Estimation of variability from tolerances

In engineering literature it is common to find variability in a design variable described in terms of a tolerance $\pm K$ or by a range r of values. If tolerance is defined as $\pm\Delta L$ on a dimension L and $(L - \Delta L, L + \Delta L)$ describes the range of values generated by a particular process, then the standard deviation σL can be estimated as

$$\sigma_L \approx 2\Delta L/6 = \Delta L/3$$

provided n is reasonably large.

Example The yield strength S_Y of a titanium alloy is reported to be in the range of 120 to 160 kpsi. Estimate the mean and standard deviation of S_Y.

solution The mean is estimated to be

$$\bar{S}_Y \approx (120 + 160)/2 = 140 \text{ kpsi}$$

[1]E. B. Haugen, *Probabilistic Approach to Design,* John Wiley & Sons, New York, 1968; E. B. Haugen and P. H. Wirschin, "Probabilistic Design Alternative to Miner's Cumulative Rule," *Proceedings of the Annual Reliability and Maintainability Symposium,* Philadelphia, 1973; H. R. Jaeckel and S. R. Swanson, "Random Lo Spectrum Test to Determine Durability of Structural Components of Automotive Vehicles," Ford Motor Company, 1970

TABLE 2.8 Elementary Statistical Analysis.

The following data were obtained from an experiment to measure the loads in a strut of an aircraft landing gear upon landing impact. Loads are given in kips. A sample of size $n = 40$ replications, under a variety of flight conditions, was taken. Data are denoted as X_i, $i = 1, 2, 3, \ldots, 40$.

Let X be a random variable denoting a landing load. X has a PDF given as $f_x(x)$ and a mean of μ_x and standard deviation of σ_X, all of which are unknown before the data are collected. The purpose of taking the data was to estimate these values.

2.88	2.41	2.42	2.72
1.79	1.97	3.20	2.61
3.02	2.86	2.59	2.68
3.12	3.05	2.27	2.46
2.40	2.80	2.19	3.10
2.79	2.52	2.82	2.07
2.37	2.48	2.63	2.71
1.90	2.10	2.57	2.32
2.74	2.90	3.14	2.81
2.80	3.32	2.70	2.63

The *sample mean* $\bar{X}$ is a measure of the central tendency of the data and is an estimate of μ_x.

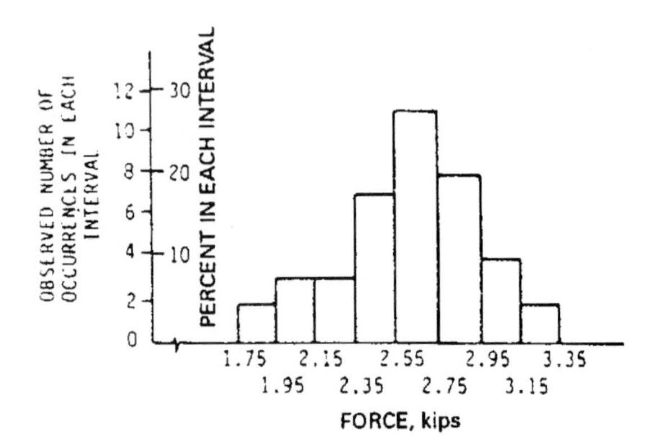

and the standard deviation is estimated as

$$\sigma_{SY} \approx (160-120)/6 = 6.67 \text{ kpsi}$$

The distributional PDF is unknown; however, the normal distribution is often an acceptable model for static material strength.

2.5.5 Functions of random variables in design for affordable automation

For design proposes, from functional expressions of unit stress σ and unit strength S, components are synthesized. Each σ and/or S is often a function of several random design variables, and thus each is a random variable. For example, peak shear stress τ in a helical spring can be written as[2]

$$\tau = (1 + d/2D)\,8FD/\pi d$$

where d = wire diameter
D = coil diameter
F = axial force

If F, D, and d are random variables and have known PDFs, mean values μ, and standard deviations σ, then the PDF, μ, and σ of τ must usually be estimated. A designer can usually make a reasonable estimate of μ and σ for each design variable by the methods of Table 2.7, but doubt remains as to the exact form of the PDF.

Example The shearing stress amplitude of a helical spring is given as

$$\tau = \frac{(1 + d/2D)dGy}{\pi D^2 N} \tag{2.1}$$

where the variables are defined in Table 2.9. (Also given in Table 2.9 are numerical values for the mean and variance of each of the variables.) Estimate the mean and standard deviation of τ.

solution The mean value of τ, denoted as μ_τ, is estimated by applying Eq. (1) using the variables of Table 2.7 to Eq. (2.1):

$$\mu_\tau = \frac{(1 + \mu_d/2\mu_D)\mu_d\mu_G\mu_y}{\pi {\mu_D}^2\mu_N} \tag{2.2}$$

Using mean-value data from Table 2.9 gives

$$\mu_\tau \approx 47{,}550 \text{ psi}$$

The variance of τ, denoted as ${\sigma_\tau}^2$, is estimated by applying Eq. (4) of Table 2.7 to Eq. (2.1):

$$\sigma_\tau^2 = (\partial\tau/\partial d)^2\sigma_d^2 + (\partial\tau/\partial D)^2\sigma_D^2 + (\partial\tau/\partial G)^2\sigma_G^2 + (\partial\tau/\partial N)^2\sigma_N^2 \tag{2.3}$$

where $\partial\tau/\partial d$ etc. are evaluated at the mean values of the variable.

Evaluating σ, using data in Table 2.9, gives a value of 1430 psi. Since the coefficients of variation of all terms were relatively small, it is known that τ will have an approximately normal distribution and can be expressed as

$$\tau \approx N(47{,}500,\ 1430) \text{ psi}$$

[2]E. B. Haugen, *Probabilistic Mechanical Design,* John Wiley & Sons, New York, 1980.

TABLE 2.9 Variability in Helical Coil Spring Parameters

Parameter	Mean μ	Standard deviation σ
Wire diameter d	0.092 in	0.399 (10^{-3}) in
Mean coil diameter D	0.658 in	0.3816 (10^{-2}) in
Shear modulus of elasticity G	11.5 (10^{6}) psi	23.0 (10^{4}) psi
Number of active coils N	14	0.0833
End contraction y	0.80 in	0.016 in

2.5.6 Probabilistic design theory for affordable automation

In classical design practice, a factor of safety $\eta>1$ is specified, and the basic design criterion uses a design factor n so that the final adequacy assessment will reveal a factor of safety equal to or exceeding that specified. The factor of safety is

$$\eta = S_0/\sigma_0$$

where σ_0 = working stress and S_0 = design strength.

Since both load-induced stress and strength are subject to statistical variability, and since it is most meaningful to specify mechanical requirements by a performance-related reliability statement, it is important that the probability of adequacy, denoted R, be specified as a basic design requirement. The following discussion explains the methodology of incorporating statistical information related to stress and strength into the design process so that the design conforms to a reliability requirement.

Distribution of design variables. The size of a part designed by probabilistic procedures depends on the statistical distribution that models the design variables. Table 2.10 shows some of the distributional models commonly used by engineers. The normal distribution plays a central role in probabilistic mechanical designs (see Table 2.11). The lognormal distribution also is of fundamental importance because of the central limit theorem for product theories.

Probabilistic design algorithms. A random variable σ denotes stress and a random variable S denotes strength. Now $(\mu_\sigma, \sigma_\sigma)$ and (μ_S, σ_S) are the mean and standard deviation of stress and strength, respectively. It is customary to use unit stress and strength expressed in pounds per square inch or megapascals. For a given component, it must be recognized that there is a finite probability that $\sigma>S$, defined as the occurrence of failure. Figure 2.20 illustrates the PDFs of σ and S (densities denoted as f_σ and f_S, respectively). Figure 2.20 also

TABLE 2.10 Basic Probabilistic Design Equations

Note that required geometric variables are contained in μ's and σ's.

If σ and S are *normally* distributed:

$$z = -\frac{\mu_S - \mu_\sigma}{(\sigma_S^2 + \sigma_\sigma^2)^{1/2}} \tag{1}$$

If σ and S are *lognormally* distributed:

$$\left(\frac{\mu_\sigma}{\mu_S}\right)^2 = \left(\frac{\mu_S^2 + \sigma_S^2}{\mu_\sigma^2 + \sigma_\sigma^2}\right)^{1/2} exp\left[z\sqrt{\left(\frac{\sigma_S}{\mu_S}\right)^2 + \left(\frac{\sigma_\sigma}{\mu_\sigma}\right)^2}\right] \tag{2}$$

P_f	$z = \phi^{-1}(1-P_f)$	P_f	$z = \phi^{-1}(1-P_f)$
10^{-2}	−3.326	10^{-7}	−5.199
10^{-3}	−3.090	10^{-8}	−5.610
10^{-4}	−3.719	10^{-9}	−5.997
10^{-5}	−4.265	10^{-10}	−6.361
10^{-6}	−4.753	10^{-11}	−6.700

where ϕ is the standard normal CDF.
Use this as an upper bound when you do not want to assume a specific distribution for σ and S:

For $\sigma_\sigma \neq \sigma_S$:

$$P_f(\sigma_\sigma - \sigma_S) = \sigma_\sigma\, exp\left[-\frac{(\mu_S + \sigma_\sigma - \mu_\sigma + \sigma_\sigma)}{\sigma_\sigma}\right] - \sigma_\sigma\, exp\left[-\frac{(\mu_S + \sigma_S - \mu_\sigma + \sigma_\sigma)}{\sigma_S}\right] \tag{3}$$

For $\sigma_\sigma = \sigma_S$:

$$P_f\sigma_\sigma = (\mu_S + 2\sigma_S - \mu_\sigma - \sigma_\sigma)\, exp\left[-\frac{(\mu_S + \sigma_S - \mu_\sigma + \sigma_\sigma)}{\sigma_\sigma}\right]$$

NOTE: If σ or S is shown to be deterministic, then $\sigma = \mu_\sigma$ and $\sigma = 0$, or $S = \mu_S$ and $\sigma_S = 0$ in Eqs. (2), (3), and (4).

provides definitions of probability of failure P_f and reliability R. Either P_f or R can be used as an index of structural performance, since they are equivalent expressions.

Basic probabilistic design formulas are presented in Table 2.10 for two possible distributions of σs and S. Haugen[3] pointed out that often σ and/or S are products of several random design variables, and therefore, one or both may approximate lognormal distributions. It

[3]E. B. Haugen, *Probabilistic Mechanical Design,* John Wiley & Sons, New York, 1980.

TABLE 2.11 Normal Distribution Cumulative Density Function

$$F(x) = \int_{-\infty}^{\infty} \frac{1}{\sqrt{2\pi}} \exp\left(-\frac{t^2}{2}\right) dt$$

x	.00	.01	.02	.03	.04	.05	.06	.07	.08	.09
.0	.5000	.5040	.5080	.5120	.5160	.5199	.5239	.5279	.5319	.5359
.1	.5398	.5438	.5478	.5517	.5557	.5596	.5636	.5675	.5714	.5753
.2	.5793	.5832	.5871	.5910	.5948	.5987	.6026	.6064	.6103	.6141
.3	.6179	.6217	.6255	.6293	.6331	.6368	.6406	.6443	.6480	.6517
.4	.6554	.6591	.6628	.6664	.6700	.6736	.6772	.6808	.6844	.6879
.5	.6915	.6950	.6985	.7019	.7054	.7088	.7123	.7157	.7190	.7221
.6	.7257	.7291	.7324	.7357	.7389	.7422	.7454	.7486	.7517	.7549
.7	.7580	.7611	.7642	.7673	.7704	.7734	.7764	.7794	.7823	.7852
.8	.7881	.7910	.7939	.7967	.7995	.8023	.8051	.8078	.8106	.8133
.9	.8159	.8186	.8212	.8238	.8264	.8289	.8315	.8340	.8365	.8389
1.0	.8413	.8438	.8461	.8485	.8508	.8531	.8554	.8577	.8599	.8621
1.1	.8643	.8665	.8686	.8708	.8729	.8749	.8770	.8790	.8810	.8830
1.2	.8849	.8869	.8888	.8907	.8925	.8944	.8962	.8980	.8997	.9015
1.3	.9032	.9049	.9066	.9082	.9099	.9115	.9131	.9147	.9162	.9177
1.4	.9192	.9207	.9222	.9236	.9251	.9265	.9279	.9202	.9306	.9319
1.5	.9332	.9345	.9357	.9370	.9382	.9394	.9406	.9418	.9429	.9441
1.6	.9452	.9463	.9474	.9484	.9495	.9505	.9515	.9525	.9535	.9545
1.7	.9554	.9564	.9573	.9582	.9591	.9599	.9608	.9616	.9625	.9633
1.8	.9641	.9649	.9656	.9664	.9671	.9678	.9686	.9693	.9699	.9706
1.9	.9713	.9719	.9726	.9732	.9738	.9744	.9750	.9756	.9761	.9767
2.0	.9772	.9778	.9783	.9788	.9793	.9798	.9803	.9808	.9812	.9817
2.1	.9821	.9826	.9830	.9834	.9838	.9842	.9846	.9850	.9854	.9857
2.2	.9861	.9864	.9868	.9871	.9875	.9878	.9881	.9884	.9887	.9890
2.3	.9893	.9896	.9898	.9901	.9904	.9906	.9909	.9911	.9913	.9916
2.4	.9918	.9920	.9922	.9925	.9927	.9929	.9931	.9932	.9934	.9936
2.5	.9938	.9940	.9941	.9943	.9945	.9946	.9948	.9949	.9951	.9952
2.6	.9953	.9955	.9956	.9957	.9959	.9960	.0061	.9962	.9963	.9964
2.7	.9965	.9966	.9967	.9968	.9969	.9970	.9971	.9972	.9973	.9974
2.8	.9974	.9975	.9976	.9977	.9977	.9978	.9979	.9979	.9980	.9981
2.9	.9981	.9982	.9982	.9983	.9984	.9984	.9985	.9985	.9986	.9986
3.0	.9987	.9987	.9987	.9988	.9988	.9989	.9989	.9989	.9990	.9990
3.1	.9990	.9991	.9991	.9991	.9992	.9992	.9992	.9992	.9993	.9993
3.2	.9993	.9993	.9994	.9994	.9994	.9994	.9994	.9995	.9995	.9995
3.3	.9995	.9995	.9995	.9996	.9996	.9996	.9996	.9996	.9996	.9997
3.4	.9997	.9997	.9997	.9997	.9997	.9997	.9997	.9997	.9997	.9998

x	1.282	1.645	1.960	2.326	2.576	3.090	3.291	3.891	4.417
$F(x$	.90	.95	.975	.99	.995	.999	.9995	.99995	.999995
$2[1-F(x)]$	.20	.10	.05	.02	.01	.002	.001	.0001	.00001

SOURCE: By permission from A. M. Mood, *Introduction to the Theory of Statistics* McGraw-Hill, 1950.

has also been shown that the normal model for both σ and S usually produces conservative designs (relative to the lognormal and Weibull σ and S). Thus it is generally recommended (unless there is compelling contrary evidence) that the designer use Eq. (1) of Table 2.10 if conservative design is acceptable.

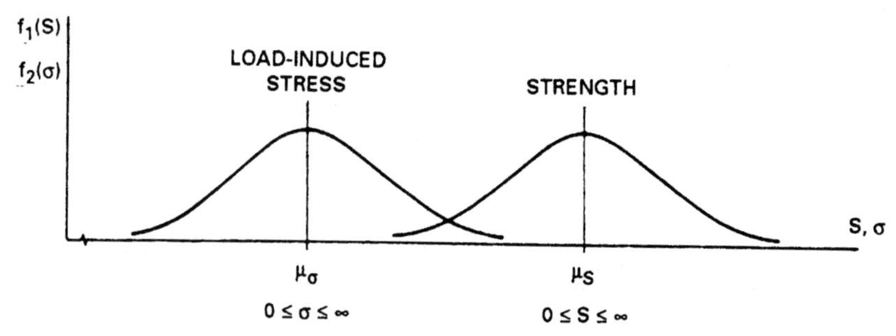

Overlap in density functions suggests that it is possible for $\sigma > S$.

Probability of failure: $P_f = P(\sigma \geq S)$

Reliability: $R = 1 - P(\sigma \geq S) = 1 - P_f$

Figure 2.20 Probability density functions of stress (f_σ) and strength (f_S).

The task of design synthesis is to size the part. The geometric dimensions calculated will be function of μ's and σ's in Eq. (1), (2), or (3) of Table 2.10. Such solutions can now be obtained very efficiently using, for example, a small desktop programmable calculator.

An acceptable P_f may depend in part on the consequences of failure, a procedure's warranty policy, or other performance requirements. However, with regard to public safety, it has been suggested that the average U.S. citizen is willing to accept fatality risks of roughly 10^{-6} per hour in commercial flying and operation of automobiles, i.e., risk roughly comparable to that of death due to common diseases in the United States. In theory, a design P_f can be established, as illustrated in Fig. 2.20. A practical problem is that of establishing anticipated profit and an economic equivalent of the consequences of failure as a function of P_f of a given system.

Example Consider a system of several components, for example, a gear, a shaft, and two bearings. If P_{fi} is the probability of failure of the ith component, then it can be shown that the probability of failure of the system is

$$P_f = \sum_{i}^{k} 1\, P_{f,i} \tag{2.4}$$

where

$$R = 1 - P_f$$

and k = number of components. An acceptable design requires that the right-hand side of Eq. (2.4) sum to less than p_f. In general, it may not be desirable to apportion all P_{fi} the same. The consequences of failure among the different components may be very different. It may cost more to replace a shaft than bearing, justifying a comparatively higher R value for the shaft.

Example When loading is repetitive, P_f can be computed as follows. After j independent applications of σ, the probability of failure of a component is

$$p_f^{(j)} = 1 - (1 - p_f)^j \tag{2.5}$$

where p_f refers to a single application of σ. Equation (2.5) is valid if fatigue failure is not assumed.

Example Calculate reliability R and probability of failure P_f, given the following normal independent random variables describing strength S and stress σ applied to a component:

$$S = (\mu_S, \sigma_S) = (27{,}000,\ 3200) \text{ psi}$$

$$\sigma = (\mu_\sigma, \sigma_\sigma) = (18{,}000,\ 1500) \text{ psi}$$

solution If the stress margin is $m = S - \sigma$, then, from Table 2.10,

$$\mu_m = 27{,}000 - 18{,}400 = 8{,}600 \text{ psi}$$

$$\sigma_m = \sqrt{3200^2 + 1500^2} \text{ psi}$$

The PDFs of S, σ, and $m = S - \sigma$ are shown in Fig. 2.21. The standard normalized random variable z_0 corresponding to $m = 0$ is

$$z_0 = \left. \frac{(m - \mu_m)}{\sigma_m} \right|_{m=0} = -(\mu_m/\sigma_m)$$

Integration from z_0 to $z = \infty$ yields reliability R.

The lower limit is

$$z_0 = -8600/3534 = -2.43$$

Utilizing standard normal area tables (Fig. 2.22),

$$R = \int_{-2.43}^{\infty} \{[\exp(z^2/2)]/\sqrt{2\pi}\}\, dz = 0.9925$$

$$P_f = 1 - R = 0.0075$$

2.5.7 The statistical nature of engineering design variables

Engineers have been slow to adopt probabilistic methods because of the scarcity of statistical data describing design parameters. Recently, however, efforts have been made to unify available statistical design information and make it available.

Examples of available information pertaining to variability in design variables follow. They provide the designer with an idea of the magnitude of the variability to expect in certain variables. However, one must also look for simplifying assumptions.

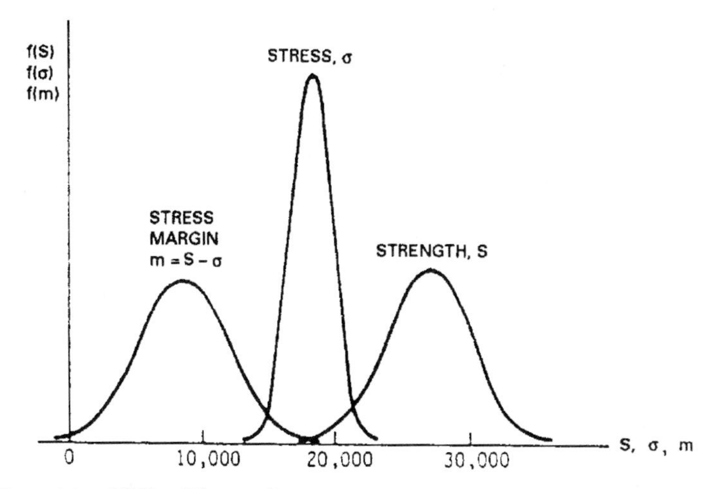

Figure 2.21 PDFs of S, σ, and m.

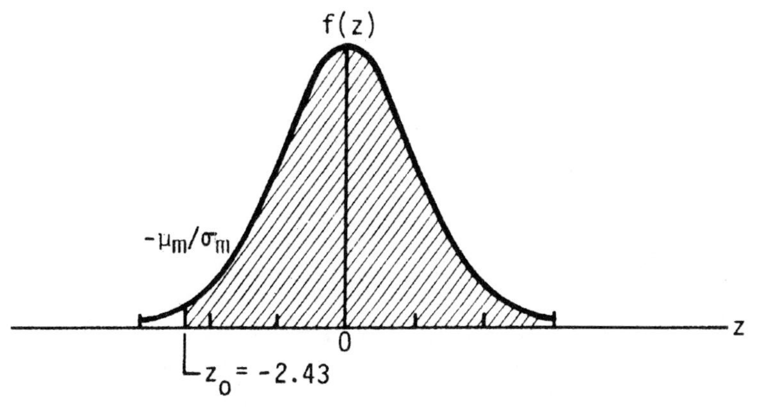

Figure 2.22 Standard normal area tables.

Variability of material strength. Extensive data have been accumulated on material strength (yield, ultimate, fatigue) and modulus of elasticity. Typical values of mean μ and standard deviation σ of static strength for widely used alloys are given in Haugen.[4]

Figure 2.23 shows the large variability in cycles to failure in a constant-amplitude, fully reversed fatigue test of 4340 steel wire. The degree of scatter in the cycles to failure data is typical of common alloys.

In design practice, the required cycle life n is often specified. It is necessary to estimate the distribution of fatigue strength. Figure 2.24

[4]E. B. Haugen, *Probabilistic Mechanical Design,* Wiley, New York, 1980.

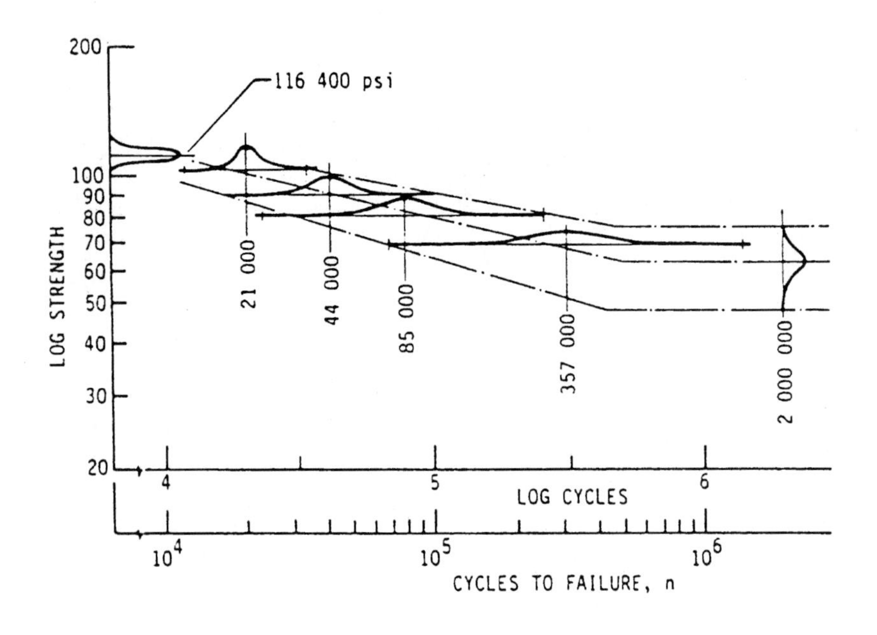

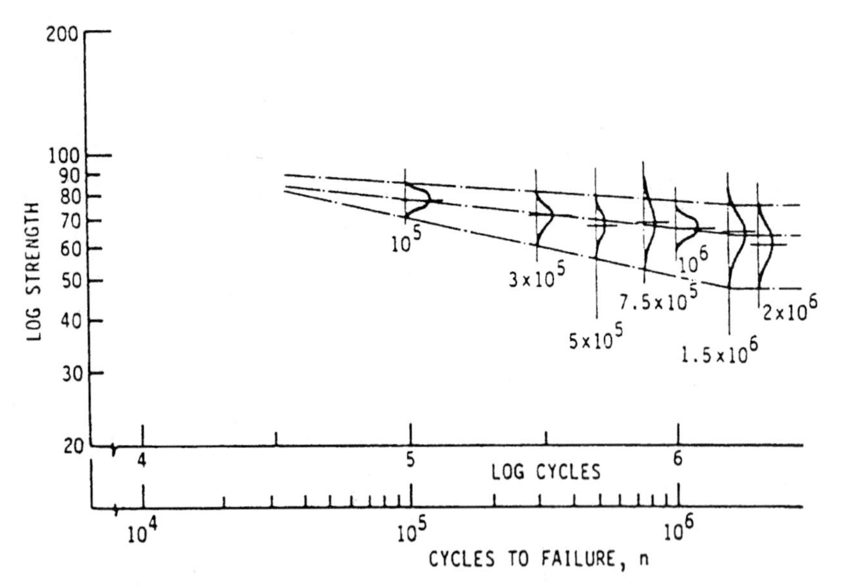

Figure 2.23 Variability in cycles to failure for 4340 steel wire.

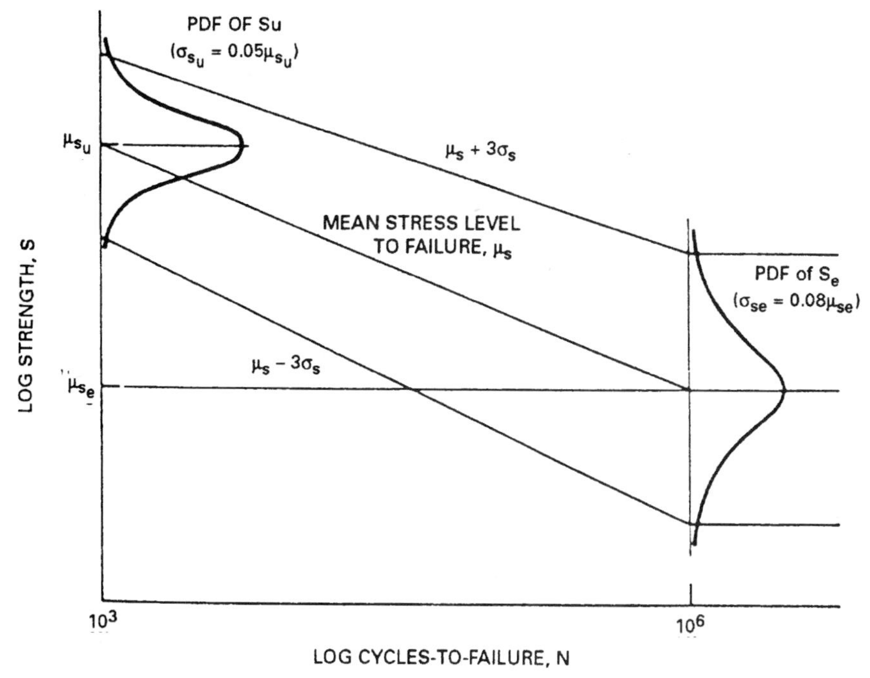

Figure 2.24 *S-n* envelope for a typical steel.

shows the usual statistical dynamic *S-n* envelope of *S* at various values of *n* for a typical steel. Note the coefficients of variation σ/μ of ultimate strength S_u and endurance limit S_e (0.05 and 0.08, respectively). This description of variability is an estimate and should be used only in the absence of specific data. A distributional Goodman diagram is shown in Fig. 2.25. A summary of typical variability in material properties to be used in the absence of data is given in Tables 2.12 and 2.13.

Material stiffness. Little published data exist describing the variability of the modulus of elasticity *E* for metallic materials. Table 2.14 suggests that the variability may not be insignificant.

Geometry. Design engineers usually describe allowable variability in production operations in terms of tolerances. Table 2.15 summarizes typical tolerance capabilities for a variety of machining operations. Table 2.16 lists typical metal sheet thickness tolerances. From tolerances, standard deviations are estimated by setting 3σ equal to tolerance *T*.

Loading. Statistical forces and moments on a component or system are usually treated on an individual basis. A wealth of statistical in-

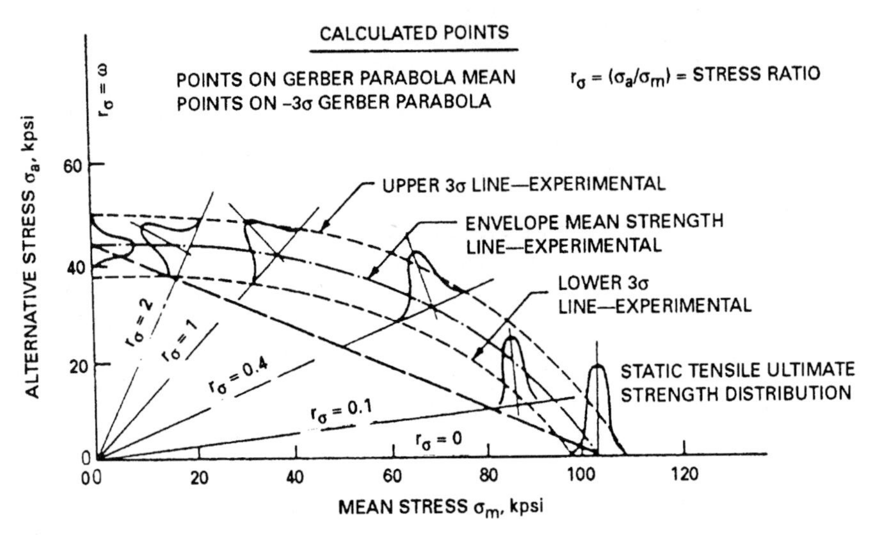

Figure 2.25 Goodman diagram.

TABLE 2.12 Variability in Materials Properties*

Parameter	Coefficient of variation
Tensile ultimate strength of metallic materials	0.05
Tensile yield strength of metallic materials	0.07
Endurance limit for steel	0.08
Brinell hardness of steel	0.05
Fracture toughness of metallic materials	0.07

*Recommended for design purposes in absence of data.

formation on various environmental phenomena, such as wind gusts, ocean wave height, and road surface roughness, is available. Often the engineer must translate statistical environmental data into forces on systems.

Mechanical components. A considerable amount of statistical data are becoming available on the statistical properties of mechanical components. Table 2.17 summarizes variability in design variables relating to helical spring performance. Table 2.18 provides data on the behavior of bolted and welded connections. It is recommended, for design purposes, that the coefficient of variation of strength of welded joints be taken as 0.15 for steel and aluminum. Wear is subject to considerable statistical variability, and rolling bearing life can be Weibull-distributed, often with a coefficient of variation approximating 0.80. Note that the tolerances given in Table 2.15 include variability arising from operator performance.

TABLE 2.13 Examples of Variability in Materials Strength

Material	Condition	Tensile ultimate strength S_u, ksi		Tensile yield strength S_y, ksi	
		μ	σ	μ	σ
2024-T3 aluminum	(Bare sheet and plate) < 0.250 in	73	3.7	54	2.7
6061-T6 aluminum	½-in sheet	46	1.9	42	2.9
7075-T6 aluminum	¼-in plate, bare	85	3.4	76	3.8
A231 B-0 magnesium	¼-in plate, longitudinal	41	3.8	24	3.8
Ti-6A1-4V titanium alloy	Sheet and bar annealed	135.	6.8	131	7.5
304 stainless steel	Round bars, annealed (0.050–4.6 in)	85.	4.1	38	3.8
ASTM-A7 steel	½-in plate	66	3.3	40	4.0
A1S1 1018	Cold drawn round bar (0.75–1.25 in)	88.	5.7	78	5.9

SOURCE: From E. B. Haugen, *Probabilistic Design*, John Wiley & Son, New York, 1980.

TABLE 2.14 Average Value of the Mean and Coefficient of Variation of Modulus of Elasticity for Metallic Materials

	Mean, Mpsi	Coefficient of variation C
Steel	30.0	~0.03
Aluminum	10.0	~0.03
Titanium	14.7	~0.09
Nodular iron cast	23.1	~0.04

SOURCE: From E. B. Haugen, *Probabilistic Design*, John Wiley & Son, New York, 1980.

Elements in nonstatistical uncertainty. In component design for affordable automation the purpose of the factor of safety in classical design practice has been to account for factors which were either nonstatistical, i.e., difficult if not impractical to observe, or for which data were lacking. The following list provides examples of uncertainty factors:

1. Uncertainty in the definition of system loads. Data in the past were often lacking.
2. If relatively complete statistical information on environmental phenomena was available, there was perhaps uncertainty in translating these data into forces. For example, although accurate wind

TABLE 2.15 Variability of Tolerances

Process	Tolerance* ±, in	Standard deviation σ, mils
Flame cutting	0.060	20.00
Sawing	0.020	6.67
Shaping	0.010	3.33
Broaching	0.005	1.67
Milling	0.005	1.67
Turning	0.005	1.67
Drilling	0.010	3.33
Reaming	0.002	.67
Hobbing	0.005	1.67
Grinding	0.001	0.33
Lapping	0.0002	.07
Stamping	0.010	3.33
Drawing	0.010	3.33

*It is assumed that tolerance equals 3σ.
SOURCE: From E. B. Haugen, *Probabilistic Design of Helical Springs*, Design Engineering Division of American Society of Mechanical Engineers, 1994.

gust velocity data were available, the designer was unable to make precise estimates regarding wind forces on a particular system.

3. Given a description of forces on a complicated mechanical or structural system, uncertainty in component member loading was introduced as a result of the static and/or dynamic analysis procedures. Contemporary static and dynamic digital computer routines generally require assumptions ranging from homogeneous material properties and linear elastic behavior to small-amplitude responses. In the case of a bar member, loading may be most seriously affected by the degree of end fixity assumed; this may involve considerable uncertainty.

4. Given the forces on a component, subsequent stress analysis may require simplifying assumptions such that the designer lacks confidence in the accuracy of the estimated peak stresses.

5. Processing operations, such as cold working, grinding, and heat treating, and assembly operations, such as fastening, welding, and shrink fitting, lead to uncertainties.

6. Material strength may be influenced by time, corrosion, and uncertain extreme thermal environments, sometimes to an unknown degree.

7. Questions of quality of workmanship may exist. Manufacturability influences the degree of quality desired and the ability to maintain the required tolerances.

TABLE 2.16 Steel Sheet, Carbon, Cold-Rolled—Thickness Tolerances (Plus or Minus Inches)

Width	Thickness, in														
	0.2299 0.1875	0.1874 0.1800	0.1799 0.1420	0. 1419 0.0972	0.0971 0.0822	0.00821 0.0710	0.0709 0 .0568	0.0567 0.0509	0.0508 0.0389	0.0388 0.0344	0.0343 0.0314	0.0313 0.0255	0.0254 0.0195	0.0194 0 .0142	0.0141 and less
< 31⁄2 incl.												0.003	0.002	0.002	
> 31⁄2–6 incl.											0.004	0.003	0.003	0.002	0.002
> 6–12							0.005	0.005	0.004	0.004	0.003	0.003	0.002	0.002	
> 12–15	0.008	0.007	0.007	0.007	0.006	0.006	0.006	0.005	0.005	0.004	0.004	0.003	0.003	0.002	
> 15–20	0.008	0.008	0.008	0.008	0.007	0.007	0.006	0.006	0.005	0.004	0.004	0.003	0.003	0. 002	
> 20–32	0.009	0.009	0.009	0.008	0.007	0.007	0.006	0.006	0.005	0.004	0.004	0.003	0.003	0. 002	
> 32–40	0.009	0.009	0.009	0.009	0.008	0.007	0.006	0.006	0.005	0.004	0.004	0.003	0.003	0. 002	0.002
> 40–48	0.010	0.010	0.010	0.010	0.008	0.007	0.006	0.006	0.005	0.004	0.004	0.003	0.003	0. 002	0.002
> 48–60			0.010	0.010	0.008	0.007	0.007	0.006	0.005	0.004	0.004				
> 60–70			0.011	0.011	0.009	0.008	0.007	0.007	0.006	0.005	0.005				
> 70–80			0.012	0.012	0.009	0.008									
> 80–90			0.012	0.012	0.010										
> 90			0.012	0.012											

TABLE 2.17 Variability in Design Variables for Helical Springs

Variable	Standard deviation	
Wire diameter d, in	$\sigma_d = 1.58 \times 10^{-3}\sqrt{d}$	(hard-drawn carbon steel)
	$\sigma_d = 1.31 \times 10^{-3}\sqrt{d}$	(music wire)
Spring diameter D, in	$\sigma_D = (48 \times 10^{-4})D$	for $d < 0.1$ in
	$\sigma_D = \dfrac{1.90 \times 10^{-3}D}{\sqrt{d}}$	for $d \geq 0.1$ in
Number of coils N	$\sigma_N = \dfrac{1}{12}$	
Shearing modulus of elasticity G	$C_G = 0.02$	

TABLE 2.18 Variability of Bolted and Welded Connections

	Mean μ	Standard deviation σ
$\dfrac{\text{Fillet weld shear strength}}{\text{Electrode tensile strength}} = (\mu, \sigma)$	0.769	0.097
Shear Strength of Longitudinal Fillet Welds		
E 60	56.4 kpsi	6.2 kpsi
E 70	63.5 kpsi	5.7 kpsi
Slip Coefficient (Bolted Joint)		
$k_S = \dfrac{\text{slip load}}{(\text{clamping force})(\text{number of surfaces})} = (\mu, \sigma)$		
No special treatment	0.336	0.070
$\dfrac{\text{Shear strength}}{\text{Tensile strength}} = (\mu, \sigma)$		
(A325 and A490 bolts)	0.625	0.033

Finally, assumptions are often required in design procedures. In probabilistic design, the best a priori statistical information is utilized and the design for affordable automation is developed from it.

Chapter

3

Part Feeding for Affordable Automation

There are distinctive principles applicable to virtually all manufacturing operations, which, if followed, will assist designers in specifying components and systems for manufacture at affordable costs.

3.1 Basic Principles

These principles may be summarized as follows:

3.1.1 Simplicity

Other factors being equal, the product with the fewest parts, the least intricate shape, the fewest precision adjustments, and the shortest manufacturing sequence will be the least costly to produce. Additionally, it will usually be the most reliable and the easiest to maintain.

3.1.2 Standard materials and components

Use of widely available materials and off-the shelf parts brings the benefits of affordable automation even in low-unit-quantity products. Use of such standard components also simplifies inventory management and purchasing, avoids extra tooling and equipment investments, and speeds the manufacturing cycles.

3.1.3 Standardized design of products

When several similar products are to be produced, it is advisable to specify the same materials, parts, and subassemblies for each as much as possible. This procedure will provide economies of scale for component production, simplifies process control and operator training, and reduces the investment required for tooling and equipment.

3.1.4 Liberal tolerances

Although the extra cost of too tight tolerances has been well-documented, it is often not appreciated by product designers. The higher costs of tight tolerances stem from factors such as

1. Extra operations such as grinding, honing, or lapping after primary machining operations
2. Higher tooling costs because of the greater precision needed initially when the tools are made and the more frequent and more careful maintenance needed as they wear
3. Longer operating cycles
4. Higher scrap and rework costs
5. Need for more skilled and highly trained workers
6. Higher materials costs
7. More sizable investments for precision equipment

Figure 3.1 illustrates how manufacturing cost increases when close tolerances are specified. Table 3.1 shows the extra cost of producing fine surface finishes. Figure 3.2 illustrates the range of surface finishes obtainable with a number of machining processes. It shows how

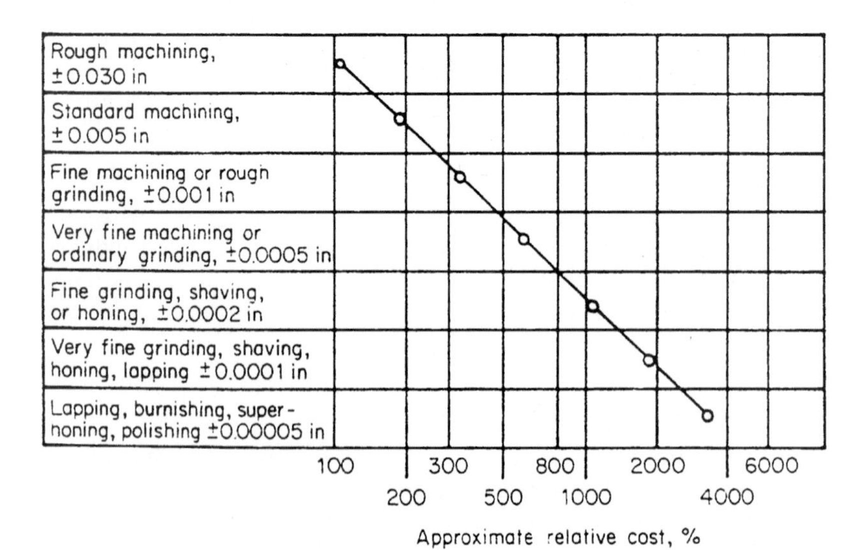

Figure 3.1 Increase of manufacturing cost with decreasing tolerance.

TABLE 3.1 Cost of Producing Surface Finishes

Surface symbol designation	Surface roughness, μin	Approximate relative cost, %
Case, rough-machined	250	100
Standard machining	125	200
Fine machining, rough-ground	63	440
Very fine machining, ordinary grinding	32	720
Fine grinding, shaving, and honing	16	1400
Very fine grinding, shaving, honing, and lapping	8	2400
Lapping, burnishing, superhoning, and polishing	2	4500

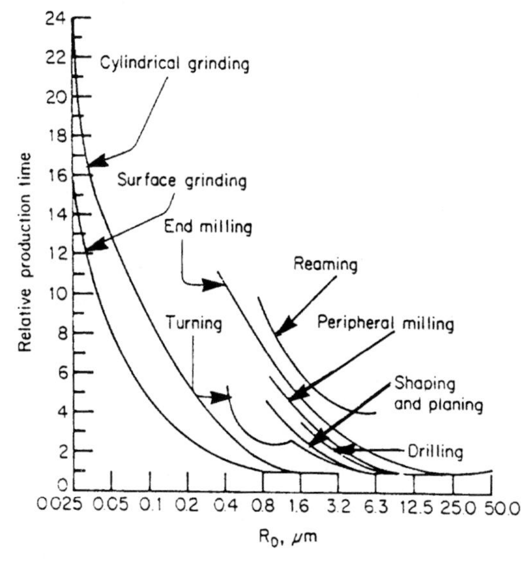

Figure 3.2 Production time versus surface roughness R_D for various machining methods.

process time for each method can increase substantially if a particularly smooth surface finish must be provided.

3.1.5 Processible materials

Use the most processible materials available as long as their functional characteristics and cost are suitable. There are often significant differences in processibility (cycle time, optimum cutting speed, flowability, etc.) between conventional material grades and those developed for easy processibility. However, in the long run the most economical material is the one with lowest combined cost of materials, processing, and warranty and service charges over the designed life of the product.

3.1.6 Collaboration with manufacturing personnel

The most producible designs are provided when the designer and manufacturing personnel, particularly manufacturing engineers, work closely together from the outset.

3.1.7 Avoid secondary operations

Consider secondary operations in design in order to eliminate or simplify them whenever possible. Such operations as deburring, inspection, plating and painting, heat treating, material handling, and others may prove to be as expensive as the primary manufacturing operation and should be considered as the design is developed. For example, firm, nonambiguous gauging points should be provided; shapes that require special protective trays for handling should be avoided.

3.1.8 Design for the expected level of production

The design should be suitable for a production method that is economical for the quantity forecast. For example, a product should not be designed for thin-walled die casting if anticipated production quantities are so low that the cost of the die cannot be amortized. Conversely, it would also be incorrect to specify sand-mold aluminum casting for a mass-produced part because this would fail to take advantage of the labor and materials savings possible with die casting.

3.1.9 Utilize special process characteristics

Wise designers will learn the special capabilities of the manufacturing processes that are applicable to their products and take advantage of them. For example, they will know that injection-molded plastics parts can have color and surface texture incorporated in them as they come from the mold, that some plastics can provide "living hinges," and that powder-metal parts normally have a porous nature that allows lubrication retention and obviates the need for separate bushing inserts. Utilizing these special capabilities can eliminate many operations and costly separate components.

3.1.10 Avoid process restrictiveness

On parts drawings, specify only the final characteristics needed, not the process to be used. Allow manufacturing engineers as much latitude as possible in choosing a process which produces the needed dimensions, surface finish, or other characteristics required.

3.2 Basic Principles of Fastener Design for Affordable Automation

The general acceptance and wide-ranging use of fasteners in high volume production is often taken for granted. In reality, though, the modern fastening system represents a complex and often critical design element that has been under continuous development and refinement. With more emphasis on reliability and long service life, the engineering community is recognizing that the so-called common fastener is usually not that common after all.

3.2.1 Advantages of threaded fasteners

As a prime joining system, standard threaded fasteners offer distinct advantages. They are commercially available in a wide range of styles, materials, and sizes, often permitting fastening where other systems are not effective or efficient. They are capable of joining the same or dissimilar materials in uniform as well as in unusual joint configurations. They can be easily installed in factories or in the field with standard tools, manual or automated, with a maximum of safety. And when disassembly or repair is required, threaded fasteners provide the same ease of removal and replacement.

Basically, the standard threaded fastening system consists of an external (male) threaded element such as a bolt or a screw, and a mating internal (female) threaded element such as a nut, insert, or tapped hole. Interestingly, the two components of the fastening system, which are so dependent on each other, are fabricated by different production methods, and often by companies specializing in the manufacture of only one of the major components. The ability to assemble and use male and female threaded elements, irrespective of the manufacturing source, emphasizes an additional advantage of fasteners as a joining technique: not only is the unit cost of the fastener itself low, but the cost of installing the fastener assembly is also low.

Threaded fasteners may be used for one function or a combination of various functions. Primarily, fasteners are intended as structural, load-carrying elements. However, they may also be selected for their nonmagnetic properties, for resistance to corrosive or other environmental exposure conditions, or even for decorative appearance. For these reasons, standard threaded fasteners are available in various combinations of materials, strength levels, and finishes.

3.2.2 Screw fasteners

Fundamental to threaded fasteners is the screw thread, which has often been described as an inclined plane wrapped around a cylinder. While there are early records of fasteners and devices employing a screw thread principle, it was characteristic that such "nuts and bolts" were generally not interchangeable and usually required specific matching or fitting. An engineering effort to establish working standards for screw threads did not fully materialize until the middle of the nineteenth century. In England, Sir Joseph Whitworth proposed a screw thread system featuring a constant thread angle of 55° which became known as the Whitworth Thread. About that same time in the United States, William Sellers developed a screw thread system based on a standard thread angle of 60° which formed the basis for the American National Thread. The eventual adoption of each system was instrumental in contributing to the success of the industrial revolution, and each system remained dominant as a national standard for about a century.

In 1948, culminating many years of intensive effort, representatives of Great Britain, Canada, and the United States signed a declaration of accord establishing the United Screw Thread system. The standards defined a thread form based on a 60° angle, tolerances and size limits, and diameter-pitch relationships for use by the three countries and most English (inch) measurement system nations. The elements of the United Screw Thread design form are shown in Fig. 3.3. In the United States, the basic documents outlining complete information on screw threads are the National Bureau of Standards Handbook H-28, *Screw-Thread Standard for Federal Services,* and American National Standard ANSI-B1.1-1974, *Unified Inch Screw Threads.*

Considerable effort is under way both in this country and through the International Standards Organization (ISO) to explore and possibly develop a uniform screw thread system and related product standards which will be accepted on a truly international basis. Should such a metric standard evolve, it is anticipated that general conversion will be phased into both production and design after adequate notice and complete engineering data are made available to the engineering community.

3.2.3 Functions of prime fasteners

Prime functions of a threaded fastener system (bolt and nut assembly) are to join and hold structures together, and to carry the applied structural loads under the specific design environment. The diversity of equipment, products, and structures using fasteners points up the fact that the design environment is not always uniform, and can even vary for different applications on the same structure. The potential design environments for a fastener system are quite broad and may involve static loading such as tension, shear, or bending, or dynamic loading such as fatigue, vibration, or shock. In elevated-temperature exposure,

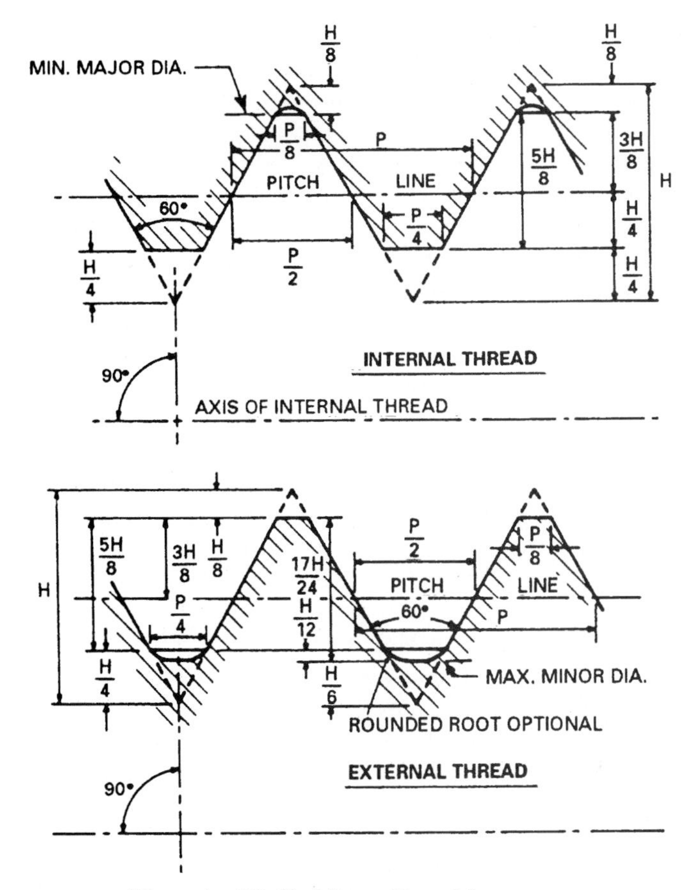

Figure 3.3 Elements of Unified Screw Thread form.

creep, relaxation, or stress rupture-strength properties may be of concern. Where the fastener system may be subjected to a corrosive environment, not only is corrosion protection a necessity, but also the stress-corrosion properties of the fastener should be taken into account to avoid premature service failure. Occasionally, a criterion other than strength may predominate, as when fasteners may be required for their nonmagnetic properties. An initial design analysis of the application should establish the most important criteria for long-life joint integrity.

3.2.4 Distinction between product and design specifications

Because of their very nature, threaded fasteners have readily lent themselves to standardization, even though design environments vary. The standards and specifications approach has permitted logical groupings of commonly used fasteners by strength level, material, finish, and function to meet the objectives imposed by actual service. Additionally, the standards have contributed to the general availability and interchangeability at competitive cost. While much deserved attention has been devoted to product specifications and standards, it is important to recognize that there is a clear distinction between the product specification for a fastener system and the design or application specification for the same fastener systems.

The *product specification* is essentially a manufacturing control document and normally identifies the specific material, finish, method of manufacture (where critical), heat treatment or strength level, and other related requirements for proper fabrication. A key feature of most product specifications is the quality assurance provision outlining test sampling to confirm that the dimensional and performance objectives for the finished fasteners have in fact been met. In essence, then, the product specification is essentially a contract to assure that the producer can furnish the finished fastener to high quality standards. However, the fastener is manufactured and supplied to the customer with no knowledge of where or how it will be used.

On the other hand, the *design specification* normally outlines the conditions for use of the fastener. The design specification is intended to ensure structural and performance integrity of the assembled joint, with the threaded fastener system considered as a critical part of the joint. In the past, design criteria have been established and applied by major industry groups on the basis of prior problems and successful solutions developed within the industry. Such criteria often reflect structural philosophy, including allowable working loads or stress levels and factors of safety, optimum or recommended fastener series for particular applications, installation and/or tightening requirements, and often special considerations requiring attention to safety, possible servicing, repair, replacement, etc. Prominent examples of design specifications in practice are the various building codes supporting design of structures and related requirements for fastening systems.

While industry-wide application criteria are gaining greater recognition, not every major industry group has adopted or established uniform design standards covering threaded fasteners. In such instances, the primary design solution has generally been identified as a fundamental company prerogative. As a result, design standards representing particular experiences within an industry are often noted in individual company specifications. These application specifications are important and constitute the basis for extensive and usually practical design use of fastening systems.

3.2.5 Product specifications

Standard finished fasteners supplied to appropriate product specifications and standards normally are designed to meet certain minimum specified performance requirements. Such requirements take into account the dramatic transformation of raw material shape into final product form and basically include dimensional tolerances, metallurgical control, and mechanical performance properties. Each characteristic should be considered important in its right and also essential to the overall quality of the finished fastener and its subsequent satisfactory functional use. Some of the more significant product criteria usually referenced in specifications are as follows:

Dimensional. Dimensions and permissible tolerances are detailed on a standard or on product drawings. Adherence to specified dimensional tolerance limits is the key to interchangeability of standard uniform series fasteners. While in-process instructions are often made during manufacture, product acceptance should be predicated on dimensional characteristics of the finished fastener. because of the large numbers of fasteners normally associated with production loss, statistical sampling is often permitted in accordance with acceptable quality levels (AQLs) established for critical and noncritical dimensional characteristics. A commonly referenced source for sampling plans is MIL-STD-105.

Threads. Inspection of fastener threads represents a special dimensional gauging requirement. Individual thread characteristics of concern to the manufacturer may include major-, minor-, and pitch-diameter tolerances, angle error, lead error, and drunken threads. However, the concern of the user is that the finished threads are functional within the established class limit, i.e., that the cumulative individual variations still fall within the acceptable tolerance range. At present, there are several prominent thread-gauging system, including factional and go/no go (or "lo") types of gauges. Details of thread gauging and proper use of gauges are summarized in American National Standard ANSI B1.2-1974, *Gauges and Gauging for Unified Screw Threads.* Applicable product specifications should be consulted to determine preferred or thread-gauging systems for specific type of fastener and class of thread fit.

Metallurgical. Metallurgical requirements, as distinguished from those for mechanical properties, are intended to control quality, uniformity, and often method of manufacture of the fastener. For example, bolts fabricated by heading have a unique grain-flow pattern that readily distinguishes them from similar-configuration bolts which have been machined. Metallurgical examination of grain-flow patterns can detect acceptable as well as poor heading practices. Marginally headed fasteners cause a problem in that service failure may occur from undesirable stress concentrations or internal effects. Conversely, threads formed by the rolling process have improved static and fatigue strength, and have work-effect patterns that can be observed in metallurgical study. Other characteristics such as discontinuities, internal defects, and grain size indicative of proper heat treatment, and excessive decarburization, can be made only by metallurgical examination.

Macro (low magnification) examination is intended for evaluation of heading and grain-flow properties. Micro (high magnification) examination is intended for evaluation of grain size, work effect, decarburization, and evidence of possible defects such as cracks, laps, etc. Where metallurgical properties are concerned, product specifications will normally identify proper etchants, locations and tolerance limits for observed discontinuities. In particular, many government specifications define permissible and nonpermissible discontinuities as illustrated in Fig. 3.4.

The photomicrographs of a steel grade 8 hexagon-head cap screw and a steel socket-head cap screw in Fig. 3.5 show representative grain-flow patterns. The photomicrographs of rolled threads in steel and a corro-

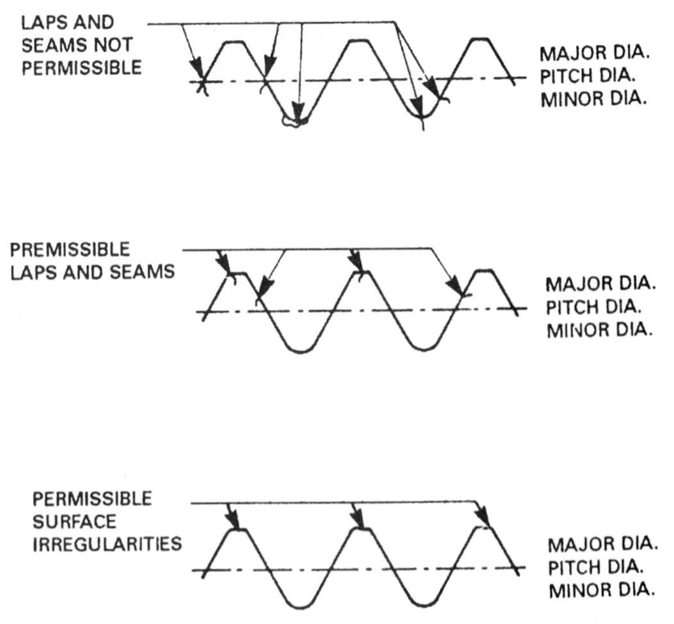

Figure 3.4 Screw thread permissible and nonpermissible discontinuity locations.

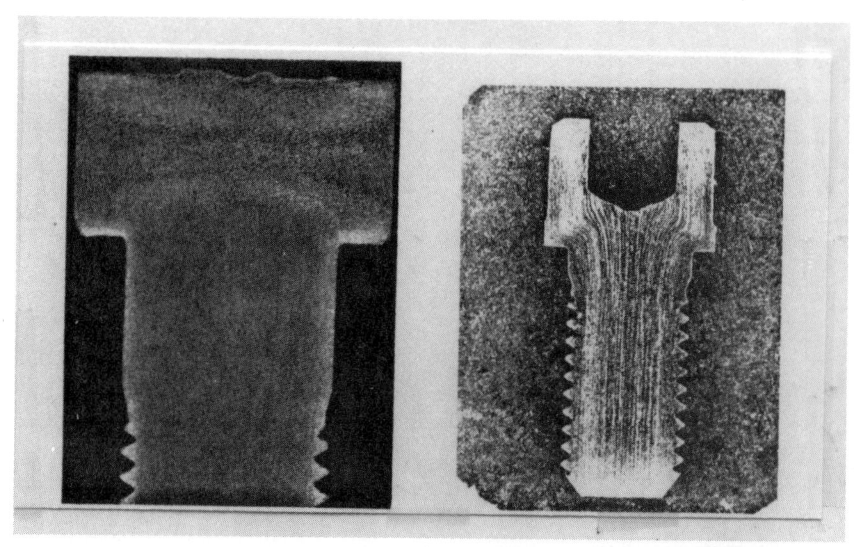

Figure 3.5 Metallurgical photomicrographs of hexagonal-head and socket-head cap screws.

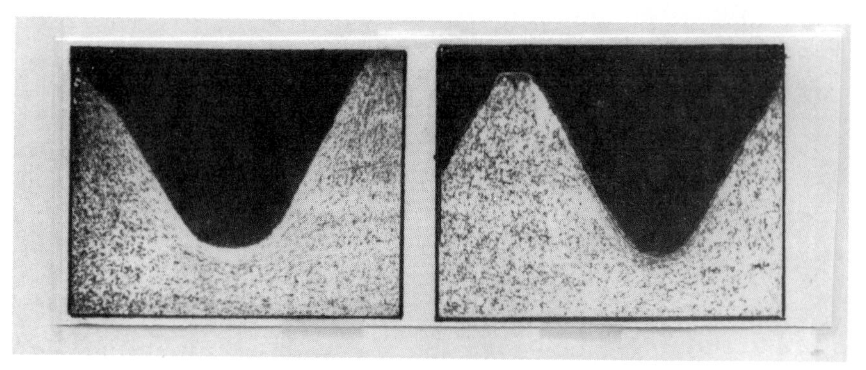

Figure 3.6 Thread sections in steel and corrosion-resistant steel bolts.

sion-resistant steel bolt in Fig. 3.6 shows work effect at the root radius. Figure 3.7 shows thread laps and cracks found by metallurgical examination. Figure 3.8 shows decarburization in a steel fastener.

Surface inspection. Surface defects in finished fasteners can occur from heat treatment and/or quenching or from surface seams present in the original material (Fig. 3.9). To detect such defects, magnetic-particle inspection can be used for ferrous materials, and fluorescent-penetrant inspection can be used for nonferrous materials. The presence of the surface indications will usually warrant further metallurgical exami-

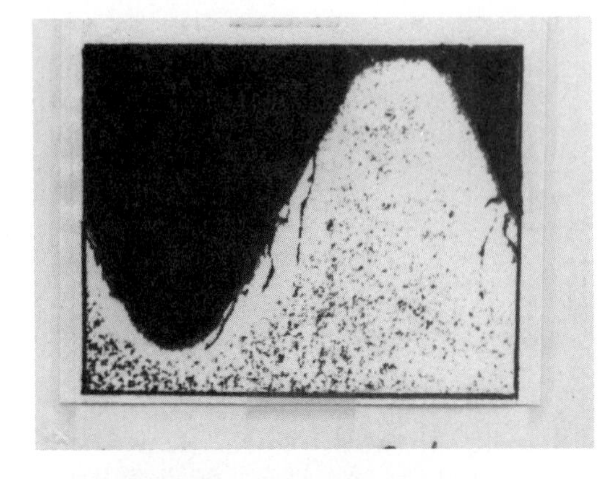

Figure 3.7 Laps and cracks in a thread.

Figure 3.8 Decarburization in a steel thread.

nation to identify the nature, depth, and severity of the defect. It should be understood that all surface indications are not necessarily cause for rejection, but rather that the surface inspection technique is a valuable tool in isolating potential metallurgical problems which could seriously affect the performance of highly stressed fasteners.

Tensile strength. The foremost mechanical property associated with standard threaded fasteners is tensile strength. It is a guide to the user of the strength class and of the ability of the fastener to carry or transfer load to specified minimum values. Processing of parent materials to

Figure 3.9 Surface defects on a fastener.

develop required strength levels may involve heat treatment, as in the case of some aluminum alloys, carbon and alloy steels, and martensitic corrosion-resistant steels. Other materials, notably austenitic stainless steel, respond to cold working (or drawing) to increase strength properties during manufacture.

The tensile strength is determined from the formula

$$S_t = P/A_s$$

where S_t = tensile strength, psi
P = tensile load, lb
A_s = tensile stress area, in^2

For any specified tensile level then, the minimum tensile load requirements for a fastener can be calculated in terms of

$$P = S_t A_s$$

For this basic relationship a significant consideration is the definition of the tensile stress area A_s for the fastener thread. Since the screw thread is formed on a helix angle, the stress area does not lend itself to absolute definition. In fact, over the years, there have been several screw thread stress areas used in calculating minimum tensile load values based on objective tensile strength criteria. At present, the commonly accepted formula for tensile stress area for static tensile strength properties of standard fasteners is

$$A_s = 3.1416 \left[\frac{E}{2} - \frac{3H}{16} \right]^2$$

or

$$A_s = 0.7854\left[D - \frac{0.9743}{n}\right]^2$$

where E = basic pitch diameter, in
D = basic major diameter, in
n = threads per inch
H = thread dimension from peak to valley (Fig. 3.3) in

In contrast, for high-strength aircraft fasteners the aerospace industry has predicated tensile strength on the pitch diameter area. For other mechanical properties, such as stress rupture and fatigue strength, the cross-sectional area at the minor diameter of the thread is often employed. The three stress areas are compared in Table 3.2. It is essential to know how the tensile stress area is defined for a specific fastener series to understand the strength of the fastener.

Several other product specification considerations are important in evaluating tensile strength properties of finished fasteners. For a standard bolt and screw, ultimate failure should occur in the threads because of the smaller sectional area. Many specifications do not permit bolt-head failures during acceptance testing, since they may be indicative of metallurgical problems or poor manufacturing quality.

Also, test experience has shown that thread engagement, hardness, and position of the mating nuts have an influence on the tensile strength of the bolt. For standard series bolts, testing requires six threads exposed above the mating test nut. Because of their shorter thread length, aerospace bolts are usually tested with two threads exposed. Interestingly, tests on grade 8 hexagon-head bolts from the same lot (Fig. 3.10), indicated a difference of approximately 15,000 psi in the tensile strength between two thread and six thread exposure. The higher breaking loads were observed with the nut positioned at two threads from the run-out. Strain hardening and notch sensitivity of the exposed threads apparently are factors contributing to this phenomenon. In view of the sensitivity of bolts to this condition, applicable specification requirements for type and position of the mating nut should be observed to accurately evaluate tensile strength properties.

A separate problem is associated with large-diameter bolts. Where testing machine capacity was not available or adequate, some early product specifications permitted machining of reduced-gauge-section coupons from the bolt in order to establish tensile strength characteristics. Typical machined test coupons are illustrated in Fig. 3.11. Caution should be exercised in evaluating the test result from machined coupons since their mechanical properties do not always corre-

TABLE 3.2 Representative Tensile Stress Areas for Fine- and Coarse-Thread Bolts

Nominal thread size	Tensile stress area,* in^2	Sectional area at minor dia.,† in^2	NAS 1348 pitch dia. stress area,‡ in^2
	Fine-Thread Series		
¼-28	0.0364	0.0326	0.0388
5⁄16-24	0.0580	0.0524	0.0614
⅜-24	0.0878	0.0809	0.0950
7⁄16-20	0.1187	0.1090	0.1288
½-20	0.1599	0.1486	0.1717
9⁄16-18	0.203	0.189	0.2176
⅝-18	0.256	0.240	0.2724
¾-16	0.373	0.351	0.3952
⅞-14	0.509	0.480	0.5392
1-12	0.663	0.625	0.7027
	Coarse-Thread Series		
¼-20	0.0318	0.0269	
5⁄16-18	0.0524	0.0454	
⅜-16	0.0775	0.0678	
7⁄16-14	0.1063	0.0933	
½-13	0.1419	0.1257	
9⁄16-12	0.182	0.162	
⅝-11	0.226	0.202	
¾-10	0.334	0.302	
⅞-9	0.462	0.419	
1-8	0.606	0.551	

*Tensile stress area per NBS Handbook H28, where $A_s = \pi (E/2-3H/16)^2$

†Minor-diameter area per NBS Handbook H28 at $D-h_o$.

‡National Aerospace standard for externally threaded fasteners in 160 through 260 ksi range with threads rolled after heat treatment.

late with the results obtained from full-scale bolt testing. Contributions of work effect in the rolled thread and in the upset or forged head are often lost in a standard machined specimen. This is particularly noticeable in corrosion-resistant steel and similar materials that have been cold-worked. For this reason, full-scale bolts should be evaluated wherever possible to assure measurement of the actual properties of the fasteners that will be used in service.

Proof load. Whereas testing to destruction is a confirmation of ultimate tensile strength, there is often an equal concern for another criterion known as *proof load* for certain classes of standard fasteners. This property, in a sense, represents the usable strength range of the

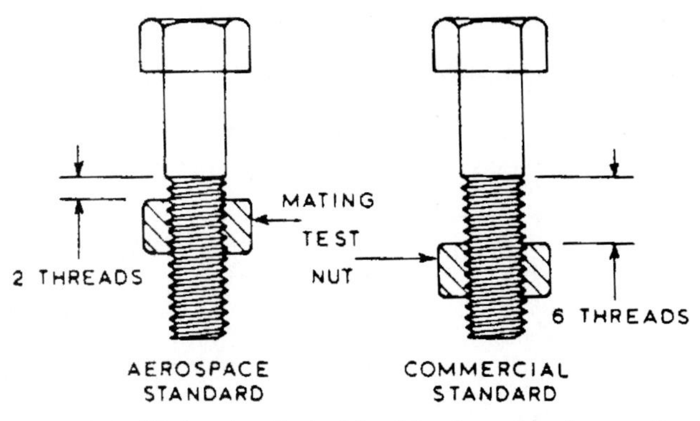

Figure 3.10 Higher tensile test load is observed when mating test nut is located at two threads from run-out.

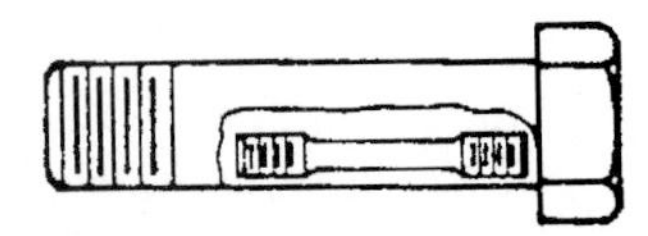

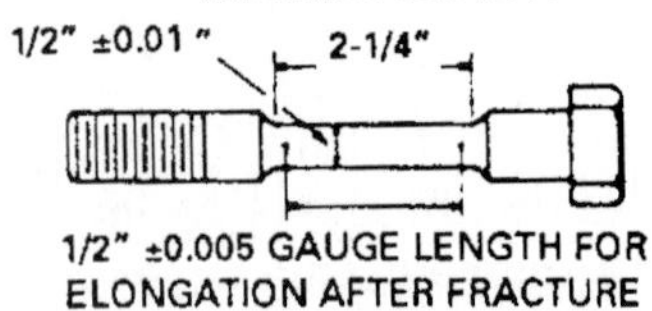

Figure 3.11 Typical reduced-gauge section tension test specimen machined from finished bolt.

fastener for many design functions. When subjected to proof-load exposure, the fastener should not exhibit permanent set as determined by length measurement prior to and after loading. As with other critical fastener characteristics, evaluation of proof-load properties, when specified, should be performed on the actual finished fasteners.

Occasionally, proof load and yield strength are interpreted as being the same. As mentioned earlier, a thread fastener is a complex component affected by changes of cross section, notches, and other regions of stress concentration, as well as by manufacturing processes. To date there is no common or verified standard for universally determining actual yield strength of a threaded fastener. The 0.2 percent offset yield strength criterion for parent material coupons should not be considered well-correlated with comparable properties for threaded fasteners fabricated from the same material.

Shear strength. Shear strength of standard threaded fasteners as a prime mechanical property has not generally been cited in industrial and commercial specifications. However, requirements for shear design applications are evolving and may be reflected in future specifications. Aerospace bolts, on the other hand, are covered by rigid requirements for shear strength as well as for tensile strength. For many of the original aerospace steel bolts heat-treated to achieve a tensile strength of 125 ksi, shear strength was estimated at 60 percent of tensile strength, and this ratio was accepted as a standard for this class of fasteners. However, the ratio is not a constant and can vary from about 54 to 65 percent. For this reason, when bolts may be subject to shear-loading service, the product specifications or specific customer requirements should therefore be reviewed and modified as necessary.

Hardness. Brinell or Rockwell hardness testing is widely used for estimating the tensile strength properties of steel fasteners. It is recognized as a valuable indicator of tensile properties, but should not be construed as a direct measure of actual tensile strength of finished fasteners. Again, true tensile strength can be determined only from full-scale fastener evaluation. Sometimes, though, production bolts are too short to permit tensile strength testing, and, in such instances, lot acceptance may be predicated on the results of hardness testing. As a manufacturing in-process control check, hardness testing is often considered as nondestructive. On finished and plated fasteners, however, hardness readings may not be valid unless the plating is removed and a flat and uniform surface prepared for the hardness indentation, and the fasteners must then be discarded after testing.

Microhardness. Microhardness testing is normally used in conjunction with metallurgical evaluation, primarily for determination of surface decarburization or carburization. Microhardnesstype instruments have the ability to make measurements 0.001 in apart. While visual metallurgical experimentation can detect obvious decarburization, specification criteria for partial decarburization are normally based on the differential hardness of three points between carburization zone and uniform parent material. Where particular limits are noted for decarburization depth, the microhardness traverse provides a means for accurate measurement. The same technique, of course, can similarly be employed for measurement of case-hardening depth. Such tests are required on prepared specimens.

Protective coatings. There are several important reasons for applying protective coatings or finishes to standard threaded fasteners. Foremost, coatings are employed as a protective barrier for the fastener base metal. In addition, they may be specified for wear resistance, decorative appearance, or thread lubrication, or as a base for subsequent painting. Quite often, these functions overlap, and selection of a coating takes into account the multiple advantages available as well as the cost. Commercial finishes include anodizing and chemical surface treatment for aluminum alloys, oxide treatment for brass, passivation for corrosion-resistant steels, and a wide spectrum of protective coatings for steel such as galvanization, electroplated zinc, electroplated cadmium, conversion coatings (phosphate and oil), electroplated nickel and chromium, and mechanical plating. From a product-specification viewpoint, the main concern is that the specific finish has been properly and fully applied to provide the required protection for the parent fastener in storage and in service.

For electrodeposited plating in particular, perhaps the foremost criteria are thickness of plating, adhesion to the base metal, and porosity. Each characteristic is critical in contributing to the overall performance of the plating. A breakdown of any individual property could result in a defective coating, even if the remaining conditions are otherwise within specification.

Evaluation of the adequacy of a coating or finish may be by a humidity test, a salt-spray exposure test, or by measurements of coating thickness, as in the case of electrodeposited plating. Thickness measurements can be made nondestructively by means of eddy-current instruments or by magnetic measuring devices capable of identifying thickness of nonmagnetic coatings on ferrous materials. Unfortunately, electroplated coatings are not uniform in thickness and have a tendency to build up on corners or at sharp changes in section. Precise measurements of thickness may therefore require sectioning and microscopic analysis.

The severest test condition for overall performance of a coating or finish is salt-spray exposure, which is intended as an accelerated test. The reliability of associating number of hours of test-chamber exposure with actual years of service life is questionable. However, the salt-spray test has been widely used to screen poor or defective plating, as evidence by the appearance of white corrosion products, red rust, pitting, flaking, or other form of attack. Hours of salt-spray exposure are generally related to the type of finish and/or thickness of coating specified for the fasteners.

3.3 Designing Fasteners for Affordable Automation

Structural design, whether it be for a massive heavy-duty structure or a miniature subsystem, involves the same basic challenge: joining the components in an economical manner with maximum joint integrity. Underlying successful design is analysis of the overall joint, since the fastener is only one vital part of the complete structural system. Obviously, it is important to start with fasteners which are of good quality and which will perform under the expected design conditions. Implementing a sound quality assurance system is well worth the effort to avoid service failures which may otherwise be attributed to fasteners of questionable quality.

The approach to joint design varies markedly from industry to industry. In part, this reflects differences in types of materials and structures being joined, service environments, the factors of safety, service life, and tooling and assembly. If there is no single set of design standards which can be set down for all standard threaded fasteners, experience nevertheless suggests certain guidelines which can contribute to sound design practice.

That valid design principles are important is borne out by industry estimates that assign as much as 50 to 60 percent of the cost of a structure to joint design, analysis, and assembly.

3.3.1 Selection of threads

Standards for the United Screw Thread form cover both coarse-thread and fine-thread series, since standard bolts in the popular nominal sizes up to 1-in diameter are available in either thread series.

The fine-thread series, by virtue of finer pitch and resultant smaller thread depth, produces a tensile area which is greater than the area for a corresponding coarse-thread fastener. The net effect is that the fine-thread series is characteristically stronger than the coarse-thread series. A stronger fastener develops higher working and clamp loads, which are certainly prime design objectives. In addition, because fine threads have a small lead angle, they are advantageous where fine adjustments may be required. Fine threads are easier to tap in harder materials but are generally limited to shorter thread engagements than coarse threads. The fine thread is preferred for thin-walled materials when tapping is necessary.

The coarse-thread series is probably the more widely used for construction and general applications, although specialized industries such as the automotive and aircraft groups tend to rely more heavily on fine-thread series fasteners. It is interesting, however, that one of the concepts being advanced in the study of the new proposed metric fasteners is a single thread series for all general commercial and industrial applications.

3.3.2 Corrosion

For many applications, the problems of corrosion create an overriding concern in design. There are several distinct types of corrosion, including galvanic corrosion, concentration-cell corrosion, stress corrosion, fretting corrosion, pitting corrosion, and oxidation. Probably the most common form of corrosion is rust associated with steel structures and steel fasteners, although the effects of corrosive attack can be observed in many other structural materials.

In the case of galvanic corrosion, the combination of two dissimilar metals with an electrolyte is all that is needed to form a battery. The use of dissimilar metals in structural design is not uncommon; fasteners are often a different material from the structure being joined. The other necessary ingredient for corrosion, an electrolyte, may be present in the form of ocean salt spray, rain, dew, snow, high humidity, or even air pollution. The anode-cathode effect of the battery couple actually results in transport of material where the metals are apart in the electromotive series.

In concentration-cell corrosion and pitting, only one metal and an electrolyte are sufficient to set up an attack system. As corrosion progresses, a differential in concentration of oxygen at the metal surface and in the electrolyte produces a highly effective localized battery with resultant corrosion and metal attack.

In stress corrosion, cracks are induced and propagated under the combined effects of stress and corrosion environments. Structures or components with high stress concentrations (such as threaded fasteners) are susceptible to this type of attack when under load. The initial corrosion may occur at a point of high stress, starting a crack, which can be either intergranular or transgranular. Continued exposure to the corrosion environment will propagate the crack, resulting in serious, if not catastrophic, failure.

Other corrosion systems can be equally severe. Alkalis, acids, marine atmosphere, industrial pollution, road salt, in addition to water products, all exert their influence. Common to all is the fact that corrosion is encountered after the structure is put into service.

Corrosion protection at design inception should be a major objective of good joint design. The first step is to identify the specific anticipated corrosion exposure in order to control or minimize its consequences in service. As noted, part of the corrosion problem arises when fasteners of one material are used to join a structure of a different material. Initially, an attempt should be made to select fasteners materials which are as compatible as practical with the structure being joined; i.e., electromotive potential should be minimized. A corollary consideration is to provide a coating or finish for the fasteners to protect the base metal and/or the joint material. Additional protection may be afforded by supplemental coatings or finishes for the entire joint, since holes drilled through a structure are bare until treated. The range of protective coatings includes primers, paints, insulation materials, inhibitors, plating, and greases. The choice depends on the severity of the application and the aesthetics of the product. As part of the design, care should be taken to prevent accumulation of corrosion elements by providing adequate draining, and also to avoid areas of high stress concentration exposed to possible corrosive environments.

3.3.3 Washers

Invariably, when standard fasteners are discussed, they are referred to as a bolt-nut system. Standard flat washers are vary rarely thought of as part of the system, and yet they often play an important role in fastening.

Washers are used to distribute bearing stresses under bolt heads and nuts, especially when joining softer structural materials. They can be used as a corrosion barrier or insulation as part of a corrosion-protection system in separating otherwise highly anodic-cathodic materials. Washers under the bolt head or nut provide a good surface for uniform torque control, such as the washers specified for use with the A325 and A490 series structural fasteners. In some instances, washers can be used to accommodate or adjust for proper grip length of bolts in an installed joint.

Flat washers are available in a wide variety of materials and finishes compatible with threaded fasteners. There are many higher-strength structural materials which do not particularly need a washer for joint structural strength, and there is some concern that the addition of a washer constitutes an added element, added expense, and added weight. However, judicious use of washers for the reasons noted may enhance overall joint efficiency and design and should be considered on that basis.

3.3.4 Joint design

There are really no standards for uniform joint configurations, and there is no limit to the variety of methods for fastening and connecting structures. The more popular types of joints include the lap joint, the flanged joint, and the butt joint. Actually, there are many variations and combinations of these joint configurations which can be assembled by through-fasteners (bolt and nut) or by securing a bolt in a tapped hole.

Modern threaded fasteners are effective as either permanent or removable connections, although each requires special considerations. Joints employing permanent fasteners can be readily and accurately assembled under shop conditions. The same joints assembled under field conditions may require additional working tolerances and may be limited by available installation tooling. Similarly, for removable fastener joints, wrench clearance, spacing, and accessibility must all be considered. Factory assembly is often geared to special tooling to expedite and facilitate fastener installation. Under field service repair or maintenance, the limiting factors may be available manual tool and working clearances.

Joint design also includes study and analysis of the forces acting on the joint. The types of loading may include tension, shear, bending, and fatigue. Even though the same fastener assembly could conceivably be used for type application, specific guidelines for structural design are applicable.

3.3.5 Tensile loading

A tensile joint is one where the applied load is in line with the fastener axis. In theory, if the tensile preload induced in the bolt system meets or exceeds the applied tensile loading on the joint, the fastener will not sense any increased tensile load and the joint will remain rigid. The key here is determining the proper tensile preload in the threaded fastener system. The action of assembling and tightening a bolt and nut produces an elongation and resultant tensile prestress in the fastener, while at the same time equal compressive stress is introduced into the material being joined. This model of the system is assumed valid as long as the applicable stresses are within the elastic limit in accordance with Hooke's law.

For most design involving static tensile loading, the assumptions are true enough. In practice, loads tend to fluctuate and tightened fasteners may actually sense tensile loading in excess of the induced preload. An increase in the tensile load on the fastener is accompanied by a decrease in the compressive load in the joint and resultant loss of clamping load effect. Where joints are designed to be rigid, excessive loss of clamp load can cause a series structural problem.

Several significant considerations are developed from the tensile joint analysis. The bolt yield strength may be a more important design criterion than the ultimate tensile strength, since yield strength indicates the load capacity of the fastener within the elastic limit of the material. Obviously, however, the higher the ultimate tensile strength of the fastener system, the higher the yield strength and clamp capability. The higher-strength fastener series with greater clamp-load ratings permit reduction in the number of fasteners, or use of smaller-diameter fasteners, to achieve the same design objective.

The foregoing discussion is applicable to rigid joints. Flexible joints with gaskets present additional problems and require additional analysis. For the gaskets to be effective, it is important for all fasteners in the joint to be uniformly loaded to the design stress condition. A gasketed joint usually requires a higher-strength or larger-diameter fastener than would be required for the same clamp-load objective in a rigid joint.

While most strength criteria are predicated in the properties of the male threaded element (bolt or screw), it should be recognized that the mating nut plays a significant role in developing the full rated strength of the fastener system. The nut should be capable of sustaining at least the ultimate tensile strength of the bolt. Generally, matching nuts require a minimum height equal to the basic diameter of the fastener in order to develop the rated tensile strength. The same minimum thread strength requirement pertains when bolts are inserted into tapped threads in high-strength steel structures. Where tapping is necessary in softer materials, thread engagement length from $1\frac{1}{2}$ to 2 times the normal fastener diameter may be needed to develop full load strength.

3.3.6 Shear loading

A shear joint is one in which the applied load is at right angles to the fastener axis, or across the bolt shank. Shear joints fall into two categories: friction and bearing. The friction connection depends on high clamp load induced in the mating threaded fasteners. As long as the mating facing surfaces are free of dirt, lubricant, paint, etc., the high clamp load contributes to high frictional forces in the joint. The fictional forces developed are normally well within the allowable design shear-stress ratings, particularly in the construction industry. Should joint shear load exceed the frictional stress, joint slipping would occur to the full tolerance of the clearance hole, resulting in shear forces

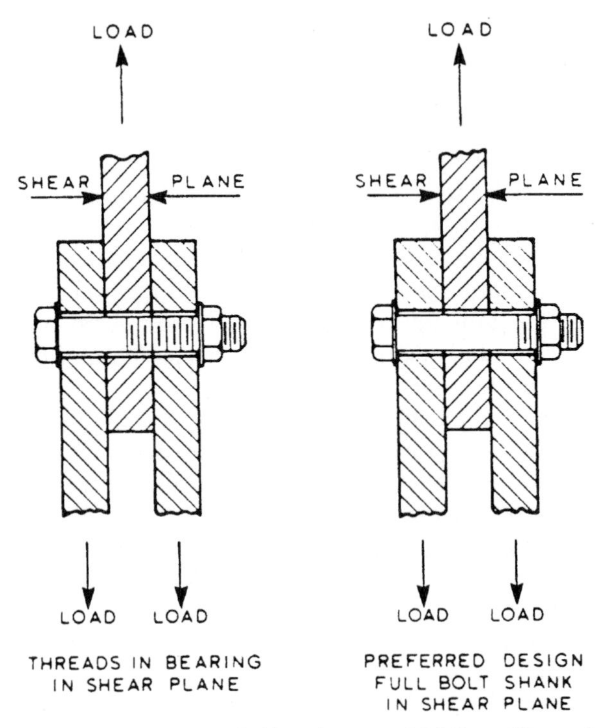

Figure 3.12 Improper (left) and proper (right) positions of threads and shank in a shear joint.

acting on the bolt and bearing forces acting on the joint materials. The bearing joint, therefore, is based on bolt shear strength and plate bearing strength properties.

For friction-type joints, it is evident that clean and flat working surfaces are necessary for high joint efficiency. Also, close control must be maintained of torque on the fastener connections to assure that the design preload is developed in the fasteners. For bearing-type joints, care should be taken to specify bolts with shanks long enough to extend through the shear planes. Occasionally, short-shank bolts are used, with portion of the bolt threads resisting shear loading, as illustrated in Fig. 3.12. Tests have confirmed that substantial loss of bolt strength is caused by the smaller net sectional area in the screw threads. As much as 20 percent of bolt shear strength may be lost when threads are in bearing. To take full advantage of strength properties, the full shank body should be positioned in the shear planes. The airframe industry, for example, maintains a strict prohibition against the use of any threads in bearing in a shear-joint connection.

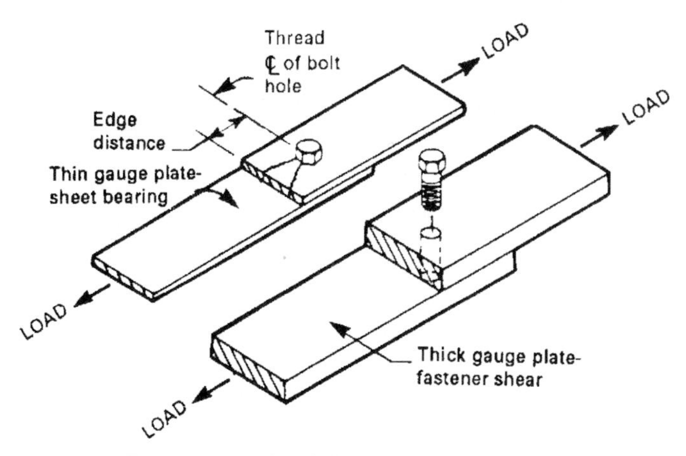

Figure 3.13 Bearing-critical and shear-critical bearing joints.

Another consideration for bearing-type shear joints is whether the bearing strength of the joint material or the shear strength of the fastener is critical. With a bolt installed in a hole acting against the plate, the bearing stress is calculated on an area equal to the thickness of the plate times the normal diameter of the bolt or fastener. In a joint where the plates are relatively thin, the plates may fail, making the joint bearing-critical. With the same diameter fastener in a joint where the plates are relatively heavy, the full strength of the fasteners can be realized, making the joint shear-critical. These conditions are illustrated in Fig. 3.13.

One of the major challenges of joint design is working within the bearing-strength rating of the joint material. Mechanical property data established for most of the common structural materials include bearing strength as a function of edge distance. Edge distance is usually defined as the distance from the centerline of the bolt hole to the edge of the plate and is expressed as a function of bolt diameter. For example, an edge distance of 2.0 for a 1-in-diameter bolt means that the centerline of the mating hole is 2.0 in from the end of the plate. Edge distances used in design range from 1.5 to 2.5, depending on the structure and material being employed. Since bearing strength properties vary with edge distance, some caution may be warranted when repairs are undertaken. To save a structure with a defective hole, or possibly to repair a damaged joint, there is a tendency to open up the hole to permit the use of the next larger size fastener. This action will change the edge-distance relationship. As long as the joint is not critical in bearing, however, it may be a perfectly acceptable approach to saving the original joint.

3.3.7 Bending loading

Unfortunately, structural joints are not always loaded in pure tension or in pure shear. Many structural applications are subjected to bending forces which effectively result in combined tension and shear loads acting simultaneously on the fastener. To establish design limits for fasteners under bending load conditions, interaction curves are usually generated for the fastener or fastener series (Fig. 3.14). Such a curve is developed from actual tests in which a number of bolts are evaluated at different combinations of applied tensile and shear loading. From the curve, it is possible to calculate maximum allowable design stresses for tension and for shear at any degree of bending. For illustrative purposes, the bolt stresses at a bending condition of 45° are shown. Interaction curves are basically empirical and are tailored to specific bolt series, taking into account strength level, material, and head design.

Even with valid interaction curves for design support, additional caution must be taken when working with joints subjected to bending. Because of the notch factor, screw threads are sensitive to applied bending stresses. Where severe or critical bending is anticipated

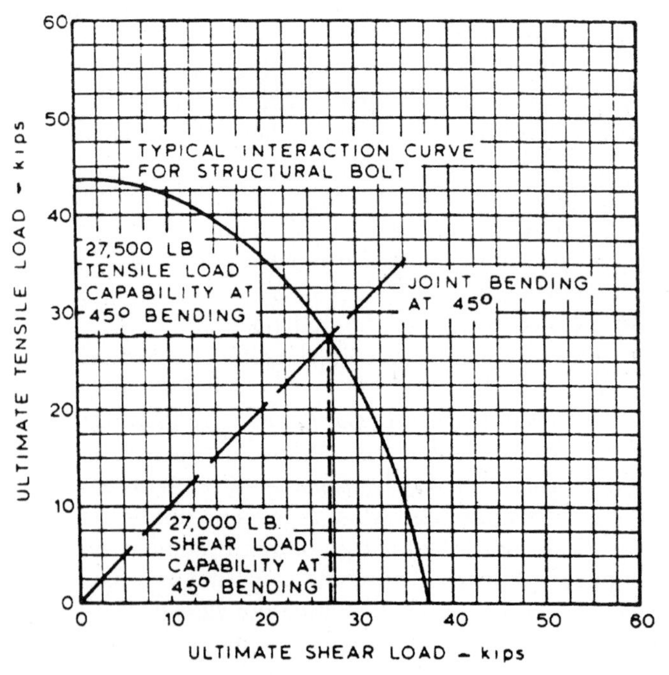

Figure 3.14 Typical interaction curve of tension versus shear.

in a joint, optimum-quality fasteners and maximum safety factors may well be advised.

3.3.8 Fatigue loading

As noted earlier, fatigue is the condition associated with repetitive or dynamic loading on a structure or component. Many types of equipment are subjected to continuous dynamic stresses during their service life. Although extensive testing and experience have contributed to the ability to control or minimize fatigue effects, the majority of service failures are still fatigue-related. This fact alone suggests the importance of good joint design.

There are two basic categories of joints subjected to fatigue loading: shear-fatigue-loaded joints and tension-fatigue-loaded joints. In the case of shear-loaded joints, fatigue failure normally occurs in the plate or shear material. Applicable dynamic stresses, hole penetration, hole clearances, and fastener preload are just some of the factors which affect shear-joint-fatigue life.

As for tension-fatigue-loaded joints, the basic principles outlined for static-tensile-loaded joints are applicable. The prime factor is the adequate preloading of the fastener to meet or exceed anticipated dynamic or cyclic loading on the joint. A properly tightened and preloaded bolt will realize only minimal external tensile forces imposed by repetitive tensile loading. The stresses expected during the service life of the equipment should be defined for accurate design. This may involve detailed experimental stress analysis and simulated service operation, and may be easier said than done.

As a result of the relationship between allowable stress and number of dynamic cycles, substantially lower design stresses have to be accommodated in fatigue joints than in comparable static-loaded joints. This is particularly true where long service life is required. Effective use of fasteners specifically designed for fatigue strength and particular attention to preload control are perhaps two of the more important steps which can be taken to overcome fatigue effects in structural joints.

3.3.9 Vibration

Fatigue loading is presumed to be relatively high repetitive or cyclic stresses in relation to the strength of the threaded fastener or the joint. On the other hand, vibratory forces are considered to be relatively low, but may be associated with various ranges of operating frequencies. Critical combinations of frequency, amplitude, and loading can force a structure into resonance with catastrophic results. However, for most structural joints the combination of these conditions is not sufficient to produce resonance. Yet, vibration is present in equipment subjected to dynamic forces and its implications are quite serious for the fastened system.

Extensive or continued vibration affects threaded fasteners in a joint, without damaging the joint material. Vibration has a tendency to loosen bolts and nuts. The mechanism is complex, but the results are serious. Under the worst condition, it is possible for a nut to literally walk off a bolt thread. In other cases, vibration can be sufficient to slightly loosen a bolted connection, with resultant loss of preload. The threaded fastener may then experience higher fatigue stresses than were originally intended for the application.

Again, initial preload control of the thread fastener assembly is vital to resisting vibration, shock, and impact loading. Where vibration is unusually severe, additional safety measures may be necessary for proper joint integrity. Such measures include using self-locking nuts, safety wire, cotter pins, or an adhesive. As with fatigue, vibration analysis is complicated and costly, but sometimes there is no alternative to assure operating performance of structural joints.

3.4 Managing Quality of Part Fastening

The continued emphasis on achieving and maintaining proper preload in threaded fasteners points up one dramatic fact: in the last analysis, the greatest influence on the structural integrity of an assembled joint is exerted by the torque control device. If a threaded fastener is torqued too high, there is a danger of the torquing tool causing failure on installation by stripping the threads, breaking the bolt, or producing excessive fastener yield. If a bolt is torqued too low, a low preload will be induced in the fastener assembly, possibly inviting fatigue and/or vibration failure. For every bolt system, there is an optimum preload objective, which is obtained by proper torquing of the bolt-nut combination.

The amount of tightening normally required to elongate the bolt and induce a desired tensile preload or clamp load in the fastener system is generally identified as the torque-tension relationship. This relationship is not always fully appreciated and is often misused. Unless the torque-tension relationship is accurately applied, there is an obvious danger of losing all of the advantages gained by good joint design.

3.4.1 Torque tension versus static tension

Threaded fasteners such as bolts and screws respond differently to application of strength tensile loading than to tightening by torquing. As illustrated by the two curves in Fig. 3.15, a bolt in which tension

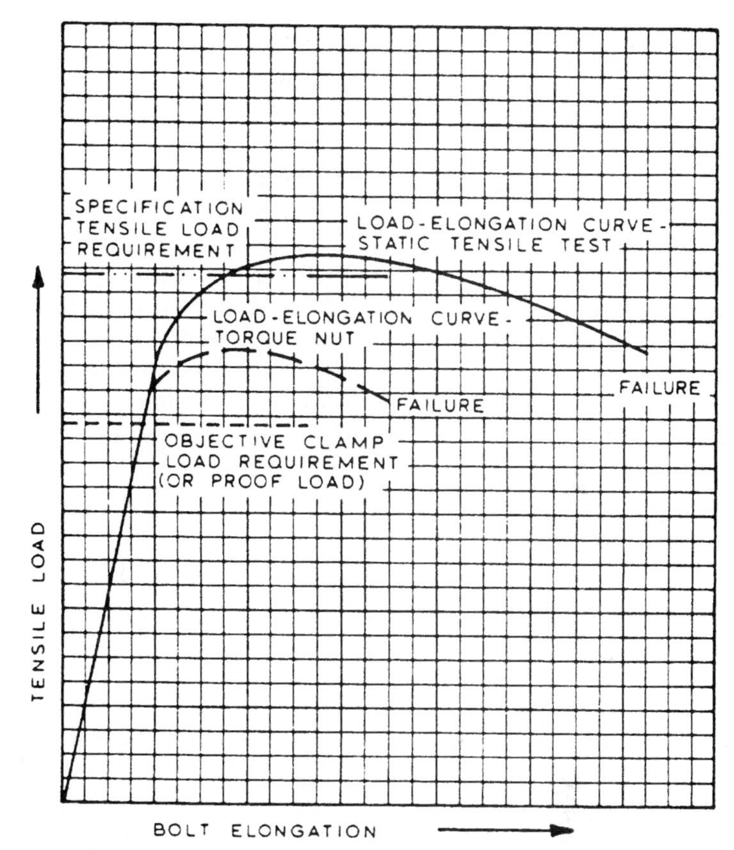

Figure 3.15 Load versus elongation curves for static tension (solid line) and tension induced by a torque nut (dashed line).

has been induced by torquing a nut does not preserve as high a tensile strength as a similar bolt that has been statically loaded. The additional torsional component induced by the torquing action on the fastener accounts for this phenomenon.

Apparent in Fig. 3.15 is the essentially straight-line relationship for both curves between increasing tensile load and bolt elongation up to the yield point of the fastener. Thus, although the two curves vary above the yield point, in practice, a bolt torqued within its yield strength will develop the full rated tensile strength when subjected to additional tensile loading in excess of the preload value, because the additional loading will be static tension.

For this reason, many designs are predicated on torquing fasteners to develop a tensile preload of 70 to 80 percent of the rated static tensile strength. Torquing in excess of this objective can result in excess bolt elongation and failure.

3.4.2 Torque tension characteristics

From the straight-line relationship of the load-elongation curve, it is evident that strict controls are imperative for developing clamp loads within design objectives. The problem of torquing to achieve tensile preload involves friction coefficients of the fastener system. Investigators have fairly well established that as much as 90 percent of applied torque is used in overcoming friction in threaded fasteners, with high points of friction located under the bolt head, under the nut, and in the actual fastener thread. This extensive work has indicated that the following formula can be used as an estimate of tension preload caused by torquing:

$$T = KDP$$

where T = installation torque, lb · in
K = torque coefficient
D = nominal bolt diameter, in
P = clamp load objective, lb

From the formula it can be seen that the prime variable is the torque coefficient K. The torque coefficient varies with different finishes, platings, and lubricant coatings normally found with standard fasteners. It was noted earlier that platings and protective coatings are used for protection of the base metal as well as for their lubrication qualities, which are sometimes vital in torque control. For example, both zinc and cadmium are popular platings. There is a noticeable difference in friction characteristics between these platings, making it necessary to rely on different installation-torque values.

Recent studies of high-strength steel, cadmium-plated bolts using mating nuts with different surface finish conditions indicated the wide spread in both torque and preload properties (Fig. 3.16). In order to maintain torque control, some fasteners are finished with supplemental zinc, oil, or molybdenum disulfide coatings applied over basic cadmium or zinc platings. Supplemental lubricants have a direct effect on torque-tension characteristics, making the use of standard tables questionable for many installation conditions.

Another factor often overlooked is that the installation-torque value normally specified for a bolt-nut combination is valid only for the initial assembly of the fastener. Continued use of the same bolt-nut assembly tends to change the coefficient of friction of the nut, resulting

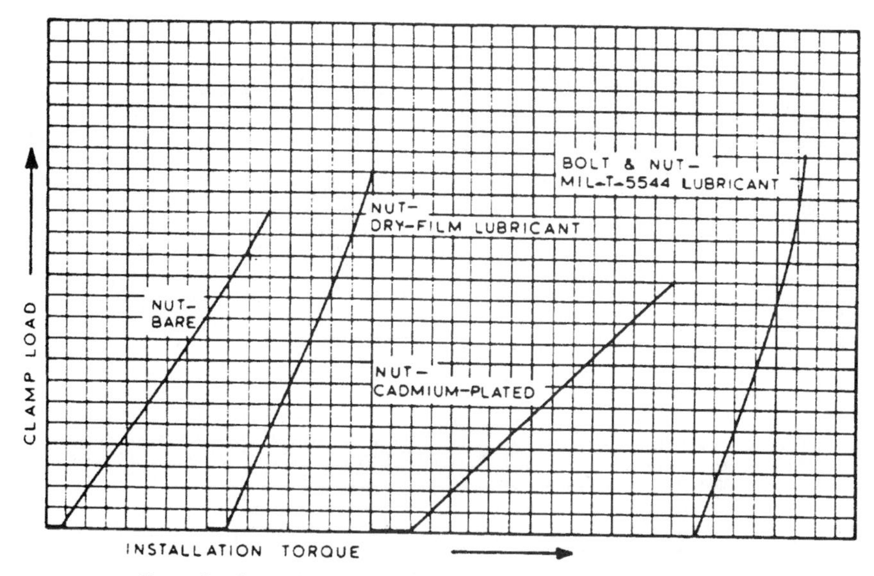

Figure 3.16 Clamp load versus torque for various surface conditions.

in lower preload in the bolt after as few as five installations. Figure 3.17 illustrates this effect with data from tests showing that preload loss for the same fastener combination can range as much as 30 to 60 percent after 10 installations.

For critical applications, the use of a calibrated torque device is a necessity. Even with a torque device, a wide spread of preload values can be observed. The effects of different torque coefficients on tightening-torque values for standard bolt series are compared in Table 3.3. A difference of 25 percent in tightening-torque requirements is noted for K factors of 0.2 and 0.15. The torque coefficient of 0.20 is an estimate for unlubricated and dry fasteners, while the coefficient of 0.15 approximates that of a plated finish. The use of additional lubricants such as oil, MIL-T-5544 graphite grease, and waxes can vary the K factor from 0.07 to 0.30. For a specific design application with exacting preload control requirements, it may be necessary to test and evaluate the bolts and nuts to define necessary installation-torque values.

Suggested tightening torque values for nonferrous threaded fasteners are listed in Table 3.4. Suggested locknut tightening-torque values for standard series bolts are listed in Table 3.5. As with all previous discussions concerning torque-tension properties, these values are intended as guidelines. It is recommended that they be substantiated for particular design joints.

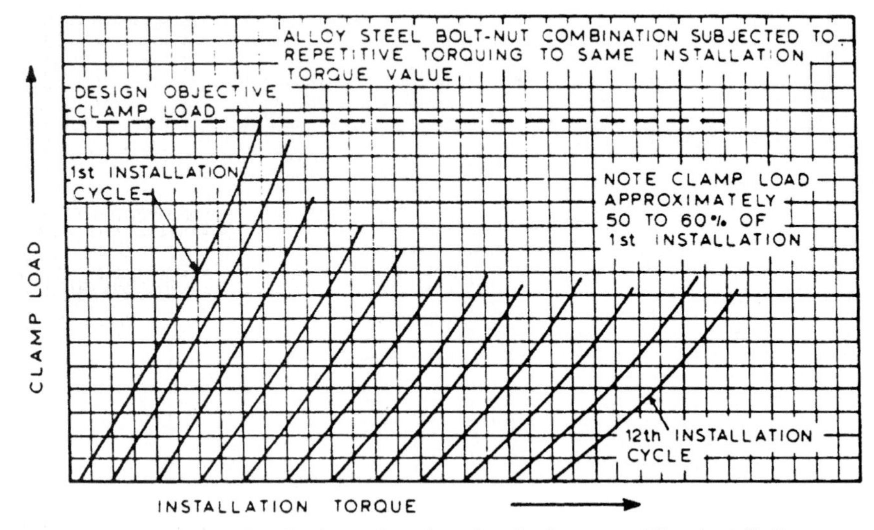

Figure 3.17 Decrease of preload as a function of preload torque with reinstallation.

3.5 Flexible Feeding Systems for Affordable Automation

In planning an assembly line or modernizing an existing one, one of the most important decisions to be made is the choice of the correct feeding system. In a fully automatic system technical considerations are at the forefront, but when the assembly is only partly automatic good work conditions for the operators carrying out manual operations must be provided. Thus, two other important criteria in evaluating a system concern (1) the equipment and (2) the personnel.

Equipment-dependent criteria can be used to evaluate the technical and the organizational aspects of the system. These criteria are

1. Flexibility in the number of operators, variety of products, mechanization, and automation.
2. Manufacturing reliability to ensure minimum faults and scrapped products.
3. High utilization, resulting from a low breakdown rate.

Personnel-dependent criteria help to judge the productivity of the operators. Among these are

1. Possibility of individual activities independent of the machine cycle and individual choice of rest periods.

TABLE 3.3 Calculated Tightening-Torque Values for Standard Bolt Series to Develop Rated Clamp Load Objectives

Nominal thread size	SAE grade 2 bolts			SAE grade 5 bolts			SAE grade 8 bolts		
	Clamp load, lb*	Tightening torque		Clamp load, lb*	Tightening torque		Clamp load, lb*	Tightening torque	
		$K = 0.15$	$K = 0.20$		$K = 0.15$	$K = 0.20$		$K = 0.15$	$K = 0.20$
Coarse-Thread Series									
¼-20	1,300	49†	65†	2,000	75†	100†	2,850	107†	143†
5⁄16-18	2,150	101†	134†	3,350	157†	210†	4,700	220†	305†
⅜-16	3,200	15	20	4,950	23	31	6,950	32.5	44
7⁄16-14	4,400	24	30	6,800	37	50	9,600	53	70
½-13	5,850	36.5	49	9,050	57	75	12,800	80	107
9⁄16-12	7,500	53	70	11,600	82	109	16,400	115	154
⅝-11	9,300	73	97	14,500	113	151	20,300	159	211
¾-10	13,800	129	173	21,300	200	266	30,100	282	376
Fine-Thread Series									
¼-28	1,500	55†	75†	2,300	85†	115†	3,250	120†	163†
5⁄16-24	2,400	112†	150†	3,700	173†	230†	5,200	245†	325†
⅜-24	3,600	17	22.5	5,600	26	35	7,900	37	50
7⁄16-20	4,900	27	36	7,550	42	55	10,700	59	78
½-20	6,600	41	55	10,200	64	85	14,400	90	120
9⁄16-18	8,400	59	79	13,000	92	122	18,300	129	172
⅝-18	10,600	83	110	16,300	128	170	23,000	180	240
¾-16	15,400	144	193	23,800	223	298	33,600	315	420

*Clamp load objective predicated on developing 75% of the proof loads specified for the respective bolt series.
†Torques values for ¼- and 5⁄16-in sizes are in lb · in. All other torques values are in lb · ft. Torque values calculated from formula $T = KDL$,

where T = tightening torque, lb · in
K = torque coefficient
D = nominal bolt diameter, in
L = clamp load objective, lb

TABLE 3.4 Suggested Tightening-Torque Values for Nonferrous Threaded Fasteners

This table is offered as the suggested maximum torquing values for threaded products and is only a guide. Actual tests were conducted on dry, or near-dry, products. Mating parts were wiped clean of chips and foreign matter. A lubricated bolt requires less torque to attain the same clamping force as a nonlubricated bolt.

All values shown on chart except for the nylon represent a safe working torque; in the case of nylon only the figures represent breaking torque.

Bolt size	18-8 stainless steel	Brass	Silicon bronze	Aluminum 2024-T4	316 stainless steel	Monel	Nylon
Values in pound-inches							
2-56	2.5	2.0	2.3	1.4	2.6	2.5	0.44
2-64	3.0	2.5	2.8	1.7	3.2	3.1	
3-48	3.9	3.2	3.6	2.1	4.0	4.0	
3-56	4.4	3.6	4.1	2.4	4.6	4.5	
4-40	5.2	4.3	4.8	2.9	5.5	5.3	1.19
4-48	6.6	5.4	6.1	3.6	6.9	6.7	
5-40	7.7	6.3	7.1	4.2	8.1	7.8	
5-44	9.4	7.7	8.7	5.1	9.8	9.6	
6-32	9.6	7.9	8.9	5.3	10.1	9.8	2.14
6-40	12.1	9.9	11.2	6.6	12.7	12.3	
8-32	19.8	16.2	18.4	10.8	20.7	20.2	4.3
8-36	22.0	18.0	20.4	12.0	23.0	22.4	
10-24	22.8	18.6	21.2	13.8	23.8	25.9	6.61
10-32	31.7	25.9	29.3	19.2	33.1	34.9	8.2
¼-20	75.2	61.5	68.8	45.6	78.8	85.3	16.0
¼-28	94.0	77.0	87.0	57.0	99.0	106.0	20.8
5⁄16-18	132	107	123	80	138	149	34.9
5⁄16-24	142	116	131	86	147	160	
⅜-16	236	192	219	143	247	266	
⅜-24	259	212	240	157	217	294	
7⁄16-14	376	317	349	228	393	427	
7⁄16-20	400	327	371	242	418	451	
½-13	517	422	480	313	542	584	
½-20	541	443	502	328	565	613	
9⁄16-12	682	558	632	413	713	774	
9⁄16-18	752	615	697	456	787	855	
⅝-11	1110	907	1030	715	1160	1330	
⅝-18	1244	1016	1154	798	1301	1482	
¾-10	1530	1249	1416	980	1582	1832	
¾-16	1490	1220	1382	958	1558	1790	
⅞-9	2328	1905	2140	1495	2430	2775	
⅞-14	2318	1895	2130	1490	2420	2755	
1-8	3440	2815	3185	2205	3595	4130	
1-14	3110	2545	2885	1995	3250	3730	
Values in pound-feet							
1⅛-7	413	337	383	265	432	499	
1⅛-12	390	318	361	251	408	470	
1¼-7	523	428	485	336	546	627	
1¼-12	480	394	447	308	504	575	
1½-6	888	727	822	570	930	1064	
1½-12	703	575	651	450	732	840	

TABLE 3.5 Suggested Locknut Tightening-Torque Values for Use with Standard Series Bolts

Locknut size and threads per inch	Steel hex locknuts						Steel hex flange locknuts					
	Grade B locknuts			Grade C locknuts			Grade F locknuts			Grade G locknuts		
	Clamp load, lb*	Locknut tightening torque‡		Clamp load, lb†	Locknut tightening torque‡		Clamp load, lb*	Locknut tightening torque‡		Clamp load, lb†	Locknut tightening torque‡	
		Max	Min		Max	Min		Max	Min		Max	Min
Coarse-Thread Series												
1/4-20	2,000	85	60	2,850	125	85	2,000	95	65	2,850	150	100
5/16-18	3,350	150	110	4,700	190	130	3,350	180	120	4,700	240	155
3/8-16	4,950	20	14.5	6,950	28	20	4,950	26	16	6,950	32	21
7/16-14	6,800	32	23	9,600	43	31	6,800	42	28	9,600	51	34
1/2-13	9,050	50	37	12,800	62.5	45	9,050	57	38	12,800	85	55
9/16-12	11,600	70	50	16,400	95	70	11,600	85	55	16,400	120	80
5/8-11	14,500	95	70	20,300	122.5	90	14,500	112	75	20,300	143	95
3/4-10	21,300	165	125	30,100	210	155	21,300	152	102	30,100	240	160
7/8-9	29,500	250	185	41,600	312.5	225						
1-8	38,700	375	275	54,600	462.5	350						
Fine-Thread Series												
1/4-28	2,300	90	65	3,250	125	85	2,300	115	75	3,250	160	105
5/16-24	3,700	160	120	5,200	200	140	3,700	200	130	5,200	230	155
3/8-24	5,600	22	16	7,900	29	21	5,600	25	17	7,900	33	22
7/16-20	7,550	34	24	10,700	43	31	7,550	45	30	10,700	60	40
1/2-20	10,200	52.5	37.5	14,400	70	50	10,200	66	44	14,400	89	59
9/16-18	13,000	77.5	57.5	18,300	95	70	13,000	94	62	18,300	132	88
5/8-18	16,300	97.5	72.5	23,000	125	90	16,300	120	80	23,000	175	115
3/4-16	23,800	165	120	33,600	210	155	23,800	192	128	33,600	270	170
7/8-14	32,400	270	200	45,800	312.5	225						
1-14	43,300	400	300	61,100	500	362.5						

*Clamp loads for grade B and grade F locknuts equal 75% of the proof loads specified for SAE J429 grade 5 and ASTM A449 bolts.
†Clamp loads for grade C and grade G locknuts equal 75% of the proof loads specified for SAE J429 grade 8 and ASTM A 354 grade BD bolts.
‡Torque values for 1/4- and 5/16-in sizes are in lb · in. All other torque values are in lb · ft.
SOURCE: The Industrial Fasteners Institute.

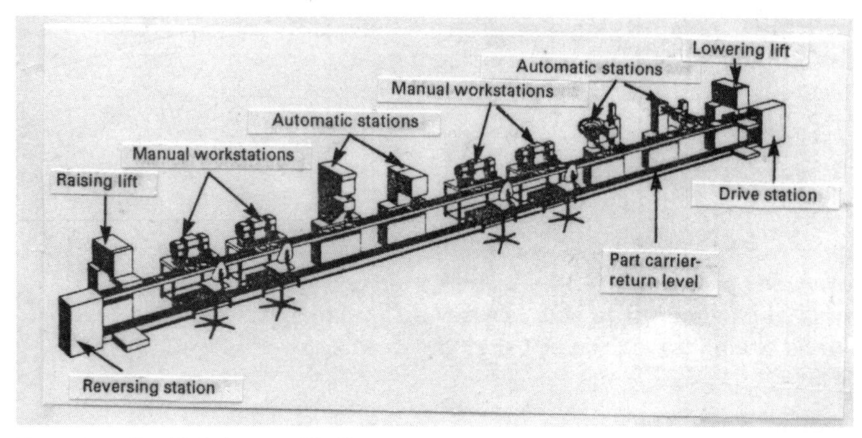

Figure 3.18 Double-belt assembly line.

2. Possibility of communication between the operators.
3. Ergonomic design of the workplace.

3.5.1 Double-belt assembly system

The needs of a flexible assembly system can readily be met by a double-belt assembly arrangement. The design of such a system on a modular line makes it possible to use standard feeding units for component transfer to manual and automatic stations arranged in-line or in a rectangle.

In the double-belt assembly line in Fig. 3.18, the parts to be assembled are transferred from one workstation to the next on part carriers which rest loosely on two constantly moving belts. The carriers are pallets, each of which contains one or more component jigs.

At each manual or automatic workstation, an escapement device stops the pallets while the belts continue moving. On completion of the operation at the station and after the escapement has been released, the carrier with its component is moved to the storage position at the following workstation by a transfer device.

At the manual stations, the escapement device is operated by a foot release. At the automatic stations, it is operated by the station's sequence controller.

On in-line assembly belts, when the pallets reach the end of the belt they are conveyed back to the starting point by means of various lowering and raising devices.

On a double-belt assembly system built in a rectangular configuration, the pallets are transferred by two double-belt conveyers at the same level, as illustrated in Fig. 3.19. Both conveyer tracks are con-

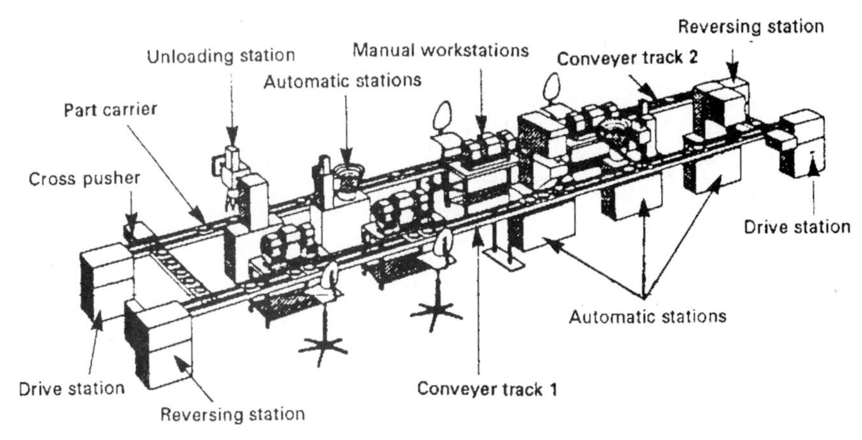

Figure 3.19 Double-belt conveyers in a rectangular configuration.

nected at the beginning and end by pneumatically or electrically driven cross-feeders. Since it is possible with this arrangement to use the belt track of the carrier return section for workstations the total layout is much shorter than for the in-line arrangement.

3.5.2 Manual arrangements in an automated environment (nonsynchronous system)

Manual workstations at double-belt assembly conveyers require certain standard equipment—work seat, foot pedal, work table, and part containers—all designed according to ergonomic principles. This arrangement is often called a *nonsynchronous* system; the cycle time of the system depends on operators integrated with the system, not on the automatic stations.

When an assembly conveyer is operated according to the principle of *work division,* each worker carries out a different operation in sequence, as illustrated in Fig. 3.20. This gives workers a degree of freedom of action with the loosely linked workstations.

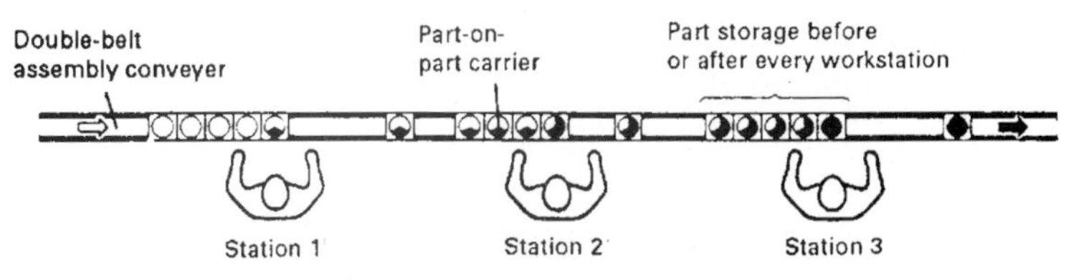

Figure 3.20 Work-division assembly on a conveyer.

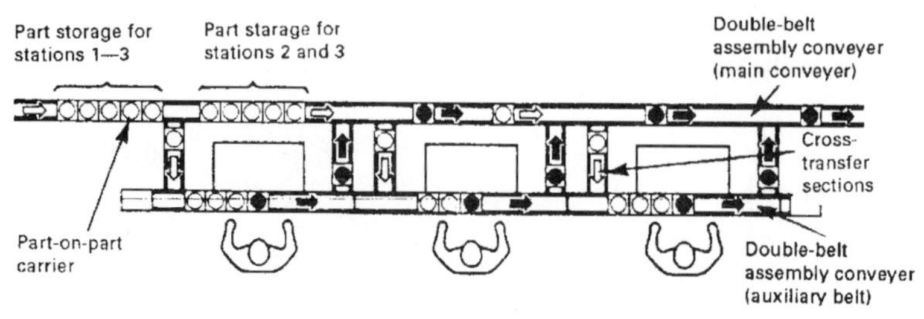

Figure 3.21 Total-work assembly on a double-belt conveyer.

If greater freedom of action is required for the operators, including the disengagement from other workstations, from each other, then so-called cycle-independent workstations can be set up. At these stations, assembly is performed according to the principle of *total work* in which each operator completes all operations in an assembly cycle. The pallets are fed from the main conveyer through cross-feed sections to an auxiliary conveyer at which similar manual workstations are situated, as illustrated in Fig. 3.21.

The command signals for a transfer to a workstation are initiated manually at each workstation; thus, the pallets are allocated according to the readiness of the operator. According to this principle, the operators can work at rates that are independent from each other.

With cycle-independent workstations it is possible to widen and enrich the operators' work.

Cycle-independent workstations also increase flexibility of an assembly line with respect to number of products assembled, variety of products, number of operators, and introduction of new manufacturing processes. Cycle-independent manual workstations with rectangular belt arrangements can be integrated with in-line automatic stations as illustrated in Fig. 3.22.

3.5.3 Automatic stations

Today, standard machines are available for assembly processes such as pressing, riveting, and screw driving, and can be included without difficulty as automatic stations in double-belt assembly lines. For many types and sizes of components, complete automatic stations for feeding and fitting can be built from standard modules.

Examples of available modules are

- Assembly frames
- Separators and escapement mechanisms for pallets

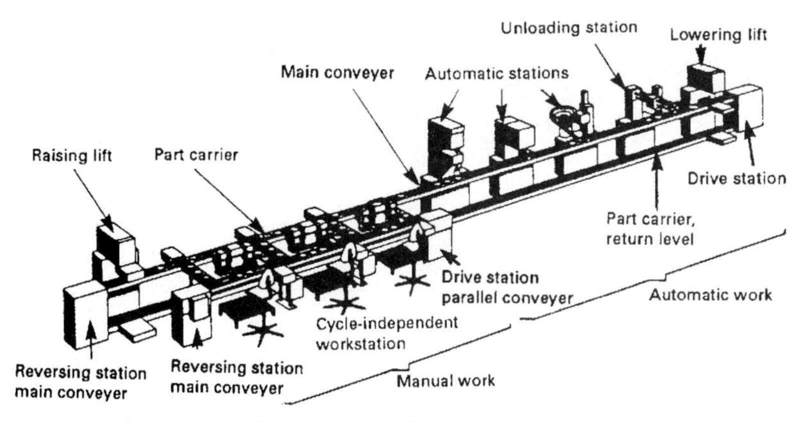

Figure 3.22 Integration of rectangular and in-line arrangements.

- Vibratory feeders
- Elevating feeders
- Belt conveyers
- Magazines
- Pick-and-place mechanisms with component grippers
- Modules for pneumatic and electrical control of automatic stations

Even for difficult assembly processes, automatic modules can be used, particularly for the following operations:

1. Assembly of such parts as sealing, insulating, and spacing washers after blanking from a strip
2. Fitting such parts as O rings, sealing rings, and slotted retaining rings with assembly sleeves
3. Assembly of spring clips and sealing pillows with expanding pliers

The growing demand for automatic stations, especially for part feeding and assembly, has lead to the development of numerical-controlled (NC) manipulating modules. These dc-motor-driven modules are in addition to traditional pneumatically operated modules. NC modules have an advantage in operations such as loading oriented pallets and magazines and assembling parts having different heights, lengths, and angular positions.

Generally, automatic stations are placed in line at the main conveyer. If, however, because of different cycle times it becomes necessary

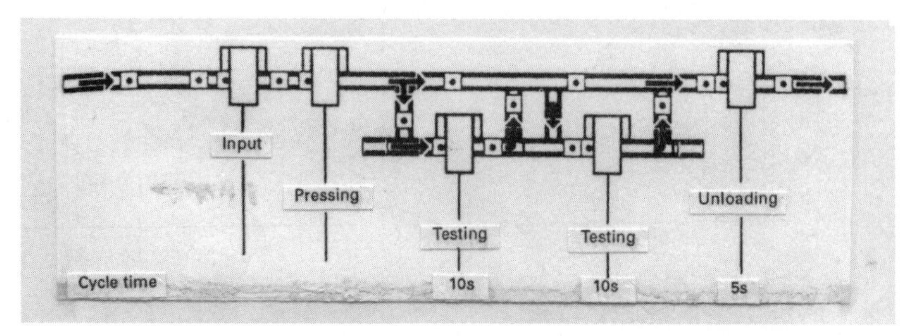

Figure 3.23 Automatic stations on an auxiliary conveyer.

to install several similar stations, these can be placed at an auxiliary conveyer dependent of the work cycle of the main conveyer, as illustrated in Fig. 3.23.

3.5.4 Component buffers

In an assembly system with component buffers, the components can be stored on pallets and double-belt conveyers. This method of storage retains the workpieces' relative positions and allows gentle handling. It also makes disengagement possible among individual manual workstations or (complete manual workstation groups) and automatic stations or between individual manual workstations and complete manual workstation groups.

With suitable configuration, any variations in cycle times at workstations and stoppages at automatic stations can largely be compensated by the buffers, as illustrated in Fig. 3.24.

For an assembly line systematically designed on modular principles, the control system is also a module that must be considered. Programmable controls with memories, used as sequence controllers

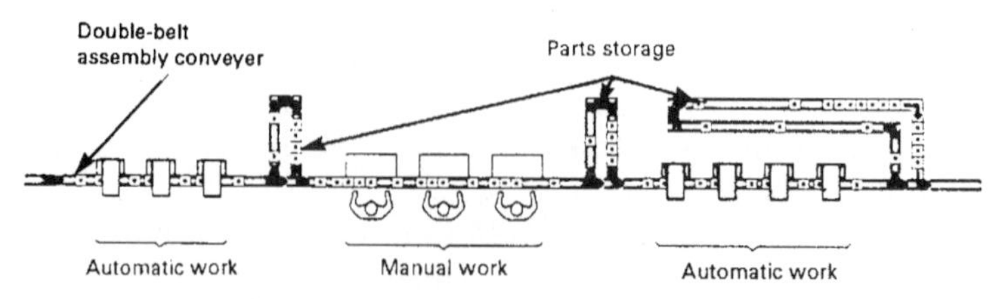

Figure 3.24 Use of buffers for temporary work storage.

for the part feeder system and automatic stations, and adaptive controls for numerical control, offer many advantages.

3.5.5 Affordable assembly lines

In planning assembly lines with a high degree of automation, it often turns out that manual stations can be placed between automatic stations for operations which cannot be automated, or are difficult to automate economically. In this case, an arrangement arises in which manual workstations (for difficult assembly and test operations) and automatic stations (for screw driving, riveting, and press operations) are in a mixed sequence, as illustrated in Fig. 3.25.

When the planning of manufacturing operations is combined with product design, then a better work structure can be obtained. By grouping all the manual operations on one side and the automatic ones on the other, manufacturing groups for manual operations and those for fully automated ones are created, as illustrated in Fig. 3.22. The formation of such groups avoids isolated manual workstations and improves communication between operators.

On an assembly line planned on the division of work principle, in which individual manual workstations and automatic stations are arranged in a varying sequence, an even work distribution between the manual workstations is often not possible. The output of such an assembly line and the wage content of the product being assembled are governed by the workstations with the greatest work content. This causes lost time for waiting at individual manual workstations.

In contrast to the principle of division of work, with its unequal work content at workstations, the principle of combining the work at cycle-independent manual stations has, in addition to the advantages already mentioned, the same work content at all workstations and therefore a lower wage content.

The most important factors that determine the formation of manual and automatic manufacturing blocks are

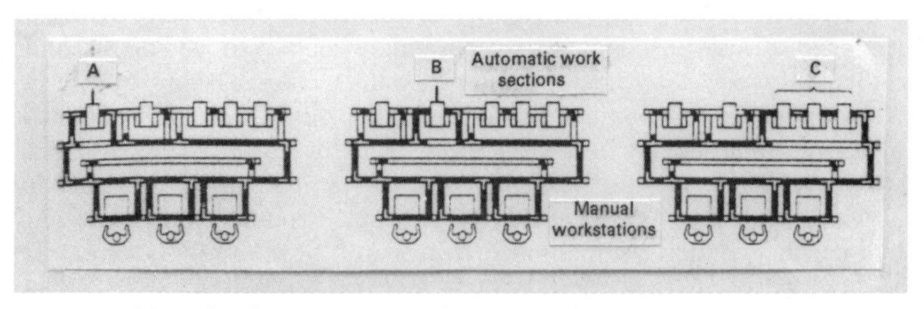

Figure 3.25 Manual and automatic operations in mixed sequence.

- Size of product
- Cycle time
- Number of workstations
- The type of work processes

The layout of an assembly line with manual and automatic work groupings can be arrived at in several ways.

For small products, the pallets can carry several parts. With multiposition pallets, all of the manual production processes at the different production stages can be completed without interruption of automatic production processes. The number of parts for each pallet must be equal to the number of the manual production stages to be combined. Every time a pallet completes a cycle, a fully assembled product is produced.

With large products, however, multiposition pallets generally are not feasible. To implement the separation of human and machine in this situation, multiple circulation of pallets may be used. A pallet may pass backward and forward between manual workstations and automatic workstations until completion of the product as required by the number of automatic production stages interspersed between the manual operations, as illustrated in Fig. 3.25.

The movements of the pallets are controlled by appropriate coding on each unit.

3.4 Affordable Cellular Manufacturing for an Unmanned Factory

Constantly rising wages have for some time been forcing the manufacturing industry into full automation. Although it has been primarily parts manufacturing that has been automated, automation measures are now being concentrated on the assembly sector. Automatic assembly, however, necessarily involves a high degree of technical complexity and high costs. It is therefore possible to apply an automated assembly system economically only if the capital-intensive means of production are used to their full capacity over a long period of time. This is made difficult, however, by more frequent product changes, and by the increasing variety of products made in smaller quantities. Moreover, it is increasingly necessary to react flexibly to customer wishes.

Conventional automatic assembly systems are mainly designed for assembling large quantities of one product. Product changes or supplementary versions, which often result in a new assembly sequence,

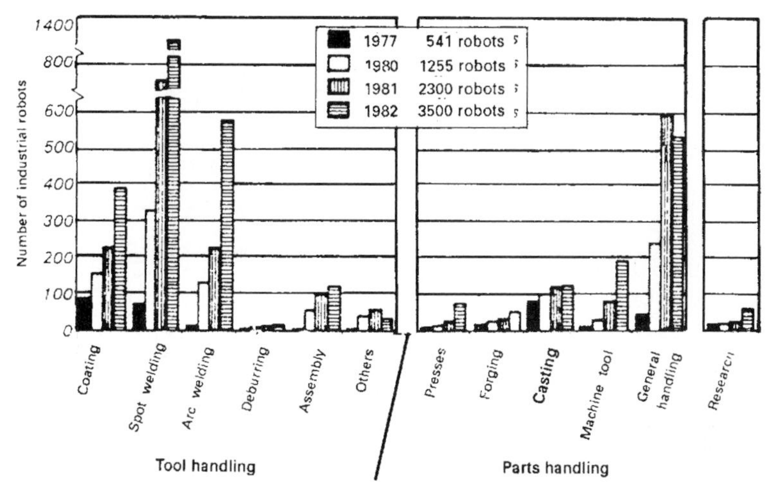

Figure 3.26 Change in use of industrial robots in Germany in various fields of application.

therefore usually necessitate complex conversion measures, insofar as it is even possible to adapt the assembly system. Successor products can only rarely be assembled automatically with an existing system.

These problems have created a demand for greater flexibility in automated assembly systems. One method of achieving more flexible automation is to use industrial robots. Compared to other manufacturing sectors, the number of industrial robots used in the assembly sector is still very low (Fig. 3.26). However, the mere use of industrial robots in assembly systems does not guarantee a flexible overall system. The assembly sector in particular requires a wide range of peripheral devices (e.g., assembly devices, tools, part feeders) in addition to the industrial robots. Both these peripheral devices and the linking system must have a flexible design to take full advantage of the flexibility of the industrial robot.

3.4.1 Cellular structure for flexible automated assembly

A concept for a flexible automated assembly system has been developed at the Fraunhofer Institute for Manufacturing Engineering and Automation (IPA) in Stuttgart. The overall concept is based on a system structure independent of specific assembly sequences. This concept is described below using the example of a system for various types of automobile assembly.

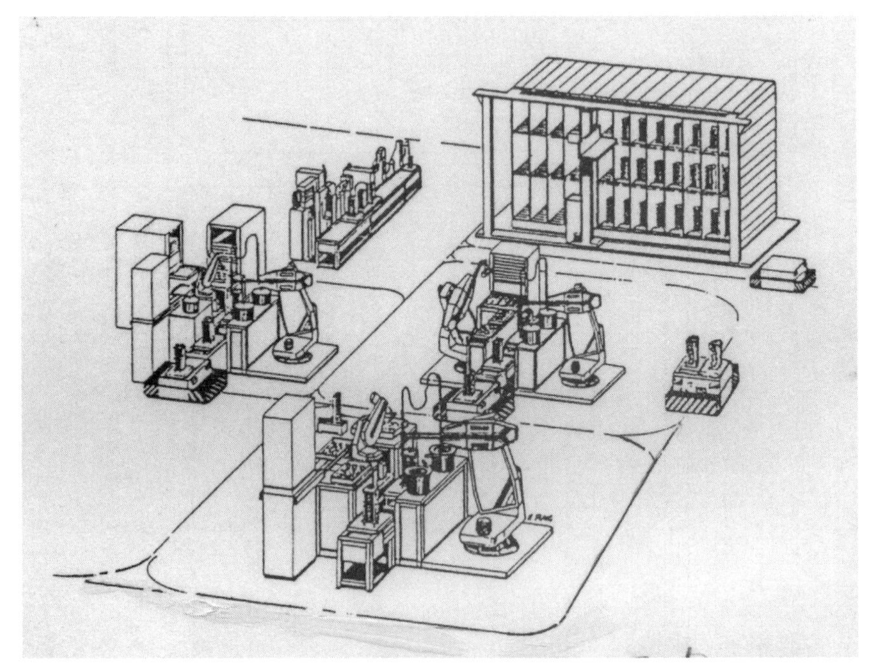

Figure 3.27 Cellular structure for automated assembly.

The assembly system shown in Fig. 3.27 consists of three parts:

1. Programmable assembly cells
2. System storage
3. Flexible conveyer system

In addition to these largely product-independent subsystems, it is possible to integrate function-specific special stations in the assembly system (e.g., test stations, cleaning station, processing stations). The central features of this assembly system are programmable assembly cells, where the actual assembly activities are performed with industrial robots.

The individual programmable assembly cells, the special stations, and the system storage are linked together by automated guided vehicles equipped with a roller conveyer for automatic acceptance and transfer of pallets with the assemblies. The automated guided vehicles are also intended to link the assembly system with the preceding and subsequent production sectors. The assemblies are conveyed to the operating rooms of the industrial robots in the assembly cells and to the special stations along friction conveyers. In conjunction with indexing equipment, these conveyers permit precise positioning of the assemblies at the assembly location and serve as buffers for the particular programmable assembly cells and special stations in case of temporary faults.

It should be possible for the assembly system to work automatically in three shifts independently of the adjacent production sectors. However, if these production sectors are manual or semiautomated, they will normally be operated in a maximum of two shifts. This is the reason for the storage subsystems, which ensures that parts are supplied to and retrieved from the assembly system, even if no production takes place for one or two shifts in the adjacent sectors. A further function of the storage is to act as a buffer for the assembly system in case of faults. The storage unit consists of shelves with programmable shelf control that gives random access to the individual storage location. The number of storage locations in the shelf depends on the shift interval which must be breached.

One of the main problems involved in flexible automation of assembly is supplying of the parts to be assembled. Since the assembly system is also intended for operation during unmanned shifts, interrupt-free supply of parts must be guaranteed for a period of several hours. This problem can be solved by supplying parts in pallet magazines. These are manufactured from plastic in a deep-drawing process, this method being extremely economical. In addition to the main task of ensuring the position and orientation of the parts, these pallet magazines also fulfill subsidiary functions, such as stacking and transporting.

Although pallet magazines are designed to accommodate particular parts, their external shape and their dimensions are standardized. This means that the same equipment can be used for storing and supplying different parts. The pallet magazines are stacked in pallet stores, from where they are conveyed directly to the working space of the industrial robot. The programmable assembly cells can thus be automatically supplied with parts in up to three shifts.

The pallet stores take the form of mobile storage devices; they can thus be loaded centrally with pallet magazines in the storage area and used to provide material flow. This concept also allows an empty store to be replenished by a full store in a very short time.

An important basic rule of automatic production is intended to prevent unnecessary repetition of organization processes: a part position which has already been defined once should not be altered before termination of all manufacturing, assembly, and packing processes without compelling reasons. The use of pallet magazines enables this requirement to be met. On termination of the preceding production process the parts can be loaded either manually or automatically into the pallet magazines. Since these magazines are ideally suited for transport because of their stacking ability, they can also be used by suppliers or in distant production sectors.

The use of industrial robots in conjunction with sensors fulfills an important requirement for unmanned operation: automatic monitoring of the assembly process.

The most important features of the cellular assembly system are

1. Minimum proportion of product-specific devices
2. Rapid changeover is possible
3. The structure is independent of assembly sequences, and thus easily adapts to new assembly sequences
4. Good expansion capacity
5. Automatic material flow
6. Low susceptibility to faults

This flexible automatic assembly system is largely independent of specific products. It can therefore be adapted to assembly of a wide range of products. The individual elements of the assembly system will then differ in detail from those described; however, the advantages of the overall concept will remain.

3.4.2 Restrictions in applying flexible assembly systems

In the short term, it will not always be possible to apply the flexible automatic assembly concept for a number of reasons, such as the often limited space available in existing production shops and the need, for financial reasons, to continue using existing production facilities. However, in the long term it will be possible to apply flexible automatic assembly concepts, especially when new production shops are built or existing ones extended.

It is, however, the aim of many enterprises to increase productivity in the short term by using a greater degree of automation in existing assembly lines. In such cases it might be possible to integrate particular assembly subsystems. In particular, it is possible to use programmable assembly cells in conjunction with parts supplied in pallet magazines as a substitute for manual workplaces.

Chapter 4

Design for Affordable Automation

In a competitive market, product design is one of the primary attributes affecting a product's market share and profitability. An example is the personal computer market, where expanding sales have created a large demand for floppy disk drives. The major suppliers of floppy disk drives are producing sufficient quantities to consider automation of manufacturing. To be profitable and competitive in the field, many manufacturers are designing their products to facilitate automation and reduce cost.

Here, four floppy disk drives are analyzed by the design for assembly (DFA) technique. The drives are 3½-in disk units with the same storage capacity, equivalent functionality, and approximately the same quality and reliability. The drives sell for about the same price. Two of the drives are reportedly designed for automatic assembly.

The major difference between the drives is the number of parts and the ease of assembly. Thus, for the same product volume, the ease with which each is assembled will determine which drive produces the most profit for its manufacturer.

4.1 Design for Assembly

Design for assembly is a disciplined technique for analysis of a product. The product is disassembled one part at a time, while part and assembly characteristics are observed. The drives are evaluated on the basis of these characteristics (ease/difficulty of feeding, handling, insertion, and assembly) and their manual and automatic assembly costs, manual assembly time, and estimated equipment costs are determined.

Characteristics of parts used in the main assembly of each of the drives are shown in Table 4.1.

TABLE 4.1 Part Characteristics

Part	Easy to feed, %	Easy to grasp and manipulate, %	Easy to orientate, %
A	59	65	63
B	63	75	69
C	80	90	55
D	80	93	60

TABLE 4.2 Assembly Characteristics

Part	Straight line from above, %	View clear, %	Access clear, %	Self aligning, %	Easy to insert, %	Self locating, %
A	69	96	94	92	82	61
B	75	98	77	40	79	71
C	77	74	68	26	84	74
D	85	100	95	20	85	65

TABLE 4.3 Main Drive Assembly Stations and Costs

	Automatic assembly			Manual assembly			
Drive	Auto work-station	Manual work-station	Cost, U.S. $	Manual work-station	Cost, U.S. $	No. of parts	No. of sub-assemblies
A	50	9	3.02	54	3.05	65	3
B	49	10	2.99	58	3.27	61	7
C	28	6	2.00	42	2.35	47	6
D	36	6	2.35	47	2.61	54	4

The characteristics of the parts from an assembly standpoint (feeding, grasping, manipulating, and orienting) are shown in Table 4.2.

The number of workstations and cost of assembly for the main assembly of the drives is shown in Table 4.3. The number of manual workstations calculated for automatic assembly indicate that it would not be practical to assemble the drive completely automatically. The estimated capital equipment costs for automatic assembly of the drives are listed in Table 4.4.

Assembly costs of the drives ranged from \$2.00 to \$3.02 per drive, a difference of 50% of one million in assembly cost based on the production volume used for the analysis (one million drives per annum). The capital equipment costs range from \$4.5 to \$6.3 million, a difference of \$1.8 million or 40 percent. These differences are quite significant and are solely a result of product design.

TABLE 4.4 Capital Equipment Costs

Drive	Automatic assembly, U.S. $	No. of parts	No. of subassemblies
A	6.3 million	65	3
B	5.9 million	61	7
C	4.5 million	47	6
D	5.2 million	54	4

Product design guidelines for assembly automation are summarized as follows:

1. Use the minimum number of parts.
2. Insert parts unidirectionally.
3. Stack inserted parts.
4. Use one screw size; minimize screw fasteners and retaining rings.
5. Avoid cables from components or provide means for handling them.
6. Insert all electrical connections in same plane as the assembly.
7. Keep maximum part weight within the limitations of the handling equipment.
8. Locate mating parts by pins or wells to align them for transport prior to fastening.
9. Ensure that parts can be assembled with one hand.
10. Chamfer mounting holes and shafts for close-fitting parts.

One message regarding product design is distinctly clear; in products with the same functionality, reliability, and price, the manufacturing cost of a product designed for assembly can be significantly less than that of its competitors. Design for assembly practices can reduce product cost and improve profitability without decreasing functionality or performance; they are a major contributor to a product's competitiveness.

4.2 Component Design for Affordable Assembly

Adapting the design of a product to match the characteristics of assembly machines is one of the essential conditions for affordable automation in assembly. An important aspect of all assembly operations is the handling of assembly parts and connecting elements. Implementing standard guidelines for handling and orienting of parts for automated assembly can best be done during the component design phase.

The present situation in assembly, in contrast to the situation of machining of parts, is not susceptible to rationalization (the process of examining alternatives and applying the principles of DFA). Small production volumes make the factory expenditures and the necessary investments too high to introduce rationalization into assembly operations. Also, the complexity and precision requirements of an assembly process cause additional restraints for material delivery, quality control, and the storage of parts. In addition there has been a lack of useful techniques for the creation of good organizational conditions to deal with integrating partial subassemblies. Moreover, the technical equipment vital to automatic assembly processes is not often available. However, the latest developments in handling techniques will make it possible to introduce innovative automated devices to rationalize the conditions in the assembly.

Experience with conventional assembly machines shows, however, that the assembly-oriented design of products is an essential precondition for an economic automation. If assembly-oriented design methods are used, the design demands of automation systems for assembly can be met.

4.3 Automated Process Demands on Product Design

Engineering attention often concentrates on product design only to achieve functional goals; little attention is given to design for product assembly processes. This tendency creates significant difficulties for production, manufacturing, and material management personnel. In addition to these internal influences, there are a number of external influences, such as the customer's requirements, which often have pivotal effects in determining product configurations, as illustrated in Fig. 4.1.

The designer is usually faced with a myriad of challenges to consider in order to satisfy the multiple demands for the product performance, qualitative and quantitative. One way to meet some of these challenges is to concentrate on product design to yield an assembly-oriented product.

4.4 Oriented Design Alternatives

It is imperative that an engineering designer extend the conceptualization of components for a product to incorporate two additional

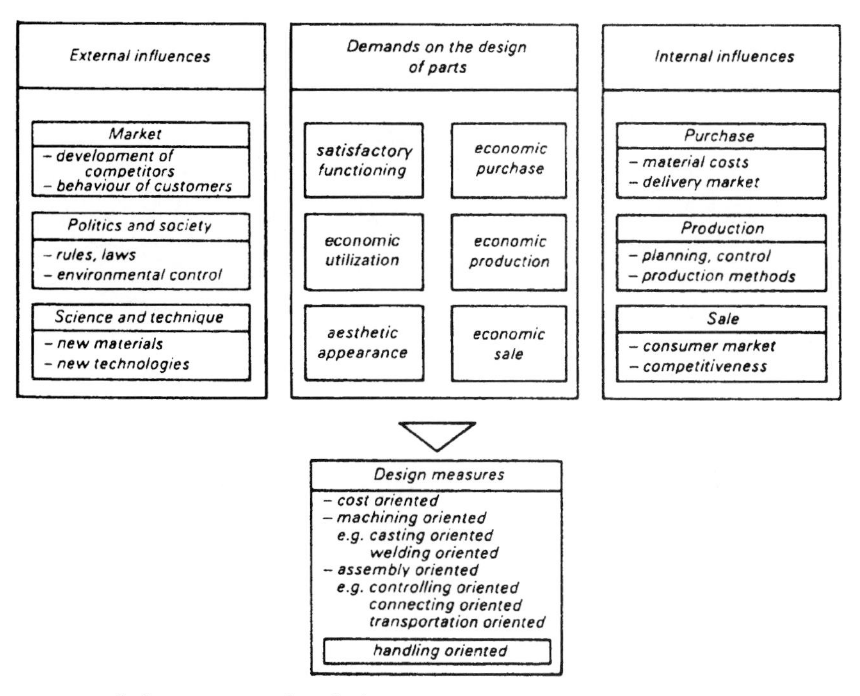

Figure 4.1 Influences on product design.

major activities: the fabrication of these components and the method of assembling and integrating them to produce the final product in the desired quantity at the highest quality possible.

The designer must also create compatibility among the component design, fabrication, assembly, packaging, storage, and delivery. This can be accomplished by applying the principle of oriented design alternatives. The principle is applied to three areas:

1. Structuring of product
2. Assembly-oriented design
3. Standardizing of parts

4.4.1 Structuring of product

The structuring of product and the standardization of parts are conducted at the rational division of assembly tasks. The frequency of parts coming in contact with each other will play an important role in developing a concept of assembly simplification. This is product structuring for alternative assembly.

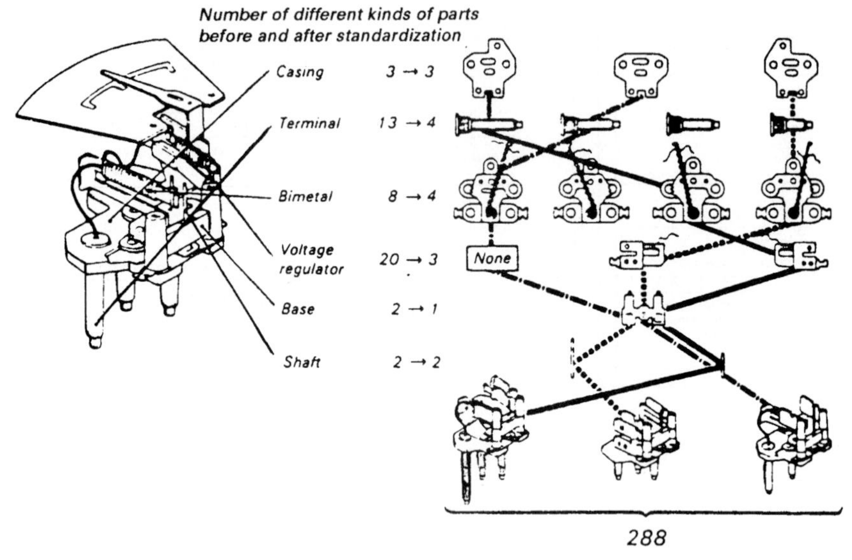

Figure 4.2 Example of systematic structuring and standardizing.

4.4.2 Assembly-oriented design

The assembly-oriented design technique helps to arrive at the simplest method of assembly. It may be assumed that a design that takes into account automatic assembly operations will also simplify its manual assembly. An example for the systematic structuring and standardizing of a product is illustrated in Fig. 4.2.

4.4.3 Standardizing of parts

Figure 4.2 shows the structure of an electric indicator instrument for the automotive industry. By standardizing on six structural levels, the number of different parts was reduced from 48 to 17. This means that storing and feeding parts for automated assembly has been simplified. In addition, expenditures for the parts supplies have been reduced. The advantages of standardizing parts in this case are evident.

If it is necessary to change the product's design to achieve an improvement, once it has been structured and standardized, it is recommended that each assembly operation be reevaluated to determine the impact of redesign.

During the execution of assembly operations, parts are manipulated. It is evident that handling and manipulating of parts is one of the most important functions of the assembly process. Sometimes it is difficult to distinguish between handling and connecting functions, since a connecting operation is often completed with the execution of a han-

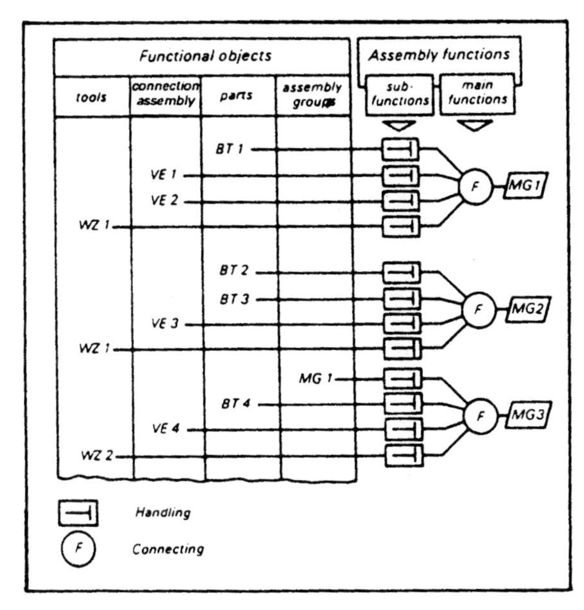

Figure 4.3 Relationship of handling and connecting in assembly.

dling function. Therefore, in many cases assembly-oriented design can be characterized by the way in which it is handled. The relationship between handling and connecting parts is illustrated in Fig. 4.3 for a simple functional process.

Rationalized reduction of assembly operations refers to a reduction of the number of assembly categories according to assembly-oriented design principles. The extent to which such measures can influence a functional process is illustrated in Fig. 4.4, which shows how the process in Fig. 4.3 can be simplified.

Repetitive assembly operations, including part handling, can be avoided by reducing the number of connection elements and by integrating different parts in a single unit. Accordingly, the assembly operations can be greatly simplified by concentrating on the simplification of the handling processes. The handling functions can be regarded as a series of elementary subfunctions, such as picking up, moving, and laying down (Fig. 4.5).

Simplification of handling functions can be approached in two basic ways, as Fig. 4.5 indicates:

1. Reducing demands that are made on the capabilities of a handling device by removing obstacles for the execution of handling functions

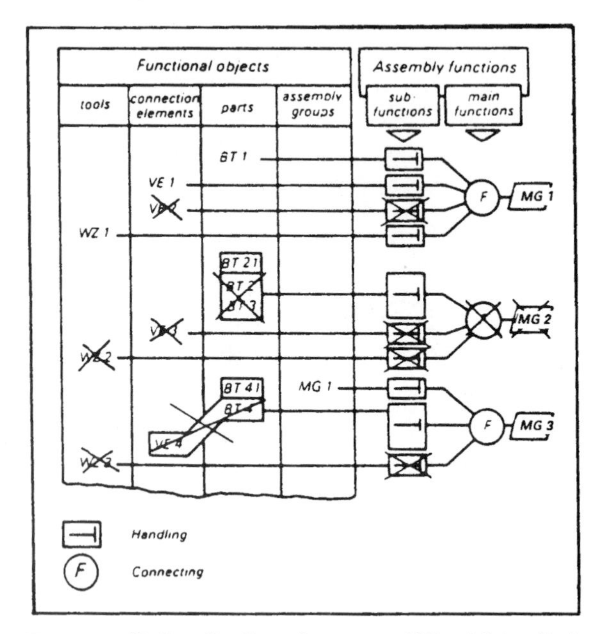

Figure 4.4 Rationalization of process of Fig. 4.3 to eliminate and consolidate functions.

		Simplification of handling functions	
		reduction of the demands on handling by the part	improvement of the suitability for handling
Picking-up	recognition of workpieces	reduction of the demand for orientation	adding of orientation aids
	transfer of forces	reduction of the demand for gripper elements	adding of gripping elements
Moving	change of workpiece position and orientation	elimination of obstacles for movements	adding of feeding aids
	change of quantities	elimination of the tendency for hooking	——
Laying down	reaching of final positions	reduction of the demand for positioning	adding of positioning aids
	releasing of forces	reduction of the demand for gripper elements	adding of suitable gripper elements

Figure 4.5 Analysis and simplification of handling subfunctions.

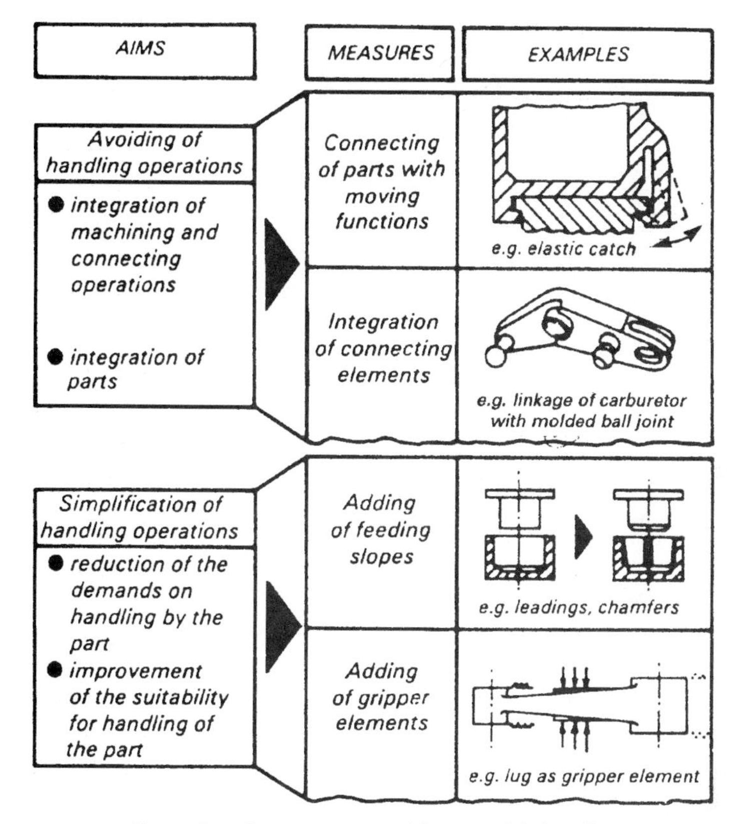

Figure 4.6 Examples of measures to avoid or simplify handling.

2. Improving the handling characteristics of a part by adding special features to it

An example is springs, which create difficulties for automated assembly because they tend to hook into each other. This tendency of tangle can be eliminated by using alternate windings and closely rolled wire ends. In the same manner, chamfers and guiding slopes can be added to make it easier to mate workpieces. Other examples are illustrated in Fig. 4.6.

4.4.4 Workpiece suitability for automatic handling

A checklist such as that in Fig. 4.7 can be a helpful means to examine the suitability of parts for automatic handling. The checklist divides workpieces into three classes:

Workpiece characteristics	Dimension of workpiece characteristics		
Form elements	orientation marks at the outer contour	same orientation by orientation marks at the outer contour	no orientation marks at the outer contour
	simple geometry of the gripper elements	complex geometry of the gripper elements	no gripper elements
	positioning aids	force-dependent positioning	no positioning aids
Relative position of the workpieces	one-sided accessibility (box construction)	accessibility from several sides (straight-lined)	limited accessibility
Form	compact	complex workpiece	dangling part
Dimensions	medium dimensions	large dimension in one direction	big or extremely small dimension
Position of gravity centre	one stable workpiece position	several stable positions	nonstable workpiece positions
Surface quality	no special demands on transfer of forces	special demands on transfer of forces	surface dirty, rusty, etc.
Weight	middle weight		heavy or extremely light
Strength	rigid	elastic	plastic
Symmetry	multiple symmetric	simple symmetric or obviously nonsymmetric	apparently symmetric
Workpiece is:	to be handled automatically	to be handled with difficulty	hardly to be handled

Figure 4.7 Checklist for workpiece suitability.

1. Suitable for automatic handling
2. Difficult to handle automatically
3. Unlikely to be handled automatically

The checklist generally does not provide a final judgment about the suitability of parts for automatic handling, because it does not consider the capabilities of the handling devices within a factory. It does, however, give essential hints on the difficulties that may be encountered when automated assembly is planned.

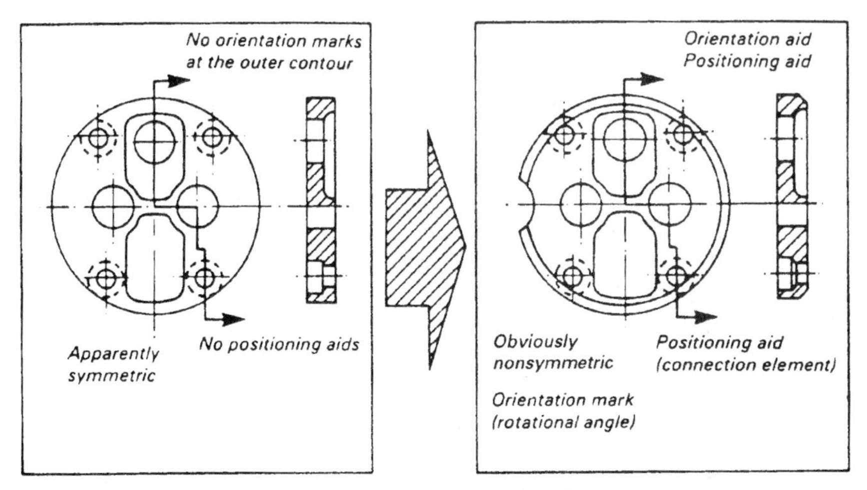

Figure 4.8 Redesign of a bearing cover for automatic handling.

Figure 4.8 illustrates an example of the redesign of a workpiece to improve handling, based on the guidelines and the checklist. The figure shows a bearing cover for a lubricant pump. The checklist revealed the weak points in the design of this part: it had no orientation marks or positioning aids and it was apparently symmetrical. The redesign added several orientation and positioning aids to the outer contour of the part that may easily be interrogated by simple mechanical sensors. The part was also made obviously nonsymmetrical.

4.5 Design Optimization for Affordable Automatic Assembly

Product design is the first of many steps in the manufacturing processes. Opportunities for production efficiency and possibilities for rationalization of assembly are established at the product design stage. However, the task of designing a new product for automated assembly is often combined with the task of designing new and specialized assembly equipment. This is because the development of commercial assembly equipment proceeds slowly, and the necessary assembly machines may not yet be available.

These tasks have to be performed simultaneously, imposing considerable demands on the designer to foresee the consequences of decisions. The designer will usually be engaged in the following activities:

1. Designing product to attain high quality
2. Establishing design parameters that directly determine the ease of assembly
3. Continuously learning of the principles of design for assembly

4.5.1 Assembly optimization

Rationalization of assembly must proceed as a total function; in other words, the assembly process must be optimized with respect to all operations of product design and production. There are four main goals that are directly linked to the optimization process:

1. Improve effectiveness of assembly; i.e., provide increased productivity in relation to labor and investment resources.
2. Improve product quality; i.e., improve product value from the buyer's viewpoint in relation to the product cost.
3. Improve profitability of assembly systems; i.e., increase utilization of equipment.
4. Improve the working environment within the assembly system.

Production systems are normally conservative. The consequences of incurring changes in product or system create various predictable and unpredictable difficulties. Therefore, any change must be carefully considered and controlled if it cannot be avoided. A reorganization of assembly should not be regarded as an end in itself, but should be regarded as a link in a total rationalization process based on the four goals mentioned above.

4.5.2 Product rationalization and DFA

The designer determines the structure of the product and of its related components together, as well as the techniques for joining these components. Additionally, the designer determines the detailed design for each of these components. This process normally results in a production process with explicit assembly steps. If the designer proposes another product structure, however, the process would have to be altered to accommodate it. Similarly, if the designer proposes a different component design, certain parts of the process would have to be modified to accommodate the changes. If, however, production is regarded as the starting point, particularly the assembly operations, then optimal assembly methods can be exploited.

A product cannot be regarded in isolation when assembly problems arise. Certain subsystems in the product can appear in other products and certain components can be applied in various subsystems.

The product rationalization concept can be summarized as follows:

- Creating degrees of freedom for product design. This encourages several alternatives to reach acceptable product assembly.
- Applying the principles of design for assembly to guarantee improvement of product assembly.

The overall goals for optimizing the assembly operation are

- High and consistent product quality
- High productivity, quantitatively and qualitatively
- High utilization of capital equipment
- High quality of work environment

The utilization of capital equipment is influenced by the product mix to be assembled by the same equipment. Obviously, if the main product design is free from variations, it undoubtedly will yield an excellent utilization factor. However, if product variations are necessary and have been considered from the inception of the main product, the utilization of capital equipment will still be good.

The productivity of a system is influenced by the frequency and duration of downtime to adjust to product variations. Thus, one must either design a system to accept many variations in products or ensure that it handles products with few variations.

Therefore, one has to decide on a policy of accepting a certain product mix in order to obtain maximum benefits of rationalization. The principles that must be considered are listed in Table 4.5.

4.5.3 Product structure

The study of the fundamental structure of a product must be geared to determine the requirements for all components so that they satisfy the demands for product performance. The designer constantly strives to achieve a fundamental structure level where problems and solutions are logically connected. Also, the designer makes decisions on shape and dimensions for components in a product group, arranging for component tolerances, positioning components in the product, and loading products into a machine for assembly. Simplicity and clarity in product design are considered important for obtaining the optimum solution for an automatic assembly problem. Also, reducing the number of parts and assemblies plays a vital role in ensuring a profitable product.

An example of a structural analysis of a product is illustrated in Fig. 4.9. Figure 4.10 illustrates three different production methods for a zip

TABLE 4.5 Survey of Main Principles for the Design of Components for Automatic Assembly

Avoid assembly operations:
Integrate component
Utilize integrating production methods
Avoid orientation operations:
Use magazines
Use components connected in bands (tapes)
Integrate the production of components into the assembly
Facilitate the orientation operations:
Avoid clamping or hooking
Put special faces on the component for orientation
Avoid components of low quality
Make the components symmetrical
Alternatively, make the components clearly asymmetrical
Facilitate transport:
Design the component for easy transport
Design a base component
Facilitate a simple pattern of movements:
Make all joins simple
Put special faces on the component for guiding purposes
Choose the right method for joining components together:
Avoid joining
Avoid separate connecting elements
Use integrating production methods

fastener. In Fig. 4.10 *a,* metal zip fastener elements produced by pressure casting are mounted one by one around a band. In Fig. 4.10*b,* the band is laid in an injection-molding machine and each element is molded around the band. In Fig. 4.10*c,* the zipper is formed from plastic cord bent and sewn into shape as the teeth of the fastener.

4.6 Basic Design of Components

The structure of a product determines the basic design of the components and the method of assembly. This means that the choice of the product structure establishes most of the assembly procedures. Considerable care must be exercised in detailing the design of components in order to ensure easy orientation in the assembly system.

Detailed design of components means specifying the basic properties, such as shape, material, dimensions, surface quality, and tolerances. The assembly quality will depend on these basic properties. The most vital one is shape, since certain surfaces on the components are utilized in the assembly process itself.

Assembly system surfaces are those used for orientating, transporting, positioning, and guiding. Designing components for assembly im-

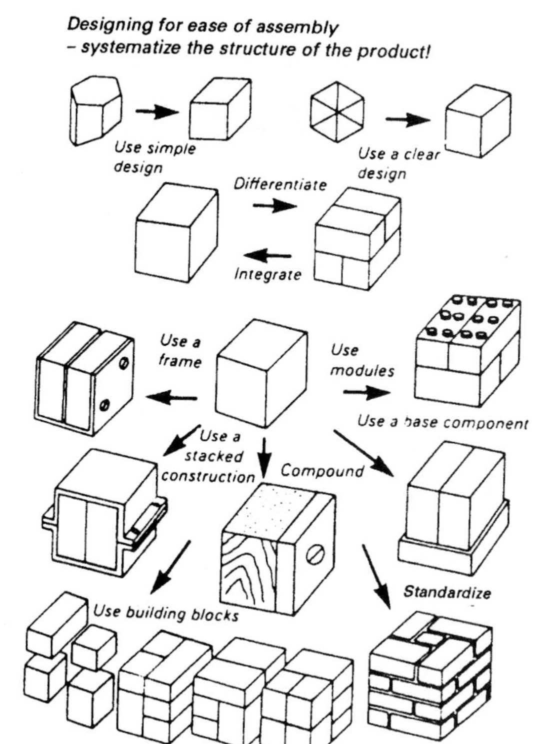

Figure 4.9 Example of structural analysis of a product.

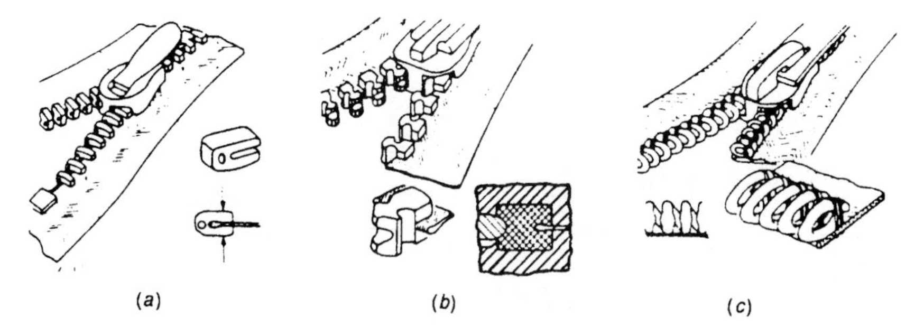

Figure 4.10 Three methods for assembling a zip fastener.

plies utilizing their functional surfaces to manipulate or change its orientation so as to create an acceptable condition for the assembly operation, as illustrated in Figs. 4.11 and 4.12.

Table 4.5 lists the main principles for designing acceptable components for assembly operations.

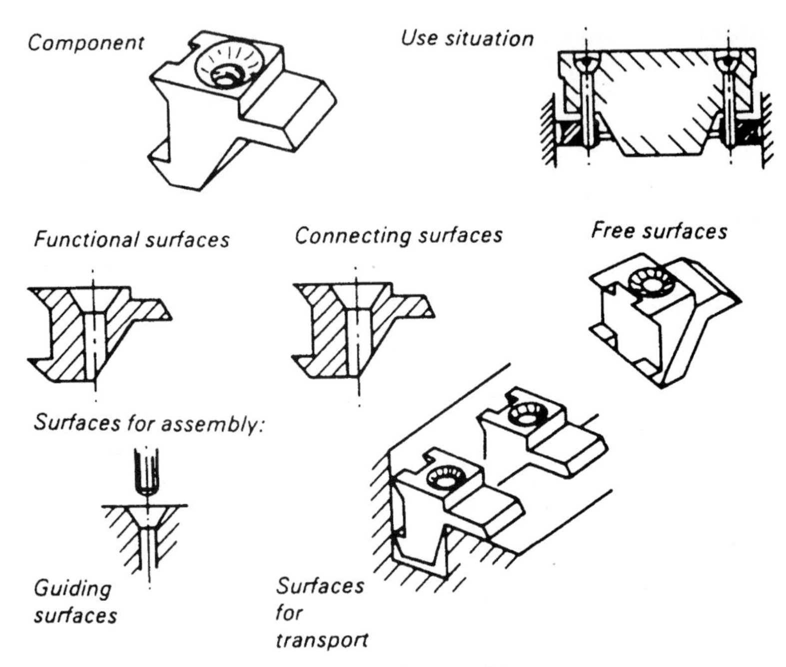

Figure 4.11 Methods of orienting parts for assembly.

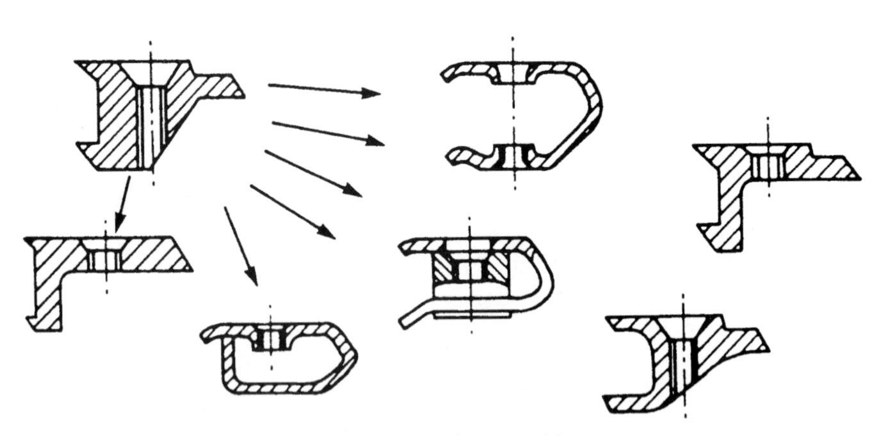

Figure 4.12 More methods of orienting parts for assembly.

4.6.1 Design strategy

The main concern in design for affordable assembly, is applying design principles during the production preparation period. It may be too late to incorporate new concepts in the assembly process when all drawings and parts lists are completed; since the designer has already consciously chosen the assembly methods. Thus, one must use a design strategy

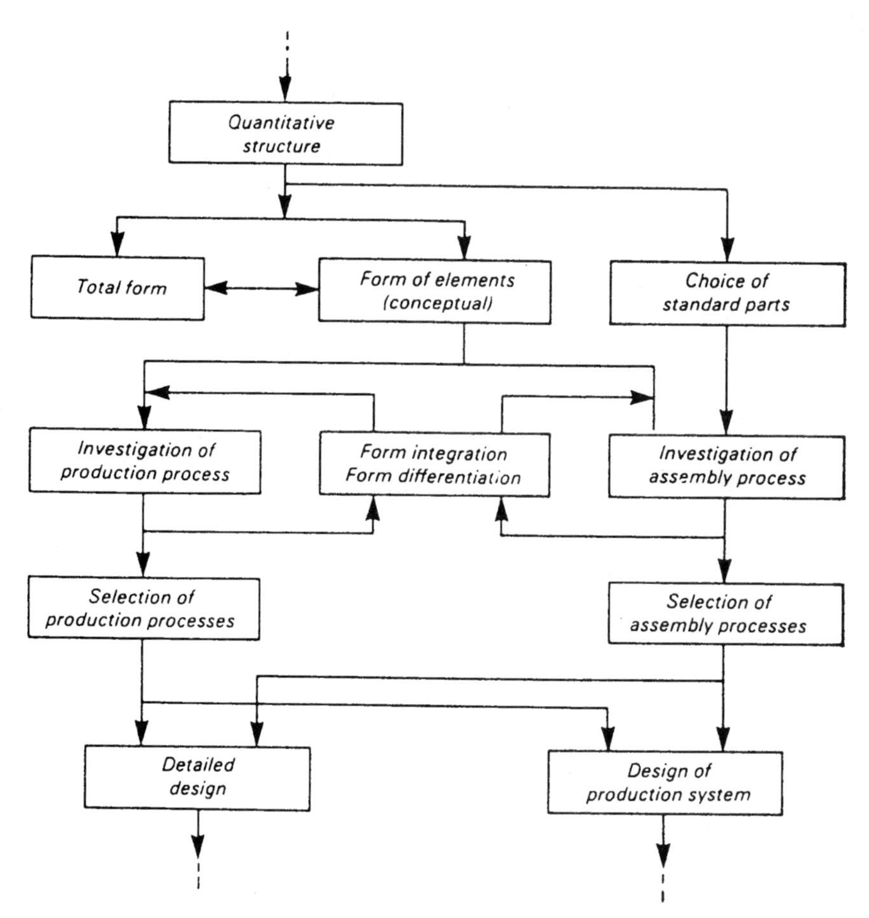

Figure 4.13 Strategy for design for assembly.

for assembly while components are being developed for the product.

The process of product design may take the form illustrated in Fig. 4.13. In this scheme it is possible to observe the effects of certain alternative solutions. Alternatives are necessary for finding the optimum assembly solution.

4.7 The Case of the Simonsen ECG Electrode

Simonsen Corporation, a Danish company, developed an electrocardiograph (ECG) electrode monitor for patients with heart conditions. An ECG electrode picks up a signal from the small electric currents produced by the patient's heart. Simonsen had purchased the electrode

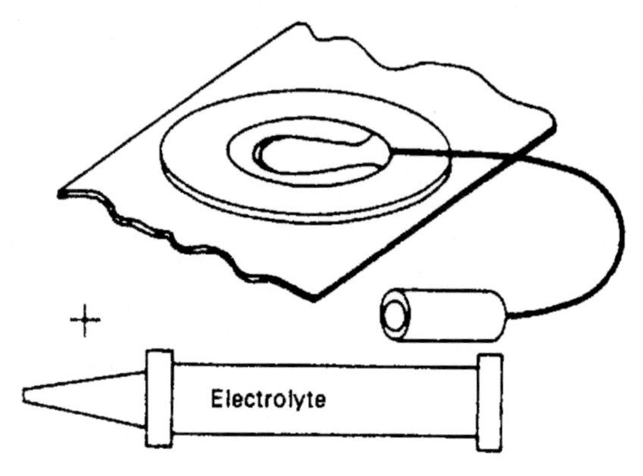

Basic specifications:

- Disposable product (lifetime max. 8 days)
- Electrode: silver–silver chloride
- Self-adhesive
- Production: 3 million/year
- Variations:
  - wire lengths, 10–100cm
  - 4 different types of plugs
  - 3 different plug colors

Figure 4.14 ECG electrode.

components from an American firm, then assembled them in Denmark, incorporating other electronic components to produce the final system. However, in order to increase the profit margin of the company, it was decided to produce the ECG electrodes in house. The electrode is illustrated in Fig. 4.14. Simonsen desired also to develop automatic equipment for production.

The company therefore instituted a parallel development project concurrent with its existing operations. The parallel development project consisted of:

1. Development of new ECG electrodes
2. Development of production technology for full automation
3. Development, debugging, and delivery of a fully automatic production system

A very simple model of the project is illustrated in Fig. 4.15. Two subprojects of very different types thus were carried out simultaneously: a horizontal project which led to the final product, and a vertical project, which led to the production system.

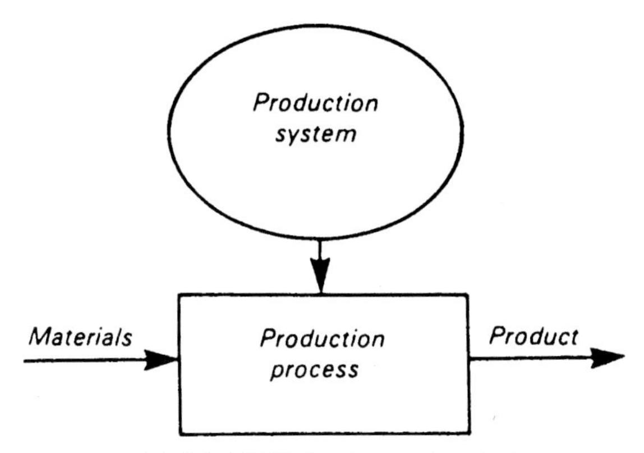

Figure 4.15 Model of ECG development project.

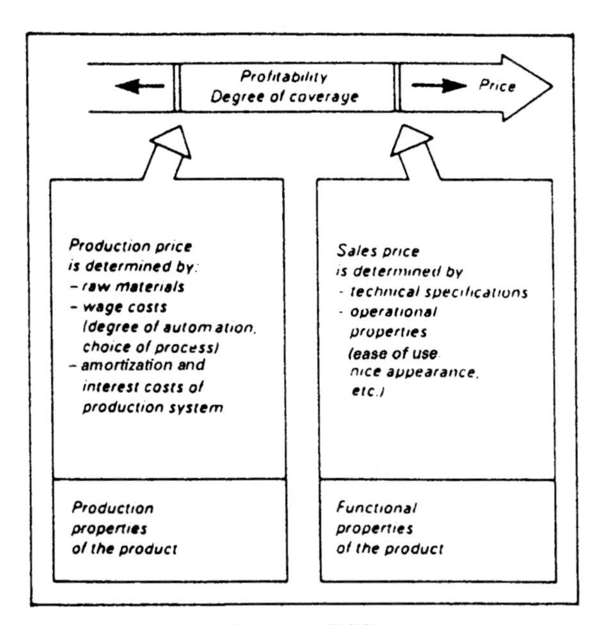

Figure 4.16 Economic factors in ECG project.

4.7.1 Considerations

The aim of the project was to create the most profitable commercial opportunity for the company. It was therefore necessary to consider the project not only as an exercise in manipulating technical factors, but also as an effort in manipulating economic factors, as illustrated in Fig. 4.16.

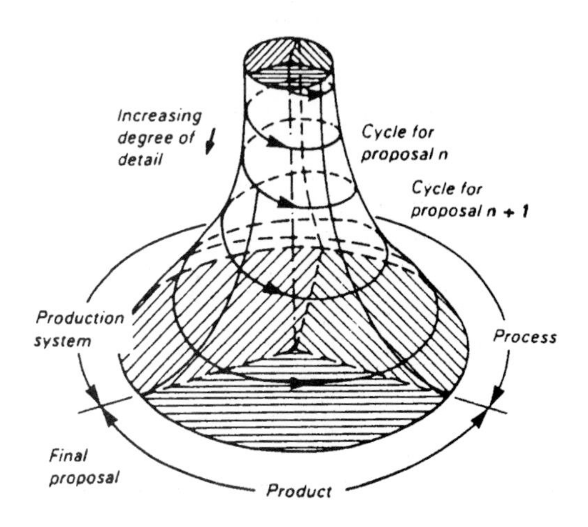

Figure 4.17 Iterative consideration of alternatives.

Each technical factor had some financial consequences. By simultaneous adjustments of the product, the production process, and the technical characteristics of the equipment, it was possible to establish the most profitable commercial opportunity.

The project followed an iterative strategy, as illustrated in Fig. 4.17. A large number of alternatives were investigated. Each alternative consisted of a proposal for the design of the product, the production technology, and the production system. The individual cycles of this process were characterized by continual rejection of proposals that were found to be less advantageous.

4.7.2 Product proposal

A product proposal primarily covered the design structure, the form of the components, the number of the components, and the choice of materials. As these features were determined, the material costs for the product were calculated by collecting estimates and testing prototypes.

It was not yet possible, however, to determine whether the product properties were optimal. If production properties could not be determined from the engineering drawings, they could nevertheless be determined at the time of pilot product run.

4.7.3 Production technology proposal

For the most promising product proposal, one or more production processes were proposed, each characterized by:

1. The manufacturing processes for the individual component
2. Assembly processes

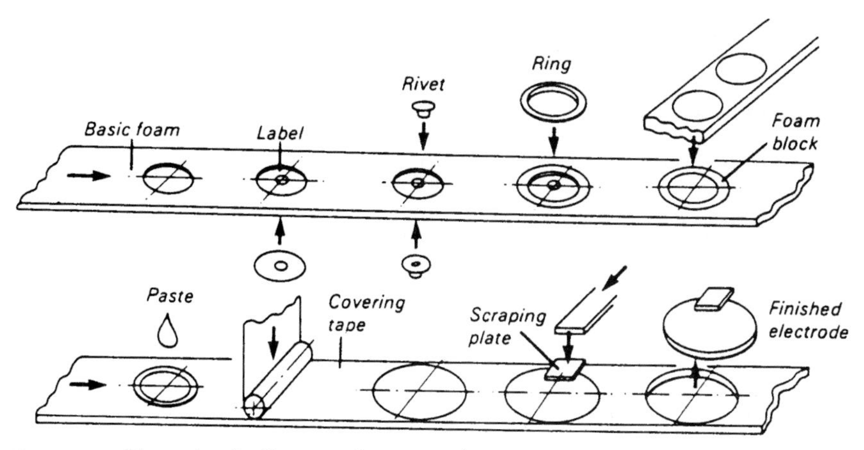

Figure 4.18 Example of a diagram of a proposed process.

3. Checking process
4. The processing sequence
5. The form of the raw material delivered (as individual component, in coils, etc.)
6. The pattern of movement of tools and other active elements
7. The timing of the processes
8. Buffer storage

The various proposals were documented and diagramed; Fig. 4.18 is an example.

4.7.4 Production system proposal (machines and operators)

Several proposals were made for various alternatives (in terms of both machines and operators) for implementing a specific production process. Each alternative system was characterized by:

- Operator tasks
- Machine tasks

It rapidly became apparent that it would be both technically possible and economically advantageous to have a fully automated production system in which the operators' tasks were reduced to:

1. Supplying raw materials

2. Removal of the finished products and packaging
3. Fault correction
4. Supervision

The automated production machine was characterized by:

1. The frame design
2. The design of tools
3. The control system
4. The checking system
5. The feeding and orientation system for raw materials
6. The storage system
7. The appearance of the system
8. The sequence of operations
9. The cycle time
10. Productivity, i.e., expected quality and quantity
11. Flexibility and readjustment to other tasks
12. The size of the system
13. The weight

Each proposed production system was analyzed in terms of capacity, cost, and direct wages. It was then possible to evaluate which combination of product design, production technology, and production system could provide Simonsen the best commercial possibilities. The chosen product design is illustrated in Fig. 4.19.

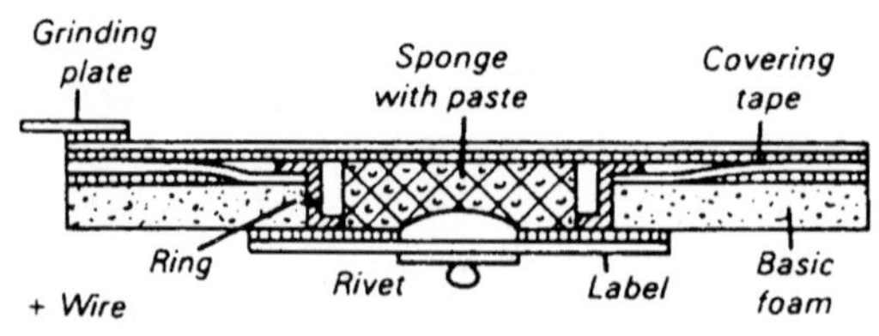

Basic specification:
- *Disposable product*
- *No variants*
- *Minimal use of silver*
- *Wire and plug are reused*

Figure 4.19 Chosen product design for ECG electrode.

4.8 High-Precision Assembly with Inexpensive Machines

The majority of assembly operations for mechanical workpieces are still carried out manually. Such operations require skilled work of a particular nature exploiting human capabilities—in particular, adaptivity related to solving different problems. Such operations need the interrelated motion and sensory perception that are subconsciously monitored and interpreted by the human nervous system.

For instance, an operator does not find particular difficulty in positioning pieces together. The operator's hands are guided by visual observation to position the pieces near the correct place for assembly. Then, when the pieces come in contact, the tactile sensation received by the hand allows the operator to correct relative position and orientation of the pieces till mating is achieved.

For repetitive assembly, the operation is performed automatically without much conscious use of the human intellect. Nevertheless, the process consists of complex interaction and feedback utilizing elements of the extremely complicated human cybernetic system.

A classic example of the difficulties involved in the insertion of a component into a close-fitting cavity is provided by a drawer that must slide into a piece of furniture. If the initial position of the drawer and/or the direction of the insertion force is incorrect, wedging occurs readily. In order to overcome wedging, appropriate maneuvers are needed, e.g., an intermittent side-to-side application of force and perhaps a sharp blow from time to time. This means that it is necessary to formulate and apply an assembly strategy that can change according to the initial conditions and to the mode of applications of the forces during the operation. Such strategies do not necessarily repeat because they are dependent on the historical pattern of the insertion.

Automatic assembly devices may be dedicated or flexible. *Dedicated devices* are specialized single-function machines for large batches. *Flexible devices* are meant to deal with smaller batches involving manipulation of different components. In both cases, however, at this time success is achieved only by replacing human skills by precision; thus high-precision machines and workpieces with strictly controlled tolerances are generally required. For inserting a peg in a hole, for example, the centering and orienting precision must be sufficient to avoid the jamming-drawer problem. Assembly robots can be used if they are programmed to use a sequence of steps according to the characteristic features of the workpiece. The use of visual sensors for assembly has advanced significantly during the past 5 years; artificial vision systems offer exceptional cost-effectiveness with acceptable speed and satisfactory reliability.

4.8.1 Active and passive assembly

Active assembly devices have sensors that provide relative positioning error measurements and control elements that provide correction. Passive assembly devices rely on contact forces to provide correction without feedback. Such devices have grippers mounted on elastic supports and use compliance to allow more deflection, thus canceling the error. With active devices, the gripping system may have high stiffness because small displacements are sufficient to measure forces, i.e., through deflection of ligaments or the output from strain gauges. Alternatively, the contact forces can be measured by potentiometers actuated by the displacement of helical springs that support the gripper.

Different sensors have been used by various industries. For example, sensitive fingertips made from silicone rubber have pressure-dependent electric resistance characteristics. Peg-in-hole insertion can be controlled by four fluidic sensors whose jets operate on the peg surface; the relative position of the peg and hole is indicated by differential pressures between diametrically opposite pairs of sensors. Though active assembly systems have theoretically great adaptivity, they also have some drawbacks. They require positioning feedback with many controlled axes; high-precision, low-friction actuators; and sensors able to detect very small forces or deflections. The systems are complex and the insertion times can be long, but they can assemble sharp-edged components into sharp-edged cavities.

Passive assembly systems are simpler and they use only contact forces to achieve the final relative setting after the first contact between the pieces. Such systems, however, need a "lead in" (a device similar to the tapered surface of a tapered pin inserted in a hole) to control the initial contact and bring the components into line. The adaptivity of the system is conditioned by the elastic support of the gripper. The joint must have at least four degrees of freedom. Furthermore, the first contact of the tip of the inserted piece must occur at the wall of the cavity of the opposite piece, or inside a chamfered zone. A chamfer may permit a large initial error, but a properly designed elastic connection between gripper and piece is essential for reliable assembly.

A remote center compliance system has been developed at Charles Stark Draper Laboratories. The gripper of this passive system is suspended by a special elastic support, designed so that its tip center coincides approximately with the instantaneous center of rotation of the part (plane of engagement). The elastic support is made so that, if a force acts in the plane of the tip center, the piece translates in the direction of the force accordingly. If a moment acts around the same point, the piece also rotates accordingly.

The system is not exceptionally sensitive to the real position of the single-piece center, but is based on the approximate assumption that a single-point contact force passes to the tip center, and a two-point contact induces a simple moment around the same center. The limits of the approximation are related to the piece shape; the approximation is good for cylindrical bodies, but not as effective for bodies with large tip faces. Furthermore, the gripper is actuated through the elastic support, so that only small insertion forces are tolerable in order to avoid large deflection.

Another passive assembly device has been constructed at the Yamanash University, Japan, called Assembly Robot Scara. The total architecture of the assembly robot is arranged to have a low resistance to movements in a direction tangent to the plane of insertion and a high resistance to rotation about an axis normal to the plane of insertion. Thus, in general, the relative angular insertion errors are limited and the centering error corrections are obtained by low contact forces.

4.8.2 Self-adaptive guided assembly

A self-adaptive guided assembly device has been developed to allow large positioning errors. Such errors arise from combinations of workpiece tolerances and fixture and positioning machine inaccuracy.

The system can overcome some problems of insertion by kinetically disconnecting the insertion elements from the gripping elements. Such a feature has been obtained with elastically supported guideways. The insertion actuator acts on the guided piece. The insertion force has no fixed direction; subsequent motion, however, must be parallel to the guideway axis. As in the robot, different compliances have been chosen for different axes of motion. The compliances affect the guideway suspension and have no influence on the insertion system. As a result, the insertion forces do not give rise to appreciable moments acting on the guideway. Rotation of the guideway is due to moments caused by contact forces, which have a straightening effect. The first device that was built according to these principles was designed to insert pegs into holes; the pegs were introduced directly into the elastically supported guideway and were pushed at their upper end by the tip of a cylinder rod.

A change in peg diameter may require a change of guideway. However, that is an inexpensive and easily interchangeable component, as illustrated in Fig. 4.20.

A guideway is made of low-friction material and is connected to a platform by a spherical joint (ball bearing). The bearing is kept in a fixed position by three symmetrical springs attached to the platform

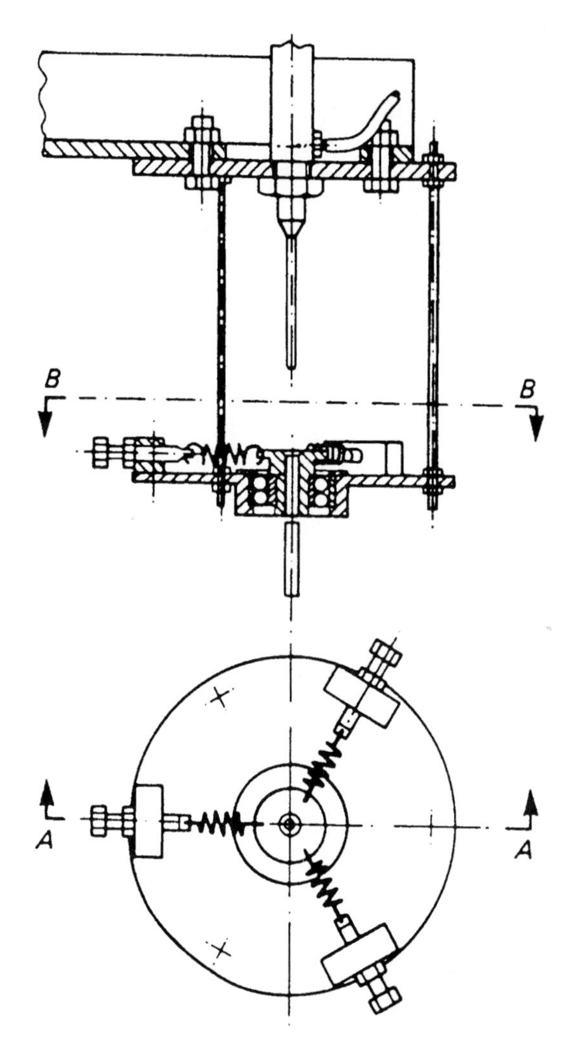

Figure 4.20 Elastically supported guideway.

(Fig. 4.21). The platform is connected to the frame by three symmetrical flexural bars that provide the guideway translational stiffness.

The insertion actuator is a pneumatic cylinder with a spherical rod tip that presses the top of the peg. Such contact allows easy sliding. Tests have shown that the friction has negligible effects on sliding; in any case, friction can be reduced to a very small value by fitting a rotating ball on the rod tip.

Peg feeding is performed in either of two ways. In one mode, illustrated in Fig. 4.22, the pegs are loaded horizontally on a slide; feeding

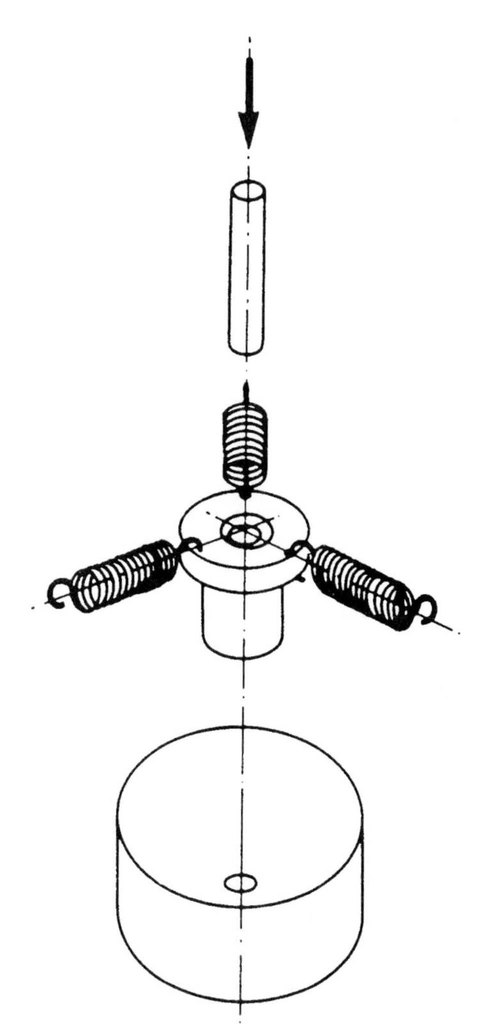

Figure 4.21 Support by three symmetrical springs.

is controlled by a ratchet actuated by a pneumatic membrane motor. When the peg is freed by the pawl, it is grasped horizontally by a two-piston pneumatic manipulator (Fig. 4.23), then it is turned to the vertical and brought over to the axis of the chamfered guideway (Fig. 4.24).

The other feeding mode uses a vertical gravity loader (Fig. 4.25). The peg is grasped by a vacuum gripper joined to a manipulator that translates between two positions, and it is brought over the guideway as illustrated in Fig. 4.26.

A few ancillary devices allow the signaling of defective assemblies, for example, those due to geometrical defects of the workpieces. A sec-

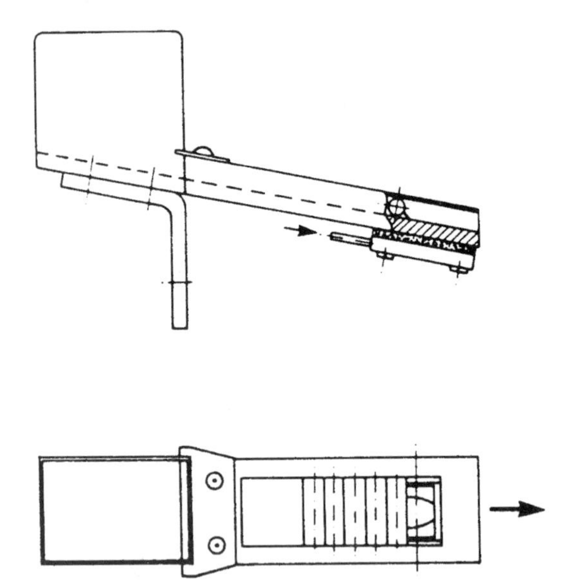

Figure 4.22 Peg feeding mode 1: horizontal ratchet feeding.

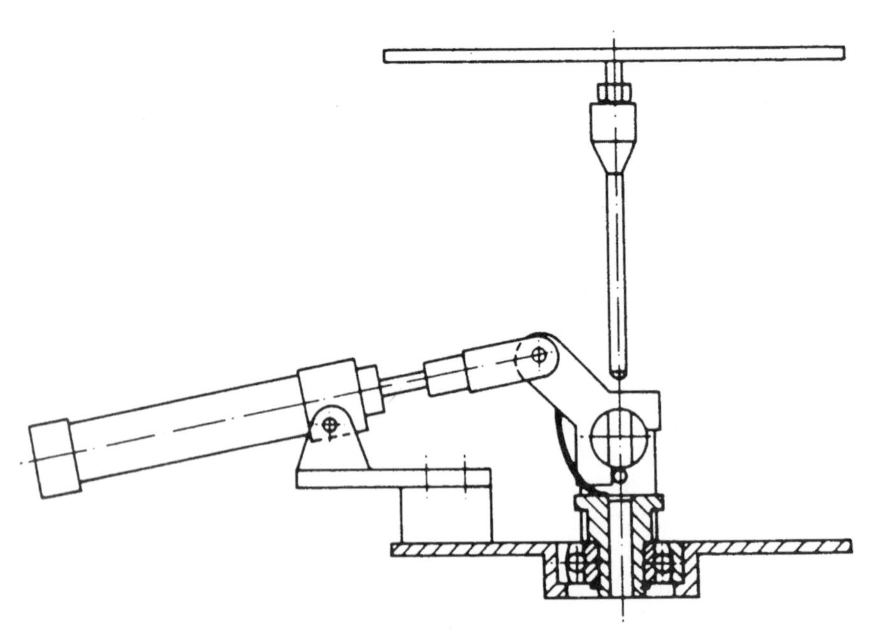

Figure 4.23 Peg is grasped by a pneumatic manipulator.

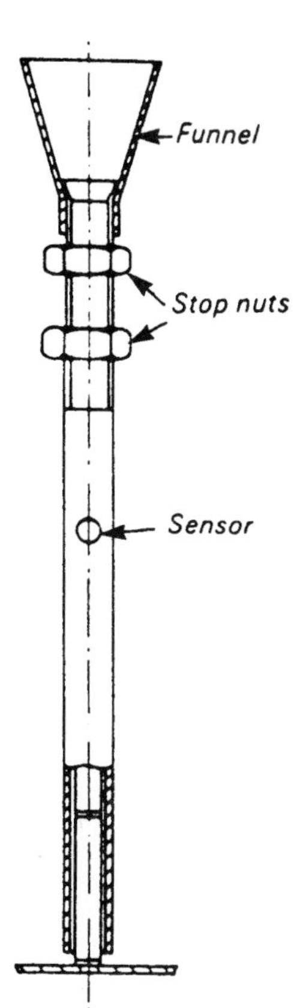

Figure 4.24 Peg is brought to a chamfered guideway.

ond assembly trial is permitted; if it is unsuccessful, the peg is extracted and discarded.

The devices described above are particularly suited for dedicated, specialized machines. Another device is able to assemble different workpieces without changing any part of its structure. The feeder has been replaced by a gripper, joined to a rod sliding inside the guideway. This rod is pushed by the actuator rod through a decoupling joint that allows small relative displacement (Fig. 4.27). Figure 4.28 shows a section of the whole device. An electromagnetic gripper is shown; a mechanical gripper may also be used. In either case, the workpiece dimensions may vary over a large range.

The hole insertion device may be used with a manipulator arm moving in a plane perpendicular to the guideway axis. The manipulator may perform workpiece gripping, transport, and insertion.

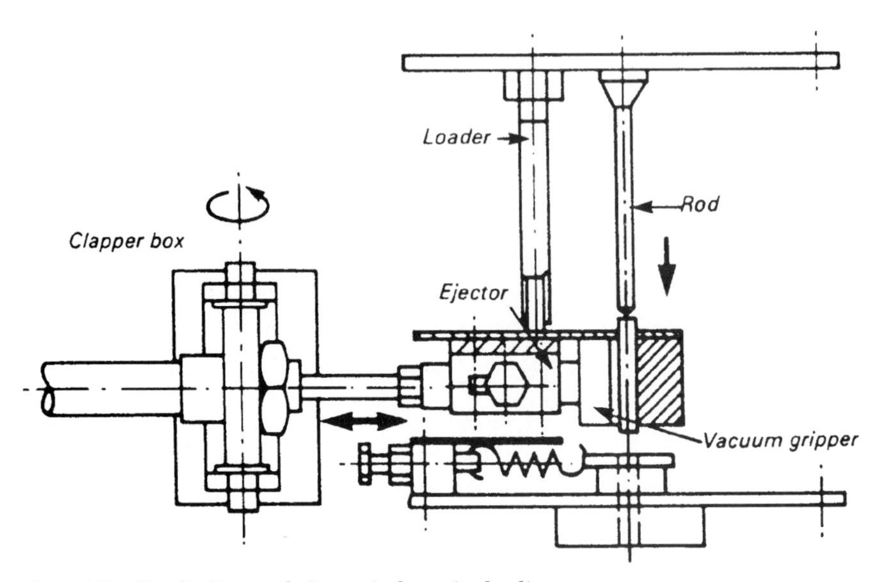

Figure 4.25 Peg feeding mode 2: vertical gravity loading.

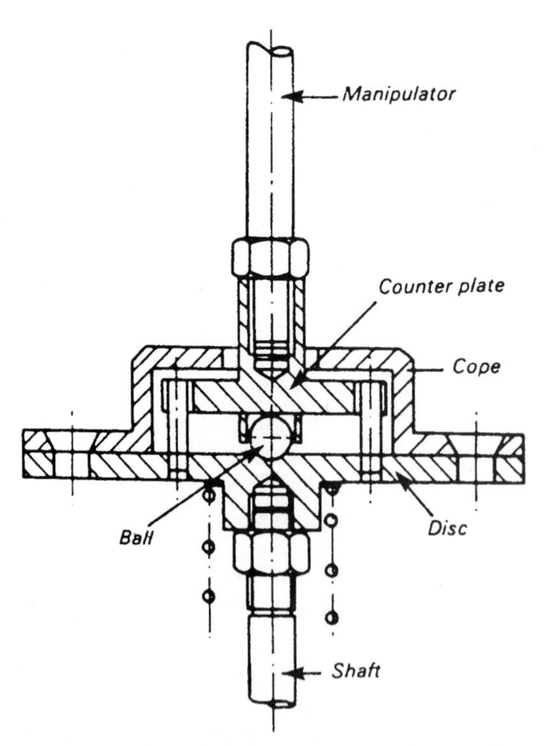

Figure 4.26 Translating manipulator. Decoupling joint between pushing and guiding devices.

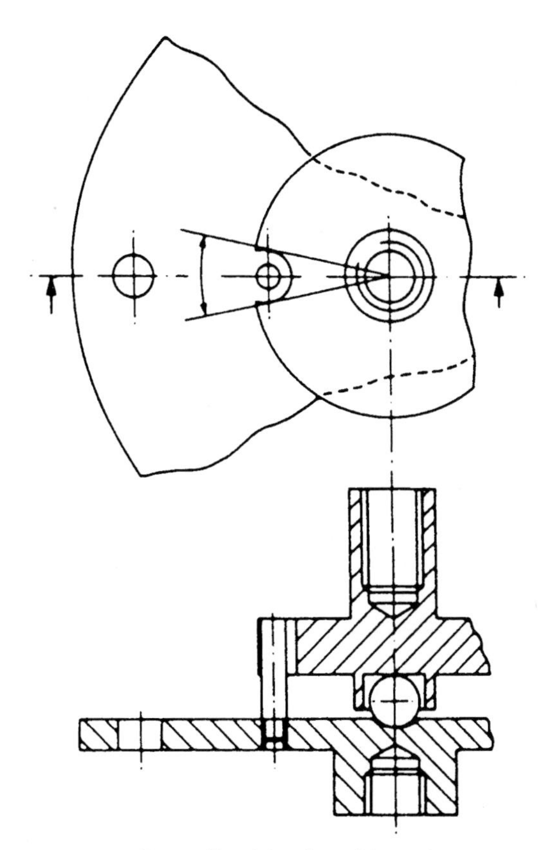

Figure 4.27 Decoupling joint for rod insertion.

During manipulator motion, large vibrations of the guideway elastic support may occur; they are avoided by allowing the gripper to be joined to the frame during motion. The part of the joint that is connected to the driven rod is shaped like a disk with vertical holes. Before manipulator arm displacement is started, the rods are lifted until the disk holes are engaged with chamfered pins mounted vertically on the frame.

Figure 4.29 shows the system control logic. The system has performed successfully on pegs and holes of various dimensions, all having the shape shown in Fig. 4.30. The pieces were axisymmetrical, with a 10-mm hole depth and a 45° chamfer with 1-mm width.

Some pieces had 20-mm diameter and 0.05-mm clearance. They were completely inserted even with initial eccentricity up to 0.8 to 0.9 mm. Other pieces had 30-mm diameter and 0.02-mm clearance, and

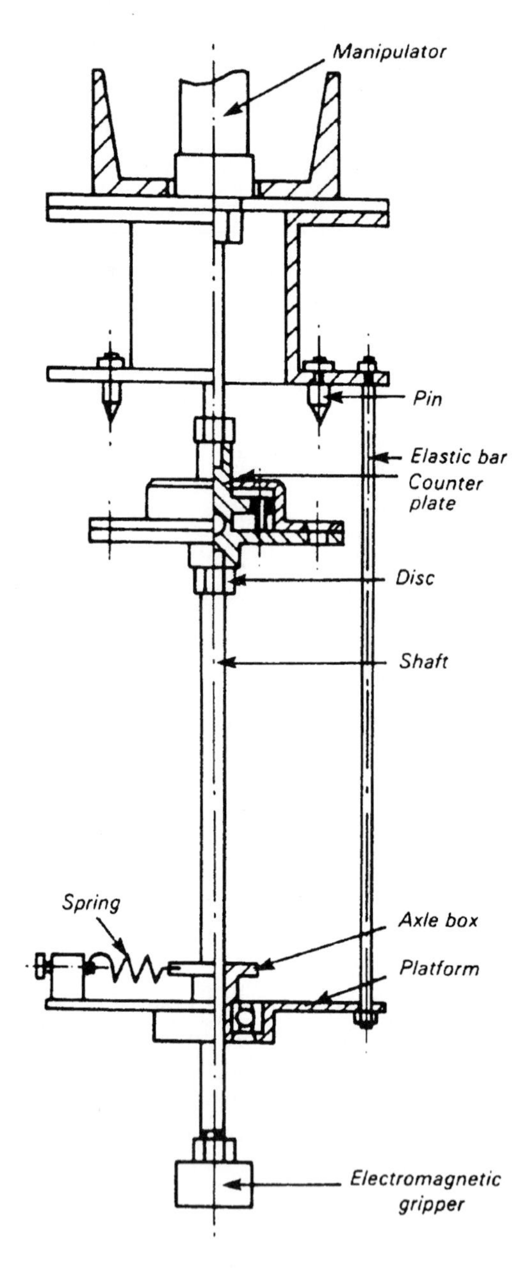

Figure 4.28 Section of assembly device for pegs of various diameters.

full insertion was possible up to eccentricity values of 0.5 to 0.6 mm. Angular errors of about 1° could be tolerated. By increasing insertion force, it was possible to form some press-fitted assemblies.

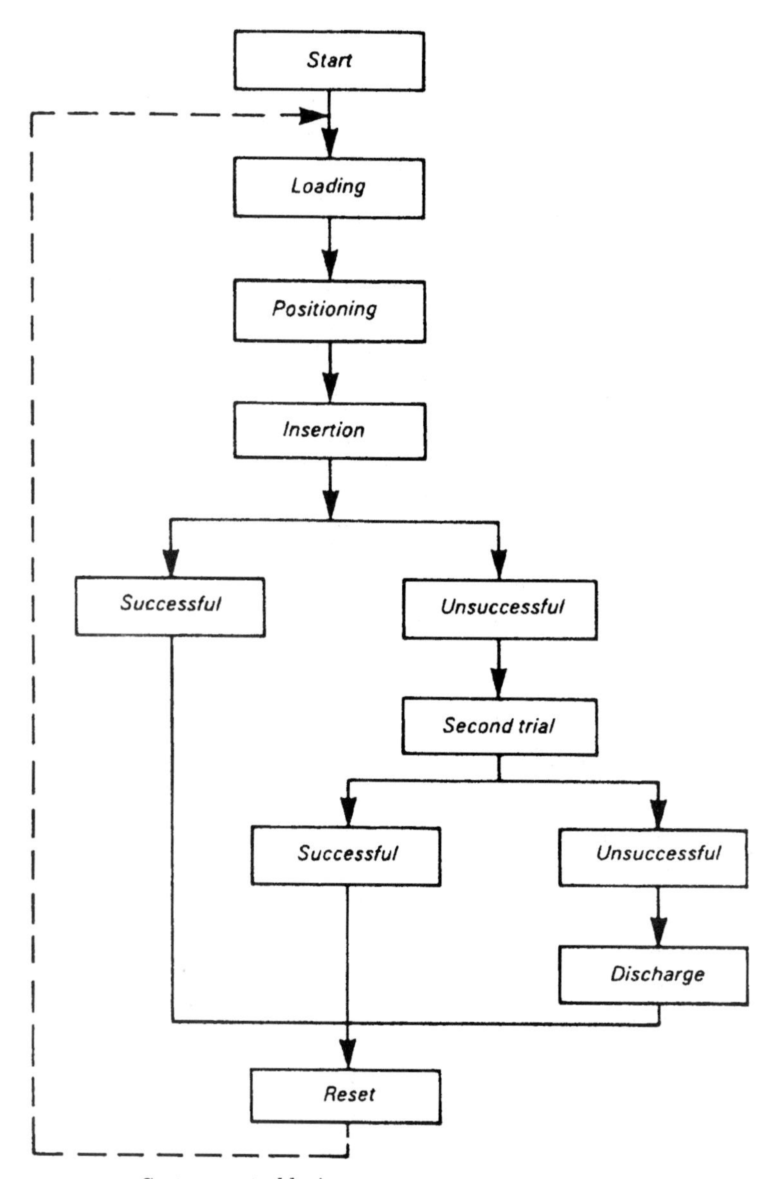

Figure 4.29 System control logic.

It appears that this very inexpensive system has the capability to extend the range of use of transfer assembly machines, and it allows them to perform precise assembly tasks at affordable cost.

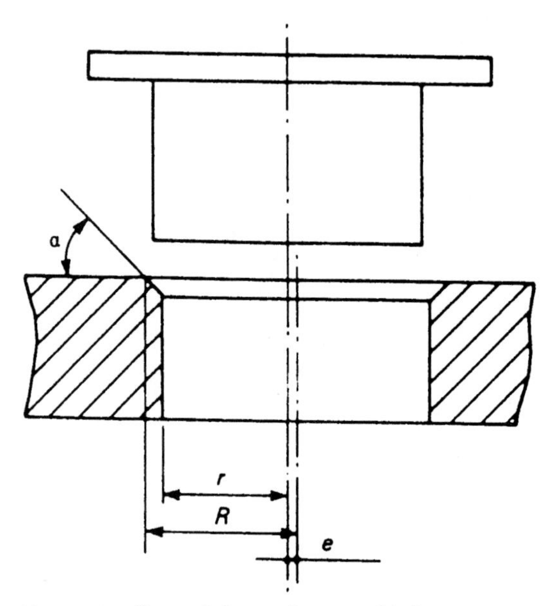

Figure 4.30 General shape of pegs and holes.

Chapter

5

Planning for Automated Assembly and Fabrication

5.1 Introduction

Affordable automation for manufacturing operations is generally implemented under either of two sets of circumstances. It may be applied to a new facility, process, or product, in which case it is incorporated into the initial plans and can be implemented routinely along with other equipment and facilities. More commonly, however, affordable automation is applied to existing processes and operations, often in response to management direction or a suggestion from a supplier of automated equipment. In this case, affordable automation concept must be integrated into ongoing operations, and the necessary changes to product, process, equipment, or facilities are often difficult to achieve.

To ensure success in either case, the application of affordable automation must be approached in a systematic manner. Launching the affordable automation concept in production systems is best accomplished in a multistep process that involves not only automation, but also the product, production equipment, layout, scheduling, material flow, and a number of other related factors.

Where affordable automation concepts are integrated into existing operations, there are five discrete steps to be taken:

1. Initial survey
2. Qualification
3. Selection
4. Engineering
5. Implementation

Where affordable automation is incorporated into a new operation, the first three steps are not explicitly followed.

5.2 Initial Survey

Manufacturing processes for affordable automation should not be selected arbitrarily. Careful identification and consideration of all potential operations is required, beginning with an initial survey of the entire manufacturing facility. The objective of the initial survey is to generate a "shopping list" of opportunities, that is, operations to which affordable automation might be applied.

During the initial survey, one should look for tasks that meet the following criteria:

1. The operation under consideration must be physically possible for affordable automated equipment.
2. An operation must not require exercise of judgment.
3. The operation must justify the use of affordable automated equipment.

At this point in the process, a detailed analysis of each operation relative to these three criteria should not be undertaken; rather, a few simple rules of thumb should be applied.

With regard to physical constraint, the following guidelines may be applied:

- Avoid the use of robots where process cycle time is less than 5 seconds. It is advisable the use automated synchronous or asynchronous dedicated systems when cycle time can be as small as 0.75 seconds.
- Avoid the use of robots where the working volume exceeds 30 m^3. It is advisable to use dedicated synchronous or asynchronous automated systems where volume can reach 200 m^3, as demonstrated in the automotive industry.
- Avoid the use of robots if the load to be handled exceeds 500 kg. It is advisable to use dedicated synchronous or asynchronous automated systems where handling weight can reach 2000 kg, as demonstrated in the aerospace and automotive industries.
- Avoid the use of robots if positioning tolerance is less than ± 0.1 mm. Automated indexing machines, either synchronous or asynchronous, are best-suited to precision applications having a tolerance of ± 0.1 mm or below.
- Avoid the use of robots where randomness in the workpiece position and orientation cannot be eliminated.

- Avoid the use of robots where randomness in the process cannot be eliminated.
- Avoid the use of robots where work lot size is typically less than 25 pieces.
- Avoid the use of robots where the number of different workpieces per process is typically greater than 10.

With regard to justification, the economic attractiveness of a potential affordable automation application, as measured by return on capital or by payback period, is usually of primary importance.

In the incorporation of affordable automation equipment or robots into existing operations, the major source of savings is the reduction of labor cost. A general estimate of these savings can be made from the labor that would be displaced. With the savings and an estimate of the cost of the equipment, a rate of return or a payback period can be calculated. For the purposes of the initial survey, however, a simple rule of thumb may suffice: a single affordable automation equipment installation can be justified by the displacement of two workers (assuming a 40-hour week for each worker).

During the initial survey, value judgments regarding the relative merits of the potential applications should be avoided. The purpose of this step is to develop objectively a list of opportunities that are technically and economically feasible and which will next be screened and prioritized.

5.3 Qualification

The second step is to determine the qualifications of the operations identified in the first step for affordable automation. There are likely to be operations on the list that, on further scrutiny, are found to be technically or economically unfeasible. Moreover, all operations on the list will not be of equal importance or complexity, and all cannot be implemented simultaneously; thus the qualification step will also involve prioritizing the qualified applications.

Qualification and prioritization is an iterative process. First, seven factors that should be considered to decide whether to apply an affordable automated system to a particular operation:

1. Complexity of the operation
2. Degree of disorder
3. Production rate
4. Production volume
5. Justification

6. Long-term potential
7. Acceptance criteria

Regarding complexity, simple affordable automation equipment exists and is well-suited to simple tasks, such as operations where an actuator, a two-way valve, and a few limit switches are sufficient. In other cases, a gravity chute may suffice to transfer and even reorient a part from one location to another.

At the other end of the scale, operations that require judgment or qualitative evaluation should be avoided. Checking and accepting or rejecting parts on the basis of a measurable standard can be achieved with an affordable automated system. For many operations, however, the only feasible measuring system is human sight or touch, and an affordable automated system such as a robot is not suitable.

In the same vein, operations that involve a combination of sensory perception and manipulation should also be avoided. An example would be a machine tool loading operation in which the part being loaded into the chuck is rotated until engagement of a notch with a key is felt, after which the part is fully inserted. While the development of a hand for a robot capable of doing this is technically feasible, the complexity of the hand and its potential unreliability will certainly reduce the overall probability of success. Operations with a greater complexity may require the actual determination of random spindle orientation and orienting the part to match. Avoidance of extreme situations like these will help to ensure successful implementation of affordable automated systems.

Affordable equipment cannot operate in a disorderly environment. Parts to be handled or worked on must be in a known place and have a known orientation. For simple equipment, parts must be always in the same position and attitude. For more complex equipment, parts must be presented in an array; however, the overall position and orientation of the array must always be the same. On a conveyer, part position and orientation must be the same, and the conveyer speed must be known.

Sensor-equipped affordable automated systems, including vision systems, can tolerate some degree of disorder; however, there are definite limitations to the adaptability of such equipment today. A vision system, for example, enables the equipment to locate a part on a conveyer belt and to position its pick-and-place mechanisms and orient its manipulators to grasp the part properly. It will not, however, enable an automated equipment to quickly remove a part, correctly oriented, from a bin of parts or from a group of overlapping parts on a conveyer belt.

A touch sensor allows affordable automated equipment or a robot to find the top part on a stack. It does not, however, direct the robot to the same place on each part if the stack is not uniform or is not always in the same position relative to the manipulator. Accordingly, repeatability is necessary to eliminate disorder and ambiguities.

As to production rate and cycle time, a small nonservo system can operate at relatively high speed, as in the case of small pick-and-place mechanisms. There is a limit, though, on the capability of even these devices. A typical pick-and-place cycle takes several seconds. A rate requiring pickup, transfer, and placement of a part in less than 0.5 s cannot be consistently supported by a robot. A synchronous or asynchronous dedicated system is better-suited to this application.

Operations that require more complex manipulation or lifting heavy parts require even more time with a robot. The larger servo-controlled equipment is able to move at speeds up to 1300 mm/s; however, as speeds are increased, positioning repeatability tends to decrease. As a rough estimate of cycle time, 1 s per move or major change of part orientation should be allowed. In addition, at least ½ s should be allowed at each end of the pass to assure repeatable positioning. For handling a part, allow another ½ s each time the manipulator gripper or handling device is actuated.

There are two factors related to production volumes. In batch manufacturing, the typical batch size must be considered. In single-part manufacturing, the overall length of the production run is important.

In small-batch manufacturing, changeover time is a major concern. Part-orienting and -locating devices may need to be changed or adjusted before each new batch is run. The robot's end-of-arm tooling and program may also have to be changed for each new batch of parts. Generally, workers do not require precise part location—their hands are instantly adaptable and their reprogramming is intuitive. Automation equipment with robots becomes impractical when its changeover time from batch to batch approaches 10 percent of the total time required to manufacture a batch.

If a single part is to be manufactured at high annual volumes for a number of years, special-purpose automation should be considered as an alternative to flexible automation such as robots. Per operation, a special-purpose device is probably less costly than the most flexible programmable automation device—that is, a robot. Single-function, special-purpose devices may also be faster and more accurate than robots. Where flexibility is required or obsolescence is likely, flexible automation or robots should be considered; where these are not factors, special-purpose automation may be more efficient and cost-effective.

Accordingly, for a very short run—about 25 pieces or less—manual operators are more effective; for a very long run—several million per year—of a single part, a special-purpose automated system is more effective; utilization of robots to complement the operations can also be achieved effectively.

The application of affordable automated systems can represent a significant investment in capital and in effort. Economic justification must therefore be carefully considered: on the balance sheet, increased productivity; reduced scrap losses or rework cost; labor cost reduction; improved quality; improvement in working conditions; avoiding human exposure to hazardous, unhealthy, or unpleasant environments; and reduction of indirect costs are among favorable factors. Offsetting factors include capital investment; facility, tooling, and rearrangement costs; operating expenses and maintenance cost; special tools, test equipment, and spare parts; and cost of downtime backup expense.

Estimates of the potential costs and savings should be made on the basis of reasonable return of investment. The savings generally can be estimated by a simple multiplication: the number of direct labor heads displaced per shift × the number of production shifts per day × the fully burdened annual wage rate. The costs can be roughly estimated by multiplying the basic cost of the automated system planned for the operation by 2.5.

Management preference and emotions are no substitutes for economic justification and, in the long run, will not support the application of automated systems. In some cases, quality issues or working conditions may override economics; however, these are usually exceptional circumstances. Accordingly, if costs do not exceed savings by more than a factor of 2, the application can probably be economically justified.

Another consideration is the potential for long-term use of affordable automated systems in the particular manufacturing facilities. Both the number of potential applications and their expected duration must be taken into account.

Because of its flexibility, an automation system can usually be used on a new application when the original operation is discontinued. Since the useful life of an automated assembly system may be as long as 10 years, several such reassignments may be made. The cost of reapplying the automated system should also be included in the justification of the initial investment. If the initial application is of significantly shorter duration than the automated system's useful life and no follow-on applications can be foreseen, it can seldom be justified.

Like any electromechanical device, an automated system requires some special knowledge and skills to program, operate, and maintain. An inventory of spare parts should be kept on hand. Auxiliary equipment for programming and maintenance or repair may also be required. Training of personnel, spare parts inventory, special tools, test equipment, and the like may represent a sizable investment. The difference between the amount invested in these items to support a single automated system and to support multiple systems is significant.

Troubleshooting and programming skills needed to solve problems tend to deteriorate without use. Few opportunities will normally arise to exercise these skills in support of a single automated system. Under these conditions, the skills may eventually be lost and any serious difficulty with the automated system may then result in its removal. Accordingly, if there is no possibility of installing more than one automated system, the single installation is seldom warranted.

Production workers are often concerned with the possible loss of jobs. Factory management is concerned with the possible loss of production. Maintenance personnel are concerned with the new technology. Company management is concerned with effects on cost and profit. Collectively, all of these concerns may be reflected in a general attitude that automated systems may be acceptable but are difficult to implement.

It is essential to know what difficulties to expect if an automated system is to be given a fair chance. Reassignment of workers displaced by an automated system can be disruptive. Training of personnel to program and maintain the automated system can upset their regular schedules, and new skills may even have to be developed. The installation and start-up can interrupt production schedules, as can breakdowns of the automated system or related equipment. Unless the personnel involved are aware of these factors and are willing to accept them, the probability of success is poor. Accordingly, an affordable automated system must be accepted by all personnel, not only in general principle, but also in their own operation.

The foregoing screening process will no doubt eliminate a number of operations from the shopping list. Those remaining should be operations that qualify as technically and economically feasible for the application of an automated system. These operations should now be prioritized, in preparation for the selection step. The prioritizing of operations and subsequent selection of an initial automated system application can be facilitated by the use of an operation scoring system. The elements of the scoring system might include:

1. Complexity of task
2. Complexity of end-of-arm tooling
3. Part orienters, feeders, fixtures, etc.
4. Changes required for facilities and related equipment
5. Changes required for product and process
6. Frequency of changeovers
7. Impact on related operations
8. Impact on work force

9. Cost and saving potential
10. Anticipated duration of the operation

For each of the elements involved in the prioritization, a set of measures and a score range is established, with the more important elements having a higher range of points than the less important elements.

With this scoring system each operation on the shopping list can be rated and prioritized; the operation with the highest score will be the prime candidate for the first application. Other factors such as timing, management preference, experience, and human relations might also be considered; however, subjectivity in establishing priorities should be minimized.

5.4 Selection

The third step in the launching of an affordable automated production system is the selection of the operation for which the system will be implemented. If the initial survey was made with care, and if the qualification and prioritization have been objectively accomplished, this step will be virtually automatic. If a scoring system is used in establishing priorities, then the selection should be made from among the two or three operations with the highest scores. Some considerations might be given to conditions and circumstances that were not measured in the second step; however, it is again important that subjectivity and arbitrariness be minimized.

It is imperative to review onsite the few top candidate operations before making the final selection to ascertain that they are, truly, technically feasible and justifiable. For the first automated system application, the critical rule is to strive for simplicity. It is prudent to forego some degree of economic or other benefit for the sake of simplicity; even the most elementary of potential applications is likely to be more complex than anticipated. Avoid the temptation to solve a difficult technological problem; those opportunities will come with later installations. Resist the pressure of uninformed suggestions from managers or others trying to be helpful. The affordable automated system should not be considered as a solution to production or manufacturing problems. If conventional approaches have not succeeded, the automated system is also likely to fail unless the proposed system is based on a technological breakthrough.

Once the operation to which the automated system will first be applied has been selected, the engineering of the application can begin. The prioritized, qualified shopping list should be retained as a source for further automated system applications.

5.5 Engineering

The fourth step in the launching of an affordable automated system is system engineering. There are a number of engineering activities involved, some of which must be accomplished sequentially and some of which can be performed concurrently. The first activity is to return to the chosen workplace and thoroughly study the task to ensure that all that must be accomplished is identified and planned. During this study phase there are a number of considerations that must be addressed:

1. *Alternatives to standard expensive robots.* Perhaps the desired result can be accomplished by a special-purpose device, by basic facility changes, or by restructuring the operation.
2. *Alternatives to robot's attitude.* There are several advantages to mounting the robot in other than the usual feet-on-the-floor attitude, such as overhead.
3. *Alternatives to existing process.* There are advantages in reversing the usual "bring the tool to the work" procedure and having the robotics system carry the work to the tool.
4. *Backup.* Arrangements must be made for equipment to back up the automated system during downtime.
5. *Environment.* The automated system usually needs special protection from excessive heat, cold, abrasive particles, shock and vibration, fire, etc.
6. *Space.* The automated system occupies significantly more space, which may cause certain difficulties.
7. *Layout.* The automated system creates problems with accessibility to itself, other equipment, and the workplace for material handling, maintenance and inspection.
8. *Safety.* The installation of the automated system must be accomplished so as to protect workers from the system and vice versa.
9. *Unusual, intermittent, random occurrences.* Contingency plans must be developed in case of possible anticipated breakdown.

The detailed study of the chosen operation is intended to provide a thorough familiarity with the operation, sequence, and space, as well as with the occasional random disruptions that seem to occur in any process. Because the automated system is not adept at handling disruptions or disturbances in its normal routine, approaches must be developed to minimize their occurrences and impact, to prevent damage when they happen, and to recover rapidly afterward.

Significant data to be gathered about the operation include the following:

- Number and description of elemental steps in the operation
- Size, shape, and weight of parts and tools
- Part orientation at delivery, acquisition, in process, and disposal
- Method and frequency of delivery and removal of parts
- In batch production, lot sizes, characteristics of all parts in family, frequency of changes, and changeover time for related equipment
- Production requirements per hour, day, week, month, and year
- Cycle times, overall and elemental
- Inspection requirements, defect disposition

At this point, a layout drawing of the installation is made. Typically this starts with a scale layout of the existing area on which the automated system and its work area are superimposed. Locations for incoming and outgoing materials, buffers, and intermediate positions of parts are determined. From this layout, potential interference points can be located and equipment relocated, if necessary. Sources and routing for utilities, such as electrical power, compressed air, and cooling water, are also shown on the layout.

Simultaneously with preparation of the layout, a detailed description of the automated system's task must be written. This task description must be broken down to a level comparable to the individual steps of the automated system's program; it will, in fact, become the basic documentation of that program. Elemental times are estimated for each step so that an approximate cycle time can be established for the entire task.

Working with a layout and the task description, the automated system's program is optimized. The objectives are to minimize the number of program steps and system movements to attain the shortest cycle time. Often, rearranging incoming and outgoing material locations or even the position of the equipment in the workstation can significantly affect the cycle time. Thus both the layout and the task description are necessary elements of this optimization process. Product and/or process changes may also be necessary or desirable.

The final selection of a specific model or configuration of the automated system should not yet be made. Ideally, several models of automated systems will be capable of performing this operation, and an alternative layout should be made for each. The task description is basic to the process and should be common for all models of automated systems considered. The selection of the desired automated system is now made, based on best fit to the layout; performance advantages, if any; price; delivery; support; and other similar considerations.

Once the model of the automated system has been chosen and the layout has been optimized, personnel and equipment access points are determined and hazard-guarding (safety barrier) locations are established. It is necessary that an area encompassing the entire automated working area be guarded against accidental intrusion by workers. Although some installations use active intrusion devices such as light curtains or safety mats interlocked with the automated system's control to stop the system when an unauthorized person enters the area, passive systems such as fences, walls, or guard rails are more dependable.

From the task description, the interlocks between the automated system and related equipment are determined. An automated system will not directly control the other equipment, that is, the automated system's controller will not directly operate other machines in the work cell. The automated system will, however, initiate other machines' cycles, and its operation will, in turn, be initiated by other machines or devices. For each of the different inputs and outputs, an input/output port on the automated system control must be hardwired to or from other devices. In more complex operations, the automated system control may not have sufficient input/output capacity. An external input/output device such as a programmable controller may be required. When determining inputs and outputs, it is also important to consider the backup method to be used. If manual backup is to be employed, then a manual control station may also have to be provided. Thus the control station must have a manual/automatic mode selector, which should be a lockable selection switch.

After the selection of the automated system and finalization of the layout, there are several engineering tasks that should be performed simultaneously because they are interdependent. These engineering tasks include the following:

1. Pick-and-place mechanism design
2. Part feeder, orienter, and positioner design
3. Equipment modifications
4. Part or product redesign
5. Process revisions

5.5.1 Pick-and-place mechanisms

Typically, a standard automated system is purchased without the special-purpose pick-and-place mechanisms it needs. The automated system supplier may furnish standard grippers or standard pick-and-place mechanisms, or a suitable device may obtained from another source; however, adaptation of a standard mechanism may still require some design effort. Similarly, a standard power tool such as a screwdriver, grinder, spray gun, or welding torch that is to be mounted on the end of the pick-up-and-place mechanisms will require specially designed mounting hardware, such as brackets and adapters. The lack of standard mechanical interfaces for tooling means that little off-the-shelf hardware is available.

Pick-and-place mechanisms lack the versatility of a human hand; thus, in the case of batch manufacturing, several interchangeable tools may be required. A multifunctional tool for such tasks must represent a practical compromise between simplicity (for reliability) and flexibility (to perform a number of functions or handle a number of different parts). Interchangeable tools should be designed for ease of removal and installation and for repeatable, precise location on the automated assembly manipulators to avoid the necessity to program the system with each tool change. In some cases, automatic exchange of tools by the automated system may be possible with quick-disconnect, collet/drawbar arrangements similar to preset machine tool holders, tool racks, and the like.

5.5.2 Part feeders

Today's automated equipment requires an ordered, repeatable environment and cannot easily acquire randomly oriented parts delivered in bulk. There are several solutions to this problem, including trays or dunnage that contain parts in fixed locations and orientations; mechanical feeder-orienter devices; manual transfer of parts from bulk containers into the feeder; and sensor-based acquisition systems with vision or tactile sensing.

The mechanically simplest approach is to use part containers that retain individual parts in specific locations. Automated equipment with computer control and sufficient memory can be programmed to move to each location in the container, in sequence, to acquire a part. Multiple layers of parts may be packed in this manner, with the automated system also programmed to remove empty trays or separators between layers. The only requirement in the workplace is to provide locators for repeatably positioning the container. The multiple pickup points, in addition to requiring computer control and large memory capacity, may increase the average cycle time for the orientation. And, if the innovative automated equipment lacks the ability to acquire parts from a matrix array, some other approach must be taken.

Another part-presentation approach is to use mechanical feeder-orienters. These, for small parts, may be centrifugal or vibratory feeders which automatically orient parts in a feeder track. Larger parts may be handled with hoppers and gravity chutes or elevating conveyers and chutes, which also present parts in proper orientation at a specific pickup point. Usually, these devices are adaptations of standard, commercially available equipment. Advantages of this approach are that the single acquisition point for each part minimizes nonproductive motions, and the orienters can often present the parts in attitudes that require little manipulation by their automated equipment after pickup. Disadvantages are difficulties in orienting and feeding some parts; relatively high cost of mechanical feeders; lack of flexibility to handle a variety of parts; potential damage to delicate, fragile, highly finished, or high-accuracy parts; and inability to handle large, heavy, or awkwardly shaped objects.

A third approach is the manual transfer of parts from bulk containers to mechanical feeders such as gravity chutes or indexing conveyers. An obvious disadvantage of this approach is the use of manual labor, especially to perform the sort of routine, nonrewarding tasks to which automated systems, such as robots or pick-and-place mechanisms, should be applied. Advantages are relatively low capital investment requirements, ability to handle difficult or critical parts, and flexibility to accommodate a variety of similar parts, as in batch manufacturing.

A fourth approach is the use of sensors such as vision or tactile feedback devices to modify the automated equipment's programmed motions, enabling it to acquire somewhat randomly oriented parts. Advantages are a reduction in the extent of mechanical orientation required and a potential to work with a variety of randomly mixed parts or to accommodate batch manufacturing lot changes with little or no physical changes required. Disadvantages are relatively high cost compared to simple mechanical feeder-orienters, the possible need for special lighting, relatively slow processing time, and difficulty with touching or overlapping parts or with three-dimensional space (such as bins).

The solutions to part presentation for the automated system often combine several of the approaches described, as well as others, such as automatic single-part-at-a-time delivery from a retention operation by means of a conveyer or shuttle device.

5.5.3 Equipment modification

Typical modifications include adding an actuator and solenoid valve to a splash guard or automatic, rather than manual, opening and closing. Other changes may be made to guards and housings for improved access by the automated system. Machine-tool chucks and collets may be modified to increase clearances or to provide leads or chamfers for easier insertion of parts. Powered clamping and shuttle devices may be substituted for manually actuated mechanisms. In machine-tool operations, coolant–cutting fluid systems may be changed or chip blow-off systems added to automatically remove cutting chips from the work and work holders. Assembly operations may require development of simple jigs and fixtures in which to place parts during the process. Likewise, manual tools such as screwdrivers and wrenches will have to be replaced with automatic power tools.

5.5.4 Product or part redesign

Part orienting and feeding and/or part handling by the automated system's pick-and-place mechanism may require some redesign of a product. Ideally, the product should be designed so that it has only one steady-state orientation, that is, it should be self-orienting. As an alternative, the product should be designed so that its orientation for acquisition is not critical; for example, it may have a flat disk or washer shape. A family of parts which are all to be handled by the automated system should have some common feature by which they are grasped; this feature should be of the same size and in the same location on all products in the family.

Vacuum pickups are simple, fast, and inexpensive. Products that incorporate surfaces or features to which a vacuum pickup can be applied are easier for the automated system to handle. Products should be designed so that the automated system's tasks, such as loading, unloading, inserting, and assembling, require a minimum of discrete motions; complex motions, especially those that require the coordinated movements in two or more axes, such as a helical movement of the part, should be avoided. Tolerances should be opened up as much as possible. Chamfers should be provided on inserted parts to aid in alignment. Parts should be self-aligning or self-locating, if possible. Parts that are to be mechanically or gravitationally oriented and fed to the automated system should be designed so that they do not jam, tangle, or overlap.

Because product redesign is costly and time-consuming, it should not be undertaken lightly, but should be considered only when its potential benefits significantly outweigh its cost. In the design of new products, however, incorporation of features that facilitate the use of automated equipment adds little or nothing to the cost and thus should be encouraged.

5.5.5 Process revisions

Another engineering requirement may be the modification of the process employed by the automatic handling and manipulating system. Examples of process revisions are

1. Changing an optional sequence so that a critical part orientation is not required.
2. Moving several machines into an area to set up a cell to take advantage of initial part orientation and to increase the automated system's utilization.
3. Linking conveyers to ensure a certain part orientation, or incorporating compartmentalized pallets or dunnage to retain orientation between operations.
4. Rescheduling batch operations to increase lot size or to minimize changeover between batches.

Process revisions can often be accomplished at minimal cost, particularly those that involve only scheduling, and can sometimes significantly increase the efficiency of the automated production system. Like product changes, process revisions should not be undertaken lightly, however, but should be carefully examined for cost-effectiveness.

The first four steps of launching an automated production system, initial survey, qualification, selection, and engineering, should, if followed carefully and thoroughly, make the fifth and last step, implementation, relatively easy and trouble-free.

5.6 Computer-Controlled Part Feeding Magazines

Magazines play an important part in the application of affordable automation. They offer these advantages:

- Careful transport of workpieces
- Minimum handling during loading and unloading
- Simultaneous processing of several workpieces

In the past, loading and unloading of magazines has been automated only in large-scale production. Now, however, industrial robots with sensors open new opportunities in the field of affordable automation. The following case study illustrates automated feeding from and to magazines.

5.6.1 Case study: System application of magazine loading

In an automated system for coating the surface of plastic parts, several magazines had to be used (Fig. 5.1). The untreated workpieces had

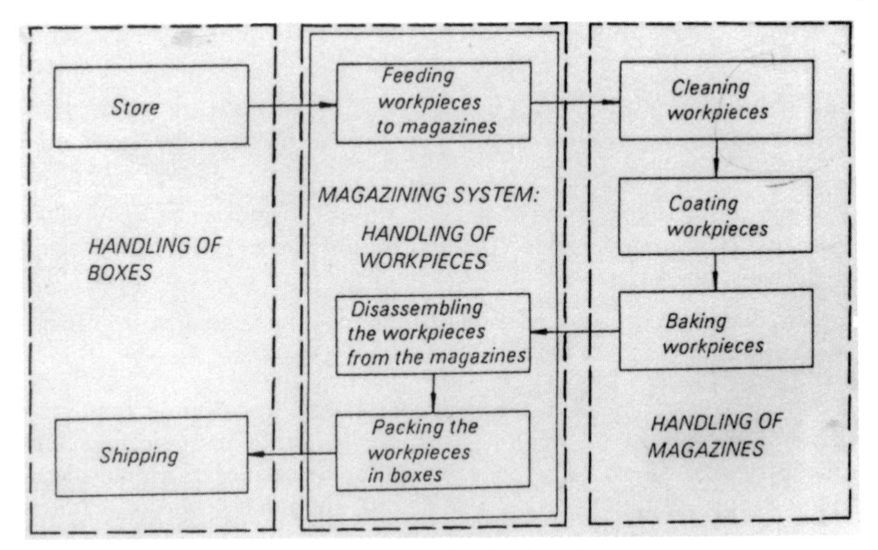

Figure 5.1 Magazine-loaded coating system for plastic parts.

to be loaded to a magazine before the coating process. The workpieces had to be transported on magazines during production. After coating, the workpieces had to be taken off the magazines and stored in palletized boxes. Once a pallet was completely loaded, it had to be immediately replaced. The manufacturer wished to adapt its automated coating system to automatic unloading, loading, and handling of the magazines.

The system handles 150 different types of plastic workpieces. The setup for selecting a workpiece for coating is illustrated in Fig. 5.2. About 70 percent of the workpieces are cylindrical with a central blind hole. Their weights range from 1 to 15 g. The diameters range from 7.5 to 100 mm, with 40 mm being the average. The average workpiece length is 26 mm, varying from 9 to 51 mm.

The magazines are partly standardized and consist of modules assembled for each workpiece type in a special combination. The number of workpieces per magazine is different for each workpiece type. The magazines are essentially the same for each workpiece type; however, they are made with large tolerances that can build up to several millimeters. In addition, the workpiece weight is not distributed symmetrically on the magazine, making the handling task more difficult. Also, because the handle is in the middle of the magazine, several workpieces are likely to be broken during transport. The design of the magazine largely depends on the type of workpieces to be carried and therefore can vary only within a small range.

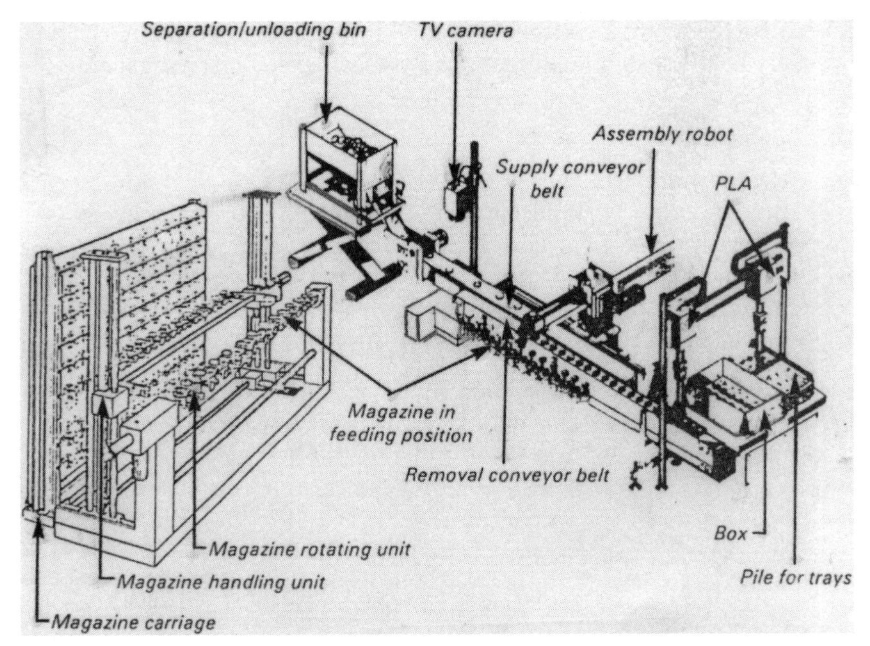

Figure 5.2 Magazine loading system (shown with vision system and assembly robot).

5.6.2 System planning

The manufacturer's production schedule imposed significant restrictions on the coating operation. Analysis identified that the workpieces had to be coated within a cycle time of 1 to 1.5 s to achieve economic goals. This time is important for planning.

In planning the automated handling system, flexibility and quality, as well as speed, were the important parameters. Flexibility means the ability to cope with the following:

- Varied workpiece types with equal handling quality. Moreover, new workpiece types are likely to be introduced.
- Magazines built to large and varied tolerances.

The following recommendations were made to balance the automated handling system with the production of workpieces:

1. Prolong the batch run time.
2. Redesign the magazines to increase quality and decrease the difficulty in automatic handling.
3. Standardize the workpiece spectrum.
4. Preorient the workpieces.

Because of the high estimated cost of implementing recommendation two, only small design modifications were achieved on the magazines.

5.6.3 System design

For the automation of the magazine feeding operation, it was necessary to orient the workpieces. Two approaches were compared: mechanical orientation by a vibratory feeder and sensing and correcting orientation by a vision system. It was found that both methods could be successfully implemented. The cost estimates are given in Table 5.1. The vision system was selected for the following reasons:

- It had a distinct break-even point at 20 different workpiece types compared to the orientation of the vibrating bowl feeder.
- It was faster.
- Its intelligent programmable sensor made it easy to install the magazining system.

5.6.4 System description

The system layout is illustrated in Fig. 5.2. The functions of the system are

1. Feeding of workpieces to the recognition area of the workpiece sensor.

TABLE 5.1 Cost Estimate of the Alternative for the Orientation of Workpieces

Sorting with Mechanical Devices	$ Cost	Sorting with TV	$ Cost
Storing bin with unloading device	3,000	TV Sensor interface	40,000
Vibratory bowl feeder: Exchangeable orienting devices, unloading device, and positioning device	8,000	Storing and separating bin	8,000
		Orienting gripper with three axes	18,000
Adjustable width gripper	5,000		
Workpiece orienting device	1,500		
Workpiece unloading device	1,500		
Workpiece unloading and positioning additional device	1,000		
Total Fixed Cost	16,000	Total Fixed Cost	66,000
Total Variable Cost, 2,500 per Workpiece			

2. Unloading workpieces from the supply conveyer belt by the assembly robot.
3. Placing workpieces on the removal conveyer belt by the assembly robot.

All functions can be achieved sequentially, with an estimated cycle time of 1.5 seconds. The system consists of the following essential components:

Separation/unloading bin. Stores workpieces for 60 minutes. It also feeds workpieces to the supply conveyer belt.

Supply conveyer. Transports the workpieces coming from the separation/unloading bin to the recognition area under the charge-coupled device (CCD) camera. It also transports workpieces to the assembly robot. Workpieces that are not taken by the assembly robot are transported to a box at the end of the supply conveyer belt. An adhesive coating on the belt prevents workpieces from rolling. The incremental motion of the conveyer belt is controlled by a rotational encoder, directed by the computer program to create synchronized displacement of the conveyer.

CCD camera. Situated above the supply conveyer, the camera senses the silhouette of the workpieces lying on this part of the belt. The information is processed by the computer.

Assembly robot. Receives the coordinates of the recognized workpieces and locates the workpieces lying on the supply conveyer belt. Using an orienting gripper, the robot picks up each workpiece from the conveyer belt and positions it to feed an empty fixture on a magazine. If a magazine full of coated workpieces is approached, first a fixture is emptied from the workpiece and placed on the removal conveyer belt, and an uncoated workpiece is fed to the empty fixture.

The assembly robot has another gripper as well: one with a vacuum suction cup that takes coated workpieces from the magazine to feed them to the removal conveyer belt.

In the magazining system, the handling system has one arm with three linear main axes, two rotating auxiliary axes, and one linear hand axis for opening and closing the gripper jaws. It is necessary to use a second arm on the assembly robot with the same features in order to decrease the cycle time. This second arm is mounted on the same way as the first arm.

Removal conveyer. Takes coated workpieces and transports them to a programmable loader workstation. The removal conveyer belt is position-controlled in the same way as the supply conveyer belt.

Programmable loader. Has two arms, each with two axes. One arm takes a row of workpieces laid in a pattern that fits the tray on the removal conveyer belt. The second arm takes trays from a pile and transports them to the box.

Magazine carriage. Stores 20 magazines in exactly defined position. It is transported manually to the magazine handling unit.

Magazine handling unit. Loads and unloads the magazine carriage; loads and unloads the magazine rotating unit. To decrease the cycle time, it has a double gripper. The vertical axis of the magazine handling unit is precision-controlled, whereas the horizontal axis is pneumatically driven and has three fixed positions.

Magazine rotating unit. Has a double-sided fixture for one magazine. The unit rotates the magazines so that the assembly robot can feed workpieces to them in the vertical direction.

5.6.5 Separation and unloading bins

The separation and unloading bin is designed to store workpieces and to feed them to the supply conveyer belt in such a way that they can be recognized by the CCD camera system. The time required to recognize a workpiece depends on how it is presented to the camera; recognition is much faster if the workpieces are not touching each other. The separation and unloading operation follows this sequence (see Fig. 5.3):

1. The bin is loaded with workpieces.
2. A signal is sent to the controller when the bin is filled.
3. The controller directs a gradual opening of the container flap.
4. The workpieces are guided to slide through the opening of the container flap.
5. A signal from the light barrier is sent to the controller, announcing that the first workpiece has slid out of the container.
6. The controller directs the container flap to close after a specific time delay.
7. The workpiece is then clamped by the container flap and the rubber stop.

By arranging the light barrier at an optimum position, the standby time from the start to end of workpiece separation is 0.5 to 1.0 s, and the number of workpieces fed simultaneously is 1 to 3 units.

To ensure that the feed rate is consistent, the container flap is removed once the light barrier has signaled the controller. This elimi-

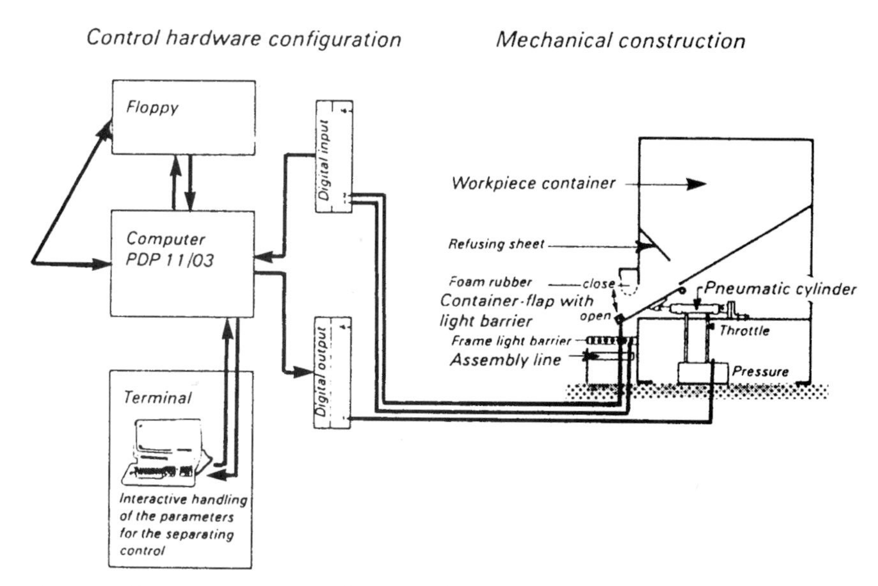

Figure 5.3 Separation and unloading operation.

nates the need for a buffer. The changeover of the separation and unloading bin to a new workpiece type requires changing only the parameters for the container movement.

5.6.6 Cost

To compare costs of in-house and outside development of the automated magazine loading system, a request for quotation was issued to three different suppliers. Their average item costs are indicated in Table 5.2 with the in-house item costs. The comparison showed that in-house development would save $87,800. The study also indicated that a savings of 2200 worker-hours per year would be realized. If the direct labor rate, including overhead and benefits, is $55 per hour, the labor cost savings is $121,000. Therefore, the total savings is $218,800.

5.7 Affordable and Adaptable Grippers for Multiple Assembly Functions

In an assembly task, a variety of objects are brought together in a predefined manner. The difference between the objects' shape and size is usually sufficient to necessitate the use of several grippers to perform the assembly satisfactorily. However, if one gripper could be designed so that it can configure itself like most or all of the grippers

TABLE 5.2 Costs of Automated Magazine Loading Equipment

Item	Suppliers' cost average, $	In-house cost, $
CCD camera and software development	45,000	12,000
Infrared reflective light barrier	4,500	450
Magnetic sensor (measuring compression)	4,500	400
Strain-gauge sensor (measuring compression)	4,500	350
Assembly robot	42,000	15,000
Grippers	15,000	3,000
Conveyers	4,500	4,500
Controller	18,000	4,800
Total	$138,000	$40,200

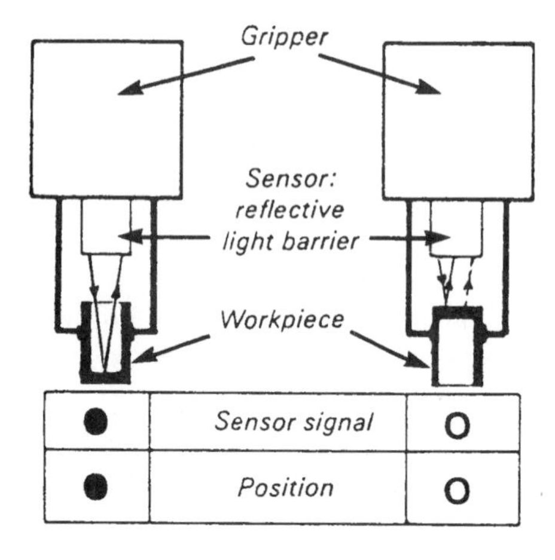

Figure 5.4 Self-configuring gripper.

required, then savings can be realized in terms of assembly time and assembly hardware.

Another major requirement for grippers is that they supply as much information about their environment as possible, so that assembly can be better controlled. If information is available about the object being gripped, where the object and gripper are in relation to the assembly site, etc., then more of the assembly subtasks can be performed under closed-loop control, leading to better quality control.

The gripper in Fig. 5.4 fulfills these requirements to a large extent. It can configure itself to accommodate dual-in-line packages having from 2 to 16 pins. Further, it provides control over the force applied to the object and feeds back information relating to this, provides facilities for testing of the package being gripped, and contains optical fibers which are used for precise positioning of the assembly site below the gripper.

5.7.1 Case study: circuit board assembly

The gripper itself and its controller form a small part of an assembly workstation which directly connects to a computer-aided design (CAD) system. The workstation comprises a network of multitasking processors, an anthropomorphic robot arm, the gripper, an x-y-z table, a set of slide feeders, and associated electronic hardware. The network can be designed so that interprocess communication is processor-independent. A process (or task) sends a message to another process requesting that one of its functions be executed using the accompanying parameters. The sending process then receives a message back at a later time that indicates success or failure and may also contain parameters relating to the performance of the function.

To perform an assembly the CAD system sends an assembly description file to the assembly scheduler process in the workstation. Among the instructions in this file are the insertion instructions; these instructions are issued to the insertion controller which coordinates all the processes necessary to insert a particular kind of pack into the circuit board at the correct position. Three key processes are:

1. The gripper control process.
2. The alignment process.
3. The chip verification process.

These correspond to use of the gripper, use of the optical fibers, and use of the insulated finger tip contacts, respectively. The following paragraphs describe the workings of these three processes, their associated hardware, and how they interact with the insertion controller.

The gripper control. The 16 fingers of the gripper are arranged in two parallel rows of eight, with 0.1-in spacing along the rows, in four blocks, having 1, 2, 4, and 1 opposing pairs per block. There are also four thumbs which separate normally to the plane of action of the fingers, one pair per block (Fig. 5.5).

The finger blocks and their thumbs are selectable by energizing the appropriate solenoids. The finger blocks and thumbs are moved by three dc motors. One motor drives the selected finger blocks, the second motor drives the unselected finger blocks and the third drives the selected thumbs. Each motor has an incremental shaft encoder with quadrature output for positional feedback.

The tips of the fingers are electrically isolated to permit separate electrical connection to each pin by the chip verification process hardware. Also along each finger is a 0.1-in-diameter plastic optical fiber that carries light from the back-illuminated circuit board to a detector. The electronics for the 16 diode detectors is also mounted on the hand

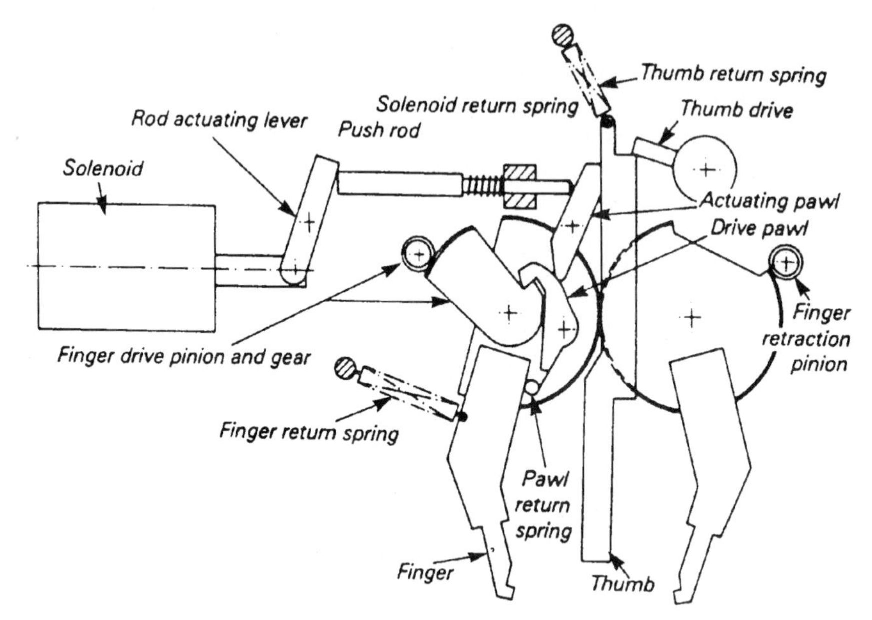

Figure 5.5 Gripper for circuit board assembly.

and the resulting sixteen ±10-V signals are available to the alignment controller through an analog-to-digital signal conditioning board.

The free end of each motor drive shaft has a differential gear with a radial slot. Absolute positioning of the motor is achieved by driving the motor until this slot is coincident with an optical proximity detector (the output is accessed by the motor controller) and using this as the datum.

The motor servo loop is implemented as a three-term controller. The new motor voltage demand to the drive is

$$V = K_1(p - d) - K_2(d - d_p) + I$$

and the new integrator value is

$$I = I_p + K_3(p - d) \qquad \text{when on}$$

$$= 0 \qquad \text{when off}$$

where p = target displacement
d = present displacement
d_p = previous displacement
I_p = previous integrator value
K_1, K_2, K_3 = constants

In addition to the volts demand, the driver circuit also receives a current limit demand from the controller. By monitoring the current taken by the motor, it ensures that this current limit is not exceeded as a primary requirement and then if possible applies the required motor volts. Thus it has the characteristic of a programmable current-limited voltage source.

The integrator is turned on toward the end of a move, if necessary, and is turned off at the start of a move. This ensures that when the move is completed, the motor displacement is that requested, or a motor current is at the requested limit value.

The servo is software-tuned to provide a minimum-duration response to a step input, with no overshoot and a programmable undershoot.

Moving the motors. The movement of the second motor, which drives the unselected finger blocks, is hidden from the user, so when the fingers and thumbs are requested to move, only parameters for motors 1 (selected fingers) and 3 (thumbs) are given. The request message to the move function of the controller contains destination position and current limits for the motors. The controller uses the destination positions in calculating the positional error, but varies the applied current limits during the movement of the motors in the following manner. For the first few servo updates, the current limit is set to the maximum value. Thereafter, when the motor volts applied are such as to move the motor toward the destination, the specified current limit is applied, otherwise the maximum current limit is applied. This both facilitates starting and stopping and limits the gripping (fingers) and pushing (thumbs) forces which the gripper can apply.

When the motors are finished moving, the positions achieved are compared with those requested, and a decision is made as to whether the move was positionally successful or not. This success or failure is conveyed to the requester, along with the actual position and final update motor currents, in the response message.

Figure 5.6 shows the actual selected thumb motor current for various limit values when the gripper grips a package at the feeders.

Reconfiguration for another package type is initiated when a request message for the configuration function is received. The number of dual-in-line pins to be configured for is included as a parameter of the message. The function controls the configuration in the following manner:

1. The motors are sequentially driven to their datum positions in accordance with the mechanical constraints imposed by the gripper design.
2. The controller uses the number-of-pins parameter as an index to a lookup table to establish the correct solenoid pattern and then en-

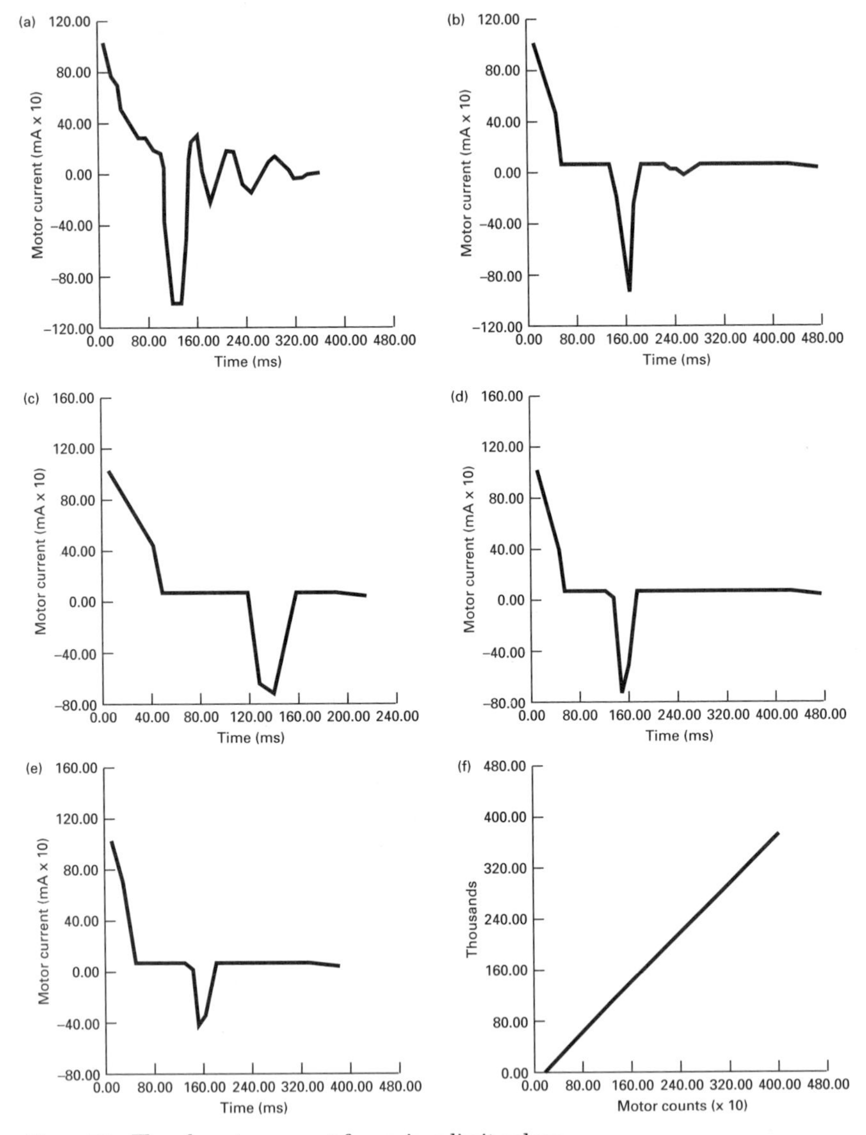

Figure 5.6 Thumb motor current for various limit values.

ergizes the appropriate solenoids. From Fig. 5.5, it can be seen that energizing a solenoid has two effects. The rod actuating lever causes the push rod to push the thumb across (via the actuating pawl), so that thumb drive will engage the flat at the top of the thumb. The actuating pawl also causes the drive pawl and the finger drive gear to engage.

3. The selected fingers are now driven inward by the finger drive pinion far enough that the finger retraction pinion no longer engages the finger cog. The selected fingers are driven upward so that they are clear of the work area.
4. If all the above steps are successfully performed, then a success response message is sent to the requester, otherwise a failure response is returned.

Reconfiguration of the gripper takes 2 to 3 seconds. However, the additional time seen by the system for reconfiguration is less than 1 second, since configuration is performed while the manipulator is moving to the slide feeder for the next component.

Gripping a package. Before the insertion controller issues a request message to the motor move function of the gripper controller, it needs to establish several parameters of the package to be inserted. It does this by sending a request message, with the package identity as a parameter, to the system database process and receiving a response message with such parameters as package width and number of pins. It uses these parameters to calculate the finger grip position and motor current limit. The thumb motor parameters are target position at the feeder surface and current limit such that the motor will stall while applying little force on the package if it is present. The move response parameters are evaluated to check that the thumb motor has stalled and is applying its current limit, signifying that the package is present, and that the finger motor has reached its target position, thereby deforming the package pins to the correct separation for insertion into the circuit board.

The package insertion is performed by the gripper itself, the manipulator being used as a stable platform during this stage of the assembly. The insertion is performed in a sequence of three operations:

1. A request is issued to the move function of the gripper controller to extend the thumbs a small distance using a low current limit. This exploratory insertion is used to confirm that all the pins are located in the holes without damaging the board or the package.
2. If the response is favorable, a second move request is sent which opens the fingers and extends the thumbs to the insertion position with a current limit sufficiently high to achieve this.
3. If this is successful, a move request is issued to extend the thumbs to the surface of the circuit board with a low current limit to avoid damaging the package or board. If the insertion has been successfully performed, then this last stage will confirm that the package

is present on the board by the motor's stalling and failing to extend thumbs fully. It is possible that the package could have been dropped between the feeder and the circuit board, in which case stages 1 and 2 would have been successful, so this final stage is a necessary check.

The alignment process. Before a package can be inserted, it must be correctly aligned over the holes in the circuit board. In a perfect system, alignment would always be correct initially, but in practice it is necessary to remove errors introduced by both the robot and the manipulator of the circuit board. This is achieved by moving the *x-y-z* table which supports the circuit board. The table movements are controlled by the inputs from the optical fibers on the gripper in order to maximize the light entering each fiber. The fiber input provides 4096 levels of light, which gives sufficient discrimination to locate the center of a hole in the circuit board. The fibers are not used merely as binary sensors. The two-dimensional response of a fiber to a hole depends on the hole diameter and the height of the end of the fiber above the board, which is a constant 1 mm; thickness of the board is a less important variable. The response to typical holes is illustrated in Fig. 5.7.

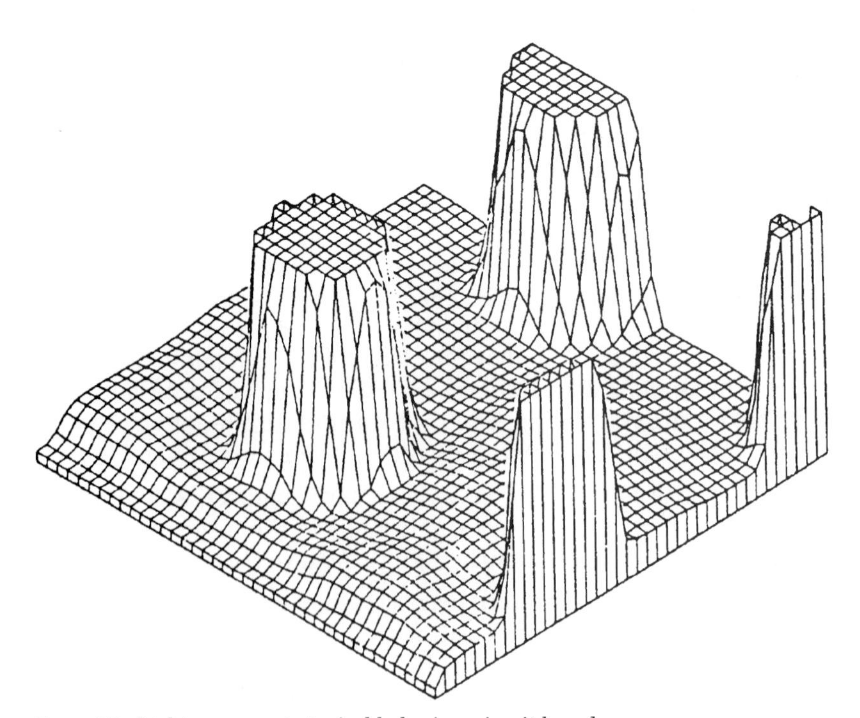

Figure 5.7 Light response to typical holes in a circuit board.

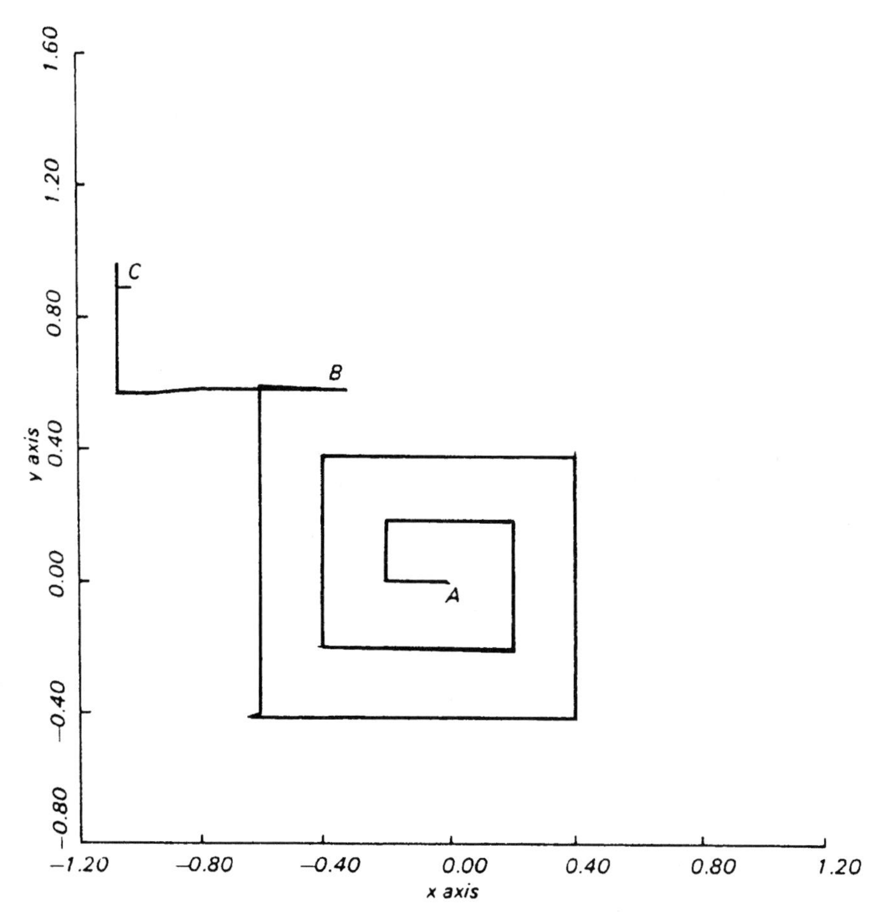

Figure 5.8 Typical path of a hole search.

The table moves incrementally under servo control, and the direction and magnitude of each step is determined by the change in light input at the fibers during the previous step. Large numbers of steps, typically 0.1 mm in length in the x or y direction, are combined to form a search pattern or "hill climb" or both to maximize the light. Incremental rotation about a vertical axis through a point on the board determined by the light input ensures that the alignment angle is correct.

A typical path followed by the table when searching for a hole and then hill climbing to its center is shown in Fig. 5.8. The process is requested to indicate success or failure. If the response is failure, insertion of the package does not proceed.

The verification process. Verification of a package begins as soon as it has been removed from the component feeder and proceeds as it is moved toward the circuit board. By applying various voltages to the appropriate pins and measuring the voltages and currents at other pins, it is possible to check:

- In which of the two possible orientations the package is held in the gripper
- That the circuits in the package are functioning satisfactorily

The result of the circuit test determines whether insertion of the package proceeds or the gripper is diverted to drop the package into a reject pin. The result of the orientation check determines the position to which the gripper is moved for the insertion stage, if the package is working properly. The two tests are performed quickly, and add nothing to the assembly time.

Experience with the gripper in an assembly station has demonstrated that it can meet its affordable automation aims. Insertion of dual-in-line packages of several sizes has been successfully achieved under controlled operation with a simple circuit board with specially toleranced holes.

Chapter

6

Planning for Affordable Automation

6.1 Introduction

When one thinks of the affordable automation concept, there is a natural tendency to place undue emphasis on the automatic assembly machine, its mechanisms, and its immediate surroundings. A better appreciation of the potential effectiveness of affordable automation is obtained if one considers the industrial handling process as a whole. This comprises many functions which are peripheral to, and just as important as, the functions performed by an innovative assembly machine. In other words, the overall marshaling and control of component parts in industry is vital at all levels and is a prerequisite to success.

Considering affordable automation as separate activities can lead to easier solutions of problems. In order to appreciate all the interactions, a problem can be divided up into various issues, such as economics, control of parts, explicit and implicit inspection, innovative mechanisms, and direct and global communications.

6.1.1 Economics

Choosing the correct system and appropriate architecture implies the optimal selection of a level of sophistication that provides reasonable financial payback. The level of sophistication (and, therefore, system cost) will depend mainly on the cost of labor, the interest rate, and the versatility required for a particular production situation.

6.1.2 Part control

Part control may be coarse or fine. Coarse part control involves scheduling, inventory, ordering, general production, and in-process handling. Fine part control involves part feeding and placement.

6.1.3 Explicit and implicit inspection

Explicit inspection ensures that a component is the correct one and carries all the required features, such as holes and screw threads. Implicit inspection ensures that all parts are linked together as planned with the tolerances attainable and component quality available.

6.1.4 Mechanisms

Mechanisms provide the motions to bring component parts together by placement devices, with jigs and fixtures for high-quantity production or “soft” tooling based on microprocessor control and sensory interaction for low-quantity batches.

6.1.5 Direct and global communication

Direct communication is used for machines requiring frequent changes to deal with smaller batches. A black-box approach with built-in microprocessor interfacing enables workers to change functions simply and quickly. Diagnosis capabilities are also necessary for fast response to malfunctions.

Global communication provides interfaces between individual control centers. It is essential for complete factory integration.

6.2 Programmable Assembly

Assembly machine technology has evolved from “hard” dedicated machinery into the more versatile systems that are currently available. Typical machines developed in the mid-1960s were in-line or rotary transfer machines, and such technologies, in the right context, are just as effective today. This equipment is mechanically actuated and built around modular elements consisting of a main chassis, a set of placement devices, and feeders. However, it is not practical on economic grounds to alter such hard automation in order to respond to product changes.

One can evaluate the adaptability of equipment by a numerical yardstick called the *versatility index*. Let P = cost of a standard modular element of the system and P_s = cost of adapting the standard system to a particular task. Then the versatility index can be expressed as

$$I = P/P_s$$

A large value of I implies high versatility, and vice versa. For the hard automation referred to above, I is in the range 0.1 to 0.2 (i.e., 80 to 90% of the installation is special-purpose); hence, there is no effective interproduct versatility (i.e., ability to change the line from one product to another).

There are very many systems available today which use a higher level of modularization, usually in the form of "free-standing" standard machines with standard control system packages, be they hardwired sequential, simple memory (i.e., plugboard), or computational. None of these machines, however, can be considered to be truly *programmable.* They have a reasonable element of "reclaimability," but are not very suitable for small-batch, variable-product production because the amount of adaptation needed is still too high. However, they exhibit extensive *interprocess versatility*; i.e., they are capable of adaptation over a wide range of relatively dedicated activities such as press feeding, in-process handling, machine serving, and assembly. It is possible in certain application areas to achieve an I of 0.5 to 1.0.

Programmable assembly machines have, as yet, made little penetration into the everyday production; this is perhaps because the philosophy of their application is so little understood. There are generally three main approaches to programmable assembly automation:

Architectural design of a machine. Machines can be architecturally designed to be capable of adaptation. This type of machine must have significant sensor capabilities, with standard functions such as part not placed, part not present, etc. Such machines will find their niches in certain activities in which they are fooled into thinking they are handling the same object when they are not. In other words, the need for special workholders and grippers and automatic tool changing is reduced to a minimum, and a large value of the factor I results. This machinery is good for dealing with families of assemblies.

Versatile machine design. Machine architecture can be designed to suit the particular task so as to create a programmable dedicated machine. The machine can be programmed to suit a particular task, but can readily be reprogrammed to suit others. The *programmable dedicated* concept becomes even more powerful when such units are integrated to form a complete line. By this means, the technology of mass production can be applicable to a batch production situation. Because a station can talk to other stations, rapid programming is possible. However, the versatility of programming of the system depends on the ease of mechanical adaptation that can be achieved, which depends once more on the degree to which family relationships can be established for the product.

Robot arm. A robot arm can be used for assembly, with extensive back-up and control. This technology carries with it a number of question marks at this time, not the least of which is whether it is really necessary to have a multiaxis complex arm carry out the majority of assembly processes in industry. Of course, with the computing power of microprocessors currently available, it is possible to provide enough computing capacity to control such an arm. But in the majority of cases, assembly insertion is a straight-line process, and to generate an accurate straight line via a six-axis robot would seem to be an unnecessary complication. With appropriately designed software, however, it is relatively easy to endow the machine with a number of routines to deal with faults.

Nevertheless, if one has continual malfunctions and the machine spends most of its time using a fault-correcting function, it is not spending enough time on production. Therefore, the need for machines that can detect an incipient malfunction and correct it before it actually occurs is very much a problem for the future.

This brings us to the very important subject of sensory interaction in assembly devices. There are around 400 machine vision systems worldwide, and it is important to appreciate just what they really achieve. On close examination, although they have been created as "universal equipment," most of the old problems of mechanical adaptation of grippers and other peripheral items to accommodate small batches of very different components remain.

There have also been significant developments in robots' "feel," or tactile sense, over the last decade. Hitachi was the first with a commercially available system, the Hi-T-Hand Expert 2 system, which is a fully reactive feedback system for plug-in-the-hole–type insertion. The Draper Laboratories in the United States has a remote center compliance (RCC) device which achieves a similar effect for relative positional errors in insertion.

Assembly machines will become economic within a total system concept for a factory in which instructions are accepted by the machine in response to input from a central source concerned with product design and component production. A computer-aided design/manufacturing (CAD/CAM) terminal is the likely medium for the input of design and functional information. However, the integrated factory concept demands a greater discipline from engineering as a whole. Unless this discipline is achieved in product design and manufacture to facilitate assembly and handling, the future for affordable automation remains uncertain.

A rule that is generally true in applying automation is that, if a machine produces more parts in a given time than its human counterpart, then it will succeed economically. This does not mean that a direct comparison should be made between the rate of manual work and the rate of machine work in terms of the cycle time to produce an assembly. The overall output over a longer period is the figure to look for. This is because, although the machine may operate on a longer cycle time than its human counterpart, so long as it only takes a fraction of a human being to supervise it and it can be left for long periods to look after itself, then daily production could be significantly in excess of that of a human work force.

Thus, a beneficial effect of increasing automation can be missed if one is preoccupied with comparing human and machine directly on the wrong basis. The machines and systems presented here, once having been "taught" the job, give their full production rate immediately, whereas a human work force has a learning curve and performance which varies throughout the working period. Clearly, to achieve the advantages that programmable assembly can offer requires a whole rethinking, reassessment, and evolutionary interaction. Success will be achieved by evolution, not revolution.

6.3 Group Technology as a Base for Affordable Automation

The amount of data that a person's mind can readily work with at one time is relatively small. A computer can manipulate considerably more data than a human mind; however, even a computer has limits. For this and other reasons it is desirable to find ways to organize data so that only pertinent items need be retrieved and analyzed at a given time. To accomplish this, methods of structuring data have been devised. Some methods are very clever, such as the data structures used in large computer databases; others are relatively simple, such as alphabetical lists of words in a dictionary. An elaborate system is the taxonomy used in biology to classify all living organisms. Taxonomy illustrates how thousands of items can be organized into small groups whose members have similar attributes. Another example is the coding and classification of books in a library catalog. Using this catalog, one can easily find all books written by an author, all books on a specific subject, or all books with a particular title.

Organizing for affordable automation in group technology is an analogous situation. A company may make thousands of different parts in an environment that is becoming more complex as lot sizes become smaller and the variety of parts increases. When they are examined closely, however, many parts are similar in some way. A design engineer faced with the task of designing a part would like to know if the same or a similar part has been designed before. Likewise, a manufacturing engineer faced with the task of determining how to manufacture a part would like to know if a similar plan already exists. It follows that there are economies to be realized from grouping parts or processes into families with similar characteristics. The resulting database would certainly make the information easier to manage; therefore, the manufacturing enterprise should be easier to manage. In 1969, V. B. Soloja defined group technology in terms of "the realization that many problems are similar, and that by grouping similar problems, a single solution can be found to a set of problems thus saving time and efforts."[1] This definition is very broad, but it is valid nonetheless. The basis for applying group technology with respect to affordable automation is coding and classification of parts.

6.3.1 Key definitions for the group technology concept

Attribute code (polycode) Each part attribute is assigned to a fixed position in a code. The meaning of each character in the code is independent of any other character value.

Average linkage clustering algorithm An algorithm for clustering objects together based on the average similarity of all pairs of objects being clustered. The similarity of each pair is measured by a similarity coefficient.

Bottleneck machine A machine in a group (cell) that is required by a large number of parts in a different group.

Cell A group of machines arranged to produce similar families of parts.

Classification The process of categorizing parts into groups, sometimes called *families,* according to a set of rules or principles.

Cluster analysis The process of sorting objects into groups so the similarities are high among members of the same group and low among members of different groups.

Coding The process of assigning symbols to a part to reflect its attributes.

Computer-aided process planning (CAPP) Use of an interactive computer system to automate some of the work involved in preparing a process plan.

Decision variables Variables in a mathematical model to which values must be assigned to optimize the model's performance. Initially, the best values for these variables are unknown.

Dendrogram A treelike graphical representation of cluster analysis results. The ordinate is a similarity coefficient scale, and the abscissa has no special meaning.

Functional layout Arrangement of machines in a factory such that machines of a specific type are grouped together.

Group layout Arrangement of machines in a factory into cells.

Group technology for affordable automation An engineering and manufacturing philosophy that groups parts together on the basis of their similarities in order to achieve economies of scale in a small-scale environment.

Group tooling Tooling designed so that a family or families of parts can be processed with one master fixture and possibly some auxiliary fixtures to accommodate differences in some of the part attributes, such as number and size of holes.

Hierarchical code (monocode) Code in which the meaning of each character is dependent on the meaning of the previous character in the code.

Hybrid (mixed) code A combination of an attribute code and a hierarchical code with the advantages of both code types.

Line layout Arrangement of machines in a factory in the sequence in which they are used. The work content at each location is balanced so that materials can flow through in a continuous manner.

Logic tree A treelike graph that represents the logic used to make a decision. This differs from a decision tree in that the branches may contain logical expressions as well as calculations, data elements, codes, and keys to other data.

Machine-component chart A matrix that denotes which machines a group of components (parts) will visit.

Part family A group of parts having some similar attributes.

Process plan The detailed instructions for making a part. It includes such items as the operations, machines, tools, feeds and speeds, tolerances, dimensions, stock removal, time standards, and inspection procedures.

Production flow analysis A structured procedure for analyzing the sequence of operations that parts go through during manufacturing. Parts that go through common operations are grouped together as a family, and the associated machines are arranged as a cell.

Rotational part A part that can be made by rotating the workpiece. It is usually symmetrical along one axis, like a gear.

Similarity coefficient A measure of how alike two machines are in terms of the number of parts visiting both machines and the number of parts visiting each machine.

Single-linkage clustering algorithm (SLCA) An algorithm for clustering together objects that have a high similarity coefficient.

Threshold value A similarity coefficient value at which clustering is to stop. That is, no more clusters are to be formed if the largest remaining similarity coefficient value is below this value.

6.3.2 History of affordable automation and group technology

The small-lot manufacturing environment has been studied extensively since World War II. This environment is very difficult to manage well. From a casual observer's viewpoint, manufacturing activities seem to occur randomly. In the 1960s and early 1970s, some professionals thought that operations research (management science) techniques, such as application of queuing theory and mathematical programming, had the potential to improve management. However, none of these techniques was practically successful. Until the late 1970s, no approach received widespread acceptance; in fact, very few new approaches were even proposed. In the late 1970s, however, a consensus emerged in the manufacturing countries that group technology provides the basis for better management of the small lot manufacturing environment. Some professionals had advocated this philosophy since the early 1950s.

People have been informally using affordable automation linked with group technology concepts for centuries; one of the first recorded applications in manufacturing was achieved by F. W. Taylor, the scientific management pioneer. In his attempts to improve productivity, he noted that there were similarities between some jobs, and he was able to categorize the similar attributes of the jobs. More recently, many companies have applied more formal group technology concepts for affordable automation, such as grouping machines into cells and establishing group tooling.

Still, no formal description of a group technology applications in manufacturing appeared until 1959, when S. P. Mitrofanov, a Russian, published his book *Scientific Principles of Group Technology.* In the editor's foreword to the English translation of Mitrofanov's book, T. J. Grason describes group technology as follows:[2]

> ...a method of manufacturing piece parts by classification of these parts into groups and subassemblies applying to each group similar technological operations. The major result of this method of manufacture is to obtain economies which are normally associated with large scale production in the small scale situation and it is therefore of fundamental importance in the batch production and jobbing section of industry.

If a company is intending to produce several million units of a specific product, large expenditures can be justified for developing methods to manufacture the item for less cost, even if the reduction per unit is only a few cents. However, if only a small number are to be produced, it is not realistic to incur such expenditures. The group technology for affordable automaton philosophy advocates combining several similar parts so that the sum is large enough to justify expenditures to reduce design and manufacture costs for the entire group of parts. In other words, any reduction in cost must be generic to the group to be cost-effective. Another premise is that group technology promotes standardization of parts, because design engineering strives to minimize new part designs by utilizing existing designs. These are some of the reasons that the acceptance of group technology for affordable automation as a manufacturing philosophy has spread throughout the manufacturing countries of the world.

In 1960, West Germany and Great Britain began serious studies into group technology techniques. Other European countries quickly followed. By 1963 the success of group technology applications in Russia prompted the government to promulgate a plan for increased implementation throughout Russian industry. By 1970 the Japanese government had begun sponsoring group technology applications. In the United States, in contrast, group technology did not receive widespread acceptance until the latter part of the 1970s.

Current trends in manufacturing have set the stage for acceptance of group technology in the United States. these trends include:

1. A rapid proliferation of number and varieties of products, resulting in smaller lot sizes.
2. A growing demand for closer dimensional tolerances, resulting in a need for more economical means of working to higher accuracy.
3. A growing need for working with increased varieties of materials, heightening the need for more economical means of manufacturing.
4. An increasing ratio of cost of materials to total product cost because of increasing labor efficiency, with reduced acceptability of scrap.
5. Pressures from the above factors to increase communication across all manufacturing functions with a goal of minimizing production costs and maximizing production rates.

It has been estimated that as much as 75 percent of all parts made in the United States will produced on a small-lot basis, from one to a few thousand. Consequently, it is not unreasonable to predict that 50 to 70 percent of American manufacturing firms will be using some form of group technology by 1995.

There have been some recent advantages in machine tool automation and metal removal technology. So, modern manufacturing is often perceived as being productive and efficient. On the contrary, studies have determined that during fabrication of an average part in an average batch-type facility, the part is on a machine for only 5 percent of the production throughput time (Fig. 6.1). Of that 5 percent, less than 3 percent is spent in metal removal. The part is being

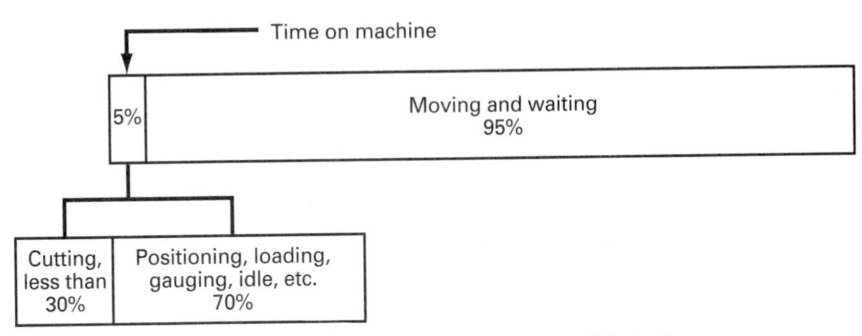

Figure 6.1 Distribution of work-in-process time during part fabrication.

moved or is waiting in a queue 95 percent of the time. These are additional reasons for batch-type manufacturing firms to consider some new concepts.

6.3.3 Affordable automation in CAD/CAM integration

As we have noted, competitive world market conditions are encouraging more and more batch-type manufacturing firms to adopt the philosophy of group technology for affordable automation. Another major contributing factor to this acceptance is the increasing integration of CAD and CAM.

It is evident that group technology is an important element of CAD and CAM. An essential aspect of CAD and CAM is the integration of information used by the engineering, manufacturing, and all other departments in a firm. Group technology provides a means to structure and save information about parts, such as design and manufacturing attributes, processes, and manufacturing capability, in a way that is a amenable to computerization and analysis. It provides a common language for the users. Integration of many types of part-related information would be virtually impossible without group technology.

Another important aspect of CAD and CAM is automation. Many manufacturing firms are automating their operations by arranging their machines into cells. The design of a cell is based on group technology. These observations reinforce the importance of group technology for affordable automation.

6.4 Methods for Developing Part Families for Affordable Automation

Group technology begins by grouping parts into families according to their attributes. Usually, these attributes are geometric and/or production process characteristics. Geometric classification of families is nor-

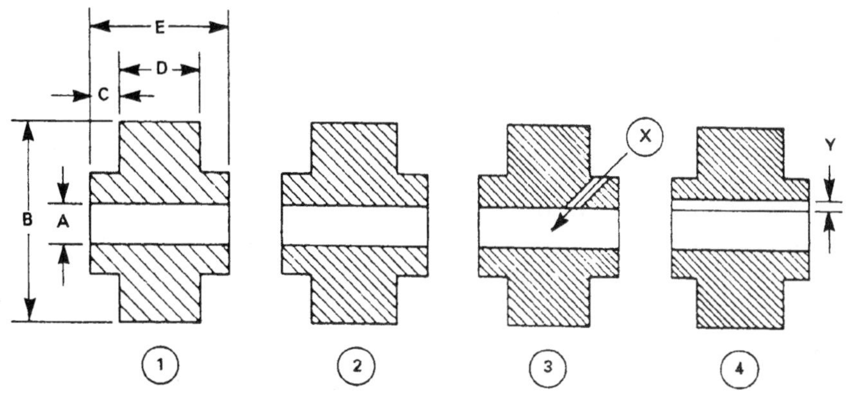

Figure 6.2 Parts grouped by geometric shape.

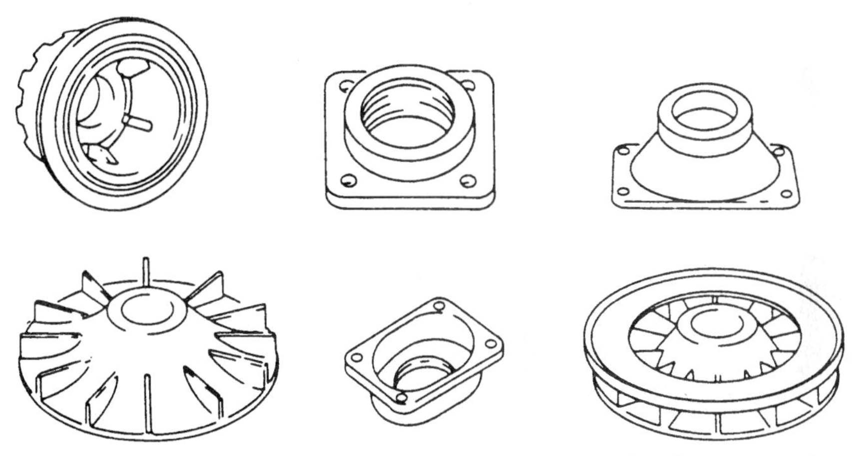

Figure 6.3 Parts grouped by manufacturing process. (*Reprinted with permission from Society of Manufacturing Engineers, Dearborn, Mich., 1984.*)

mally based on size and shape, while production process classification is based on the type, sequence, and number of operations. The type of operation is determined by the method of processing, the method of holding the part, the tooling, and the conditions of processing. For example, Figs. 6.2 and 6.3 show families of parts grouped by geometric shape and by production process. The identification of a family of parts that has similarities permits the economies of scale normally associated with mass production to be applied to small-lot batch production. Therefore, successful grouping of related parts into families is a key to implementation of group technology for affordable automation.

There are at least three basic methods that can be used to form part families for affordable automation:

1. Manual visual search
2. Production flow analysis
3. Classification and coding

Manual visual search is not often used in formal group technology for affordable automation applications. It leads to very inconsistent results because seldom will two people group sets of parts into the same families. There are many reasons for this, such as the fact that each person will have a different knowledge of the processing capabilities of the factory, recognition of the significant part attributes may differ, and many different tools and machines can be used to perform a specific function though the costs may be significantly different.

Production flow analysis (PFA) is a structured technique developed by Burbidge[3] for analyzing the sequence of operations (routings) that parts go through during fabrication. Parts that go through common operations are grouped into part families. Similarly, the machines used to perform these common operations may be grouped as a cell; consequently, this technique can be used in facility layout. Initially, a machine-component chart must be formed. This is an $M_i x N_j$ matrix, where M_i = machine number, N_j = part number, and $x = 1$ if part j has an operation on machine i; 0 otherwise.

If the machine-component chart is small, parts with similar operations might be grouped together by manually sorting the rows and columns. However, a more appealing method is to use a computer procedure to perform this work.

Figure 6.4 illustrates the use of PFA to form part families. For this technique to be successful, accurate and efficient routings must exist for each part. In many companies these routings do not exist. If routings exist, they are often inaccurate from lack of maintenance or they may be very inconsistent. The latter situation will occur if routings are established without using a coding and classification system.

Using PFA involves judgment, because some parts may not appear to fit into a family when one or more unique operations are required. Furthermore, additional analysis is required to determine when a particular machine should be duplicated in another group. In Fig. 6.4, for example, machine D is in groups 1 and 2. In this case, since machine D was visited by almost all of the parts, it was duplicated to keep the groups small. Otherwise, groups 1 and 2 might have been combined into one group having several parts with dissimilar routings. One cannot determine how many machines of type D are required without evaluating demands and machine capacities. In addition, PFA does not consider the part features as functional capabilities. Therefore, this technique should be used for part families for design engineering.

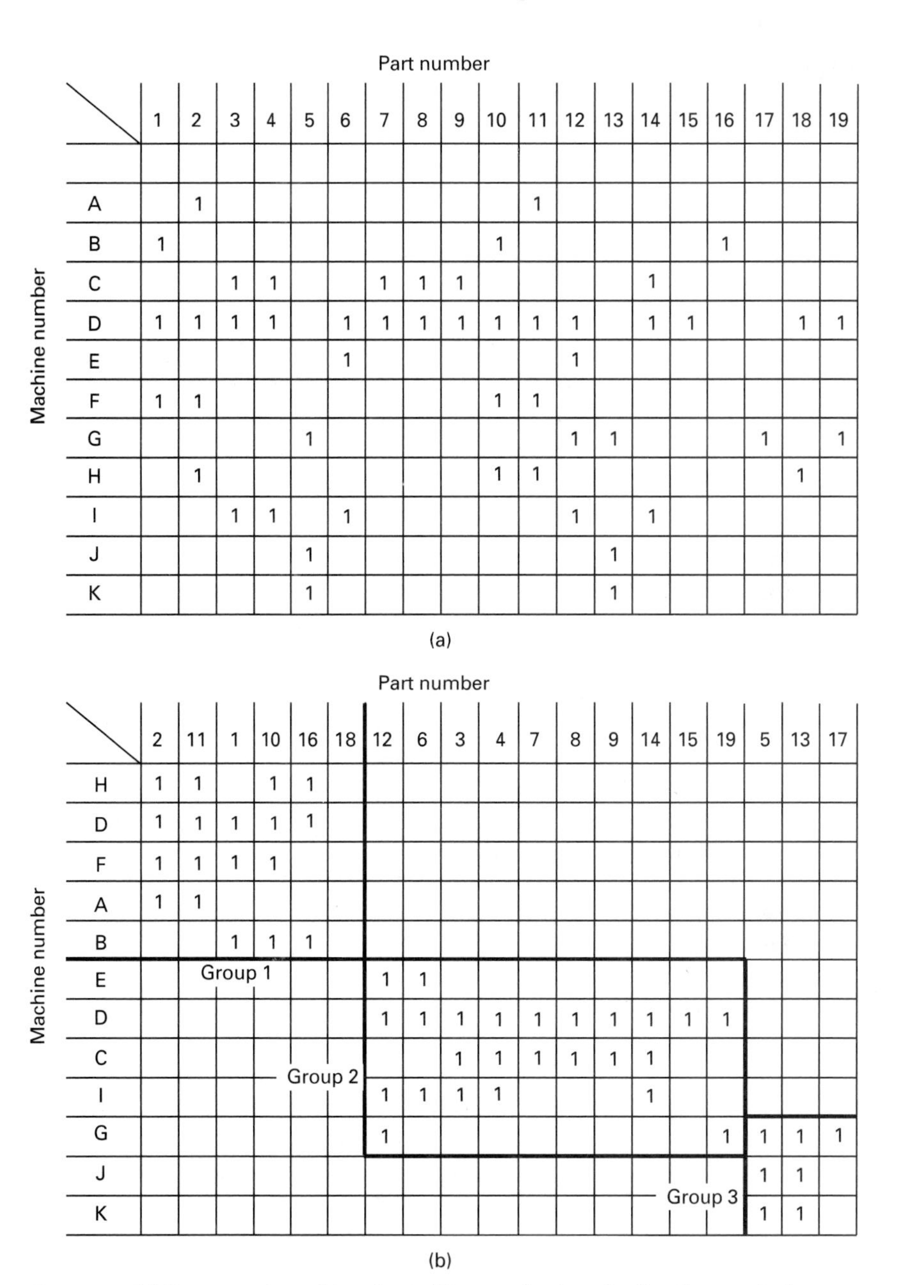

(a)

Machine number	1	2	3	4	5	6	7	8	9	10	11	12	13	14	15	16	17	18	19
A		1									1								
B	1									1						1			
C			1	1			1	1	1					1					
D	1	1	1	1		1	1	1	1	1	1	1		1	1			1	1
E						1						1							
F	1	1								1	1								
G					1							1	1				1		1
H		1								1	1							1	
I			1	1		1						1		1					
J					1								1						
K					1								1						

(b)

Machine number	2	11	1	10	16	18	12	6	3	4	7	8	9	14	15	19	5	13	17
H	1	1		1	1														
D	1	1	1	1	1														
F	1	1	1	1															
A	1	1																	
B			1	1	1														
E							1	1											
D							1	1	1	1	1	1	1	1	1	1			
C									1	1	1	1	1	1					
I							1	1	1	1				1					
G							1									1	1	1	1
J																	1	1	
K																	1	1	

Figure 6.4 (*a*) Component-machine chart; (*b*) example of production flow analysis.

One advantage of using production flow analysis compared to a coding and classification system is that part families can be formed with much less effort.

If the coding and classification technique is used, parts are examined and codes are assigned to each part according to its attributes. These codes can then be sorted so that parts with similar codes are grouped as a part family. Because these codes are assigned in a manner that does not require much judgment, the part families developed by this technique do not suffer from inconsistencies. A disadvantage of using the coding and classification technique is that a large amount of time may be required to develop and tailor a code to meet the needs of a specific company. Afterward, coding the parts will take an even larger amount of time. However, when properly applied, the results are much better than those of other techniques. Consequently, coding and classification is the preferred approach.

6.4.1 Part classification and coding

Classification of parts is the process of categorizing parts into groups, sometimes called families, according to a set of rules or principles. The objectives are to group together similar parts and to differentiate among dissimilar parts. Coding of a part is the process of assigning symbols to the part. These symbols should have meanings that reflect attributes of the part, thereby facilitating analysis (information processing). Although this does not sound very difficult, classification and coding are very complex problems.

Several classifications and coding systems have been developed. Many professionals have attempted to improve them. No system has yet received universal acceptance, however, because the information that is to be represented in the classification and coding system varies from one company to another. This is understandable because the two greatest uses of group technology for affordable automation are design retrieval and group (cell) production, and each company has some unique needs for these functions. Therefore, even though classification and coding systems can be purchased, a good rule of thumb is that 40 percent of a purchased system must be tailored to the specific needs of the particular company.

One reason that a design engineer classifies and codes parts is to reduce design effort by identifying similar parts that already exist. Some of the most significant attributes on which identification can be made are shape, material, and size. If the coding classification system is to be used successfully in manufacturing, it must be capable of identifying some additional attributes, such as tolerances, machinability of materials, processes, and machine requirements. In many companies the design department does not exchange very much information with the manufacturing department. The expression "Design engineering throws the design over the wall for manufacturing" is often used to describe the lack of communication between these departments. The classification and coding system selected by a company should meet the needs of both design engineering and manufacturing. A system that meets these combined needs will improve communication between departments and facilitate an affordable automation strategy. Although well over 100 classification and coding systems have been developed for group technology, all of them can be grouped into three basic types.

1. Hierarchical or monocode
2. Attribute or polycode
3. Hybrid or mixed code

6.4.2 Hierarchical code

As noted, manufacturing professionals have been classifying and coding parts for a long time. The hierarchical coding system supports this statement. It was originally developed for biological classification by Linnaeus in the 1700s. In this type of code, the meaning of each character is dependent on the meaning of the previous character; that is, each character amplifies the information of the previous character. Such a coding system can be depicted as a tree structure as illustrated in Fig. 6.5, which represents a simple scheme for coding a spur gear. Using this scheme, one can assign the code A11B2 to the spur gear.

A hierarchical code provides a large amount of information in a relatively small number of digits. This advantage will become more apparent when we look at an attribute coding system. Defining the meaning for each digit in a hierarchical system can be difficult, although application of the defined system is relatively simple. Starting at the main trunk of the tree, one needs to answer a series of questions about the item being coded. Continuing in this manner, one can find a way through the tree to a termination branch. By recording each choice in answering each question, an appropriate code number will be built. However, determining the meaning of each digit in the code is complicated, because each preceding digit must first be decoded. For example, in the code development for the spur gear in Fig. 6.5*b,* a 1 in the second position means *round with deviations* because there is an A in the first position of the code. However, if there had

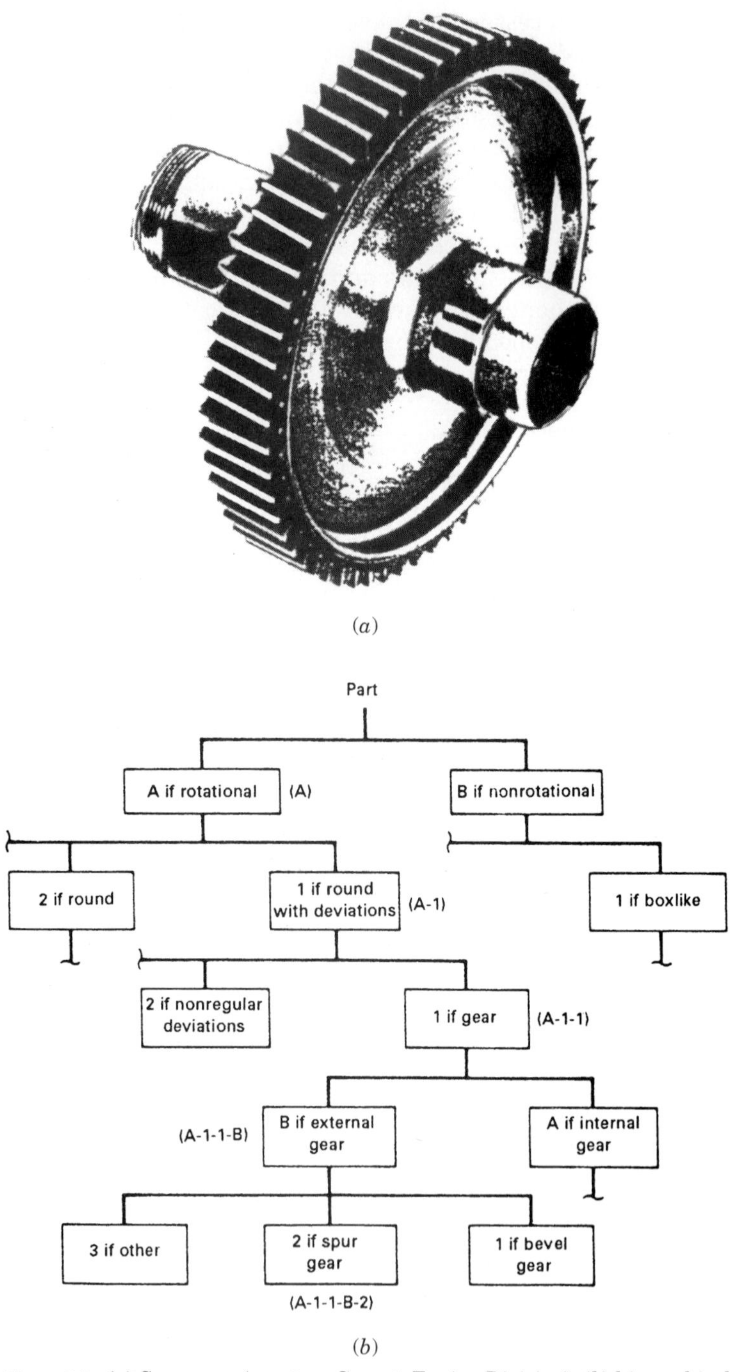

Figure 6.5 (*a*) Spur gear (*courtesy Garrett Engine Division*); (*b*) hierarchical code for the spur gear.

been a B in the first position, a 1 in the second position would have meant *boxlike.*

Design departments frequently use hierarchical coding systems for part retrieval because this type of system is very effective for capturing shape, material, and size information. Manufacturing departments, on the other hand, have different needs which are often based on process requirements. It is difficult to retrieve and analyze process-related information when it is in a hierarchical structure.

6.4.3 Attribute code

An attribute code is also called a *polycode,* a *chain code,* a *discrete code,* or a *fixed-digit code.* The meaning of each character in an attribute code is independent of any other character; thus, each attribute of a part can be assigned a specific position in an attribute code. Figure 6.6 shows an example of an attribute code. By using Fig. 6.6 to code the spur gear in Fig. 6.5*a,* the code 22213 can be generated. Referring to Fig. 6.6, note that a 3 in position 5 means that the part is a spur gear, regardless of the values of digits in any other position.

If one had used this attribute code to code several parts and wanted to retrieve all spur gears, it would be necessary only to identify all parts with a 3 in position 5 of their code. This becomes a simple task if a computer is used. Consequently, an attribute code system is popular with manufacturing organizations because it makes it easy to identify parts that have similar features that require similar process-

Digit	Class of feature	Possible value of digits			
		1	2	3	4
1	External shape	Cylindrical without deviations	Cylindrical with deviations	Boxlike	• • •
2	Internal shape	None	Center hole	Brind center hole	• • •
3	Number of holes	0	1-2	3-5	• • •
4	Type of holes	Axial	Cross	Axial cross	• • •
5	Gear teeth	Worm	Internal spur	External spur	• • •
⋮	⋮	⋮	⋮	⋮	⋮

Figure 6.6 Attribute code example.

ing. One disadvantage of an attribute code is that a position in the code must be reserved for each different part attribute; therefore, the resulting code may become very long.

6.4.4 Hybrid code

In reality, most coding systems use a hybrid (mixed) code so that the advantages of each type of system can be utilized. The first digit, for example, might be used to denote a type of part, such as a gear. The next five positions might be reserved for a short attribute code that would describe the attributes of the gear. The next (seventh) position might be used to designate another subgroup, such as material, and be followed by another attribute code. In this manner a hybrid code can be created that is more compact than a pure attribute code but still makes it easy to identify parts with a specific common characteristic.

6.4.5 Selecting a coding system

Because well over 100 coding systems have been developed, selecting the best one for a particular application can be a difficult and time-consuming task. The following factors should be kept in mind when a system is selected.

Objectives. The objective of installing the system will vary depending on the user—engineering, manufacturing, or both. Some typical engineering objectives are

1. Provide an efficient retrieval system for similar parts
2. Provide part information in a standard form
3. Provide an efficient means to determine the manufacturing capabilities and producibility

From the manufacturing viewpoint, some typical objectives are

1. Provide information required to form part families
2. Provide efficient retrieval of process plans
3. Provide an efficient means to form machine groups or cells for part families

Robust design. The system selected should be capable of handling all parts being sold or planned by the firm. This analysis will involve looking at planned group technology for affordable automation and the part attributes that might be needed. Table 6.1 provides some example applications and the associated attributes.

TABLE 6.1 Applications versus Part Attributes

	Part attributes									
Applications	Shape	Form features	Treat-ments	Functions	Size envelopes	Tolerances	Surface finish	Material type/ condition	Quantity	Next assembly	Raw material form
Design retrieval	×	×		×	×			×			
Generative process planning	×	×	×	×	×	×	×	×	×		×
Equipment selection	×	×	×		×	×	×	×	×		×
Tool design	×	×	×		×	×	×	×	×		
Time/cost estimation	×	×	×		×	×	×	×	×		×
Assembly planning	×	×			×	×	×	×	×	×	
Quality planning	×	×	×	×	×	×	×	×	×	×	×
Production scheduling	×	×	×		×	×	×	×	×	×	×
Parametric part programming	×	×	×		×	×	×	×			×

Expandability. Because it is very difficult, if not impossible, to define everything that a coding system must be capable of handling during some indefinite future time period, ease of expanding the code is a very important characteristic.

Differentiation. The amount of differentiation varies a great deal among codes. Taking an extreme case, after all parts manufactured by a company are coded, they might be classified as being in one family. At the other extreme, after all parts are coded, each part might represent a distinct family. In the former case, the coding system did not provide enough differentiation; in the latter case, it provided too much.

Affordable automation in coding systems. Most coding and classification systems in use today must be implemented on a computer. Therefore, when a potential system is evaluated, sufficient attention should be given to determining how it can be automated. This evaluation should not be restricted to the coding and classification capabilities; the associated methodology and retrieval and analysis functions should also be considered.

Efficiency. The code efficiency—the number of digits required to code a typical part—should be evaluated. If the number of available digits is too small, determine whether this number can be increased.

Cost. Cost includes several facets: the initial cost of the system, the cost of modifying the system to meet the particular needs of a specific company, the cost of interfacing the system to existing computer systems, and the cost of using the system.

Simplicity. Ease of use is important. Many of the manufacturing personnel who must use the system will not be very familiar with computer systems. Therefore, simplicity and user-friendliness gain user acceptance, make training less difficult, and reduce the cost of use.

Other concerns. For a particular firm there may be other considerations that could be added to this list. Even if none are added, selecting a coding and classification system may not be easy. Consequently, before a particular system is selected, a thorough evaluation should be performed. Because the problem is so complex, before a final decision is made, other companies that are using the same system should be visited.

Developing a coding system in house. Many companies have opted to develop their own coding system. This is not a simple project that can be done in a short period of time, even though it is not difficult to describe a series of steps to follow.

Initially a sample of parts should be selected. The coding system should accommodate purchased parts as well as those manufactured in house; consequently, purchased parts should be included in the sample. The size of the sample will vary from several hundred (25% of the active parts out of a database of 3000) parts to several thousand if the active database is large.

The next step is to assemble drawings of the parts in the sample. Then these drawings can be sorted into families by manually examining each drawing and grouping together those that have similar features and therefore require similar processing. This procedure will identify the part features having a high frequency of occurrence. Once this step is completed, it is relatively easy to identify the machine tool requirements for making these parts and link the groups of parts to existing machines in the factory. One may not be able to link some of the purchased parts to in-house abilities. However, one should consider these parts, because sometime in the future one may purchase the equipment to make these parts. In addition, the coding system can be used to select vendors when the machining capabilities of the vendors are known.

Once the part features are identified, a hierarchy of these features should be established with the objective of minimizing the time required to code a part. As one codes parts, one will quickly learn that a part can be coded in less time if certain attributes are identified before others. For example, if one identifies the part as rotational, then all attributes that apply only to nonrotational parts can be ignored.

The next step is to test the coding system that has been developed. This may be done by coding the sample of parts. When the coding is completed, the results should be analyzed to determine how well parts can be grouped into families. This process may involve several iterations before the coding system is satisfactory. Because this process requires so much effort, many companies purchase a coding system and then modify it to meet their special needs.

Sample coding systems. The codes that would be obtained by applying four representative coding systems to the bushing in Fig. 6.7 are depicted in Figs. 6.8 through 6.11. The DCLASS code (Fig. 6.8) is an 8-digit hybrid code. CODE (Fig. 6.9), from Manufacturing Data Systems (MDS) Inc., is an 8-digit hybrid code that can be expanded to 12 digits. The 8 basic positions of the code can have any one of 16 characters: 0 through 9 and A through F. When CODE is expanded to 12 digits, the additional digits may be used for attributes such as heat treatment, hardness, finish, material, production cost, and time standards. The MICLASS code (Fig. 6.10) by The Netherlands Organization for Applied Scientific Research (TNO) is a 12-digit hybrid code. An op-

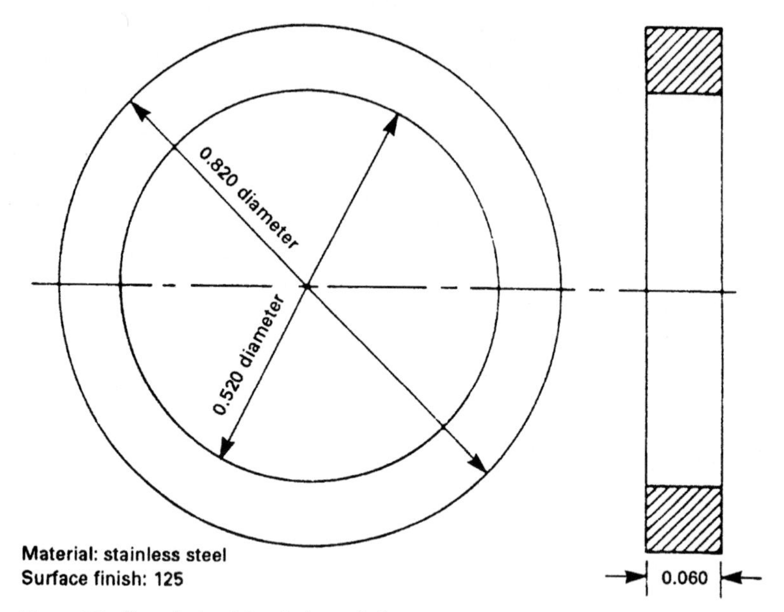

Figure 6.7 Sample bushing to be coded.

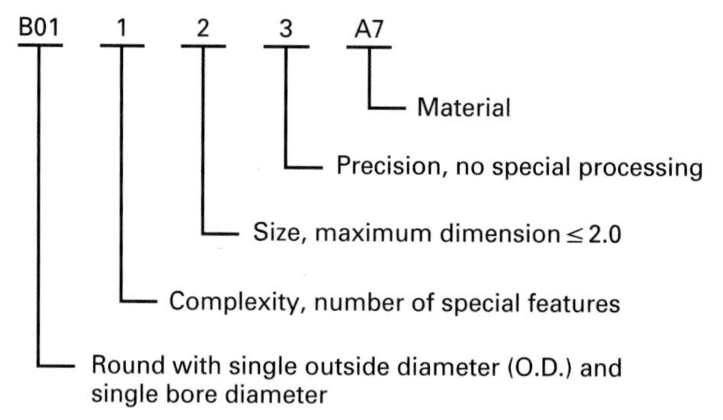

Figure 6.8 DCLASS 8-digit hybrid code for the bushing in Fig. 6.7.

tional 18-digit expansion may be used to capture company-specific information. The OPITZ code (Fig. 6.11), developed in Germany by H. Opitz, is a 9-digit hybrid code that can be expanded by adding 4 digits. The first 9 digits present design and manufacturing data. The extended 4 digits are referred to as the *secondary* and are intended to represent manufacturing data. However, how these digits are used is up to the particular firm.

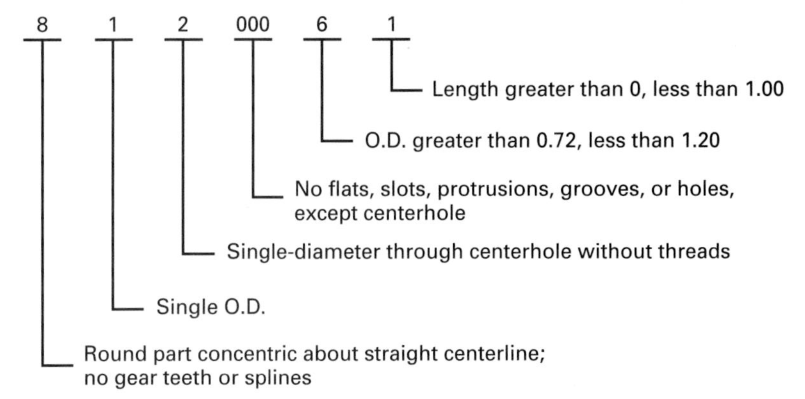

Figure 6.9 CODE 8-digit hybrid code for the bushing in Fig. 6.7.

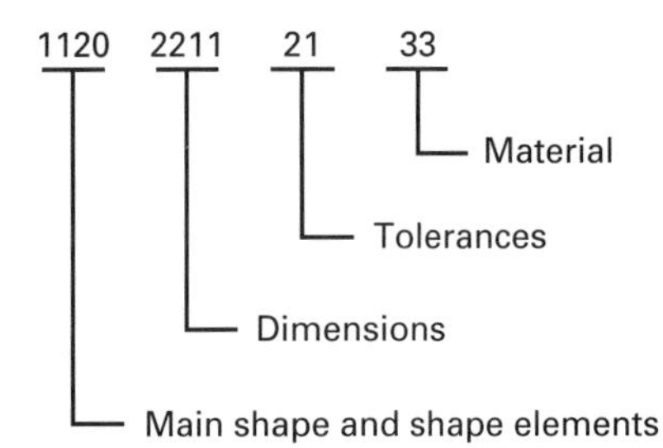

Round part with single O.D. and I.D. without faces, threads, slots, grooves, splines, or additional holes. O.D. and length are within certain size ranges.

Figure 6.10 MICLASS 12-digit hybrid code for the bushing in Fig. 6.7.

There are several reasons why so many coding systems have been developed. For example, the existing systems may not have captured all the information that the developer thought was important, the manner in which a code had to be developed may have been inconvenient, or the code may have been difficult to understand. Table 6.2 represents the information content obtained from each of the four coding systems used to code the bushing. Note that for this very simple part the information content differs for each of the systems.

Table 6.2 also illustrates the different types of attributes that a firm may need to represent when a part is coded. As already noted, the number of required attributes varies with the application. In general, as the number of applications increases, the number of required

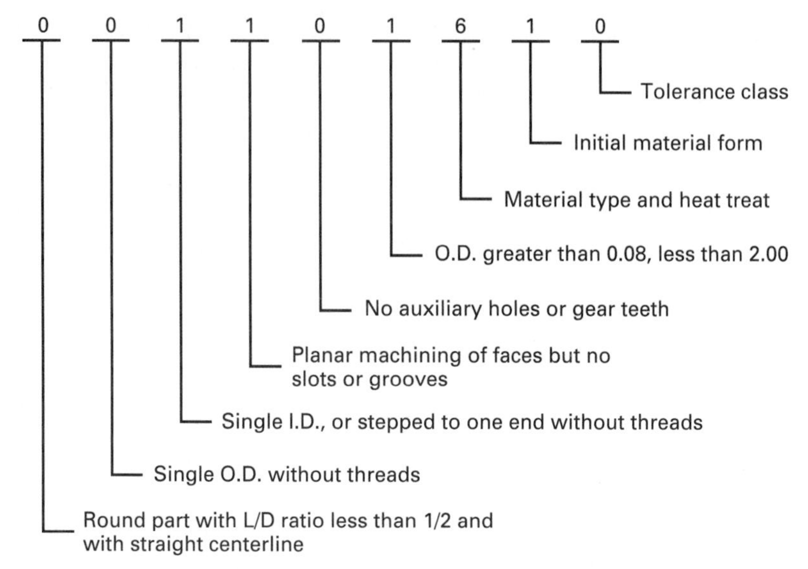

Figure 6.11 OPITZ 9-digit hybrid code for the bushing in Fig. 6.7.

TABLE 6.2 Information Content of Group Technology Codes

Information content	DCLASS	CODE	MICLASS	OPITZ
End shape			×	×
Outside shape	×	×	×	×
Inside shape	×	×	×	×
Protrusions	×	×	×	×
Additional holes	×	×	×	×
Threads	×		×	×
Grooves or slots	×	×	×	×
Flats		×	×	
Gear teeth or splines	×	×	×	×
Splits, keyways, knurls, or swages	×			
O.D. range		×	×	×
I.D. range			×	×
Length range		×	×	×
Size ratios	×	×	×	×
Tolerance	×		×	×
Heat treat	×		×	×
Material form				×
Material type	×		×	×
Finish	×		×	

attributes increases. Consequently, the amount of information that a coding system can represent may be a very important consideration.

6.4.6 DCLASS coding system

The Design and Classification Information System (DCLASS) was developed at Brigham Young University. The part code portion of this system was developed because no commercial vendor was willing to provide such a system for educational and research purposes. Although its primary use to date has been in the university environment, many companies are using it for prototype development.

Several premises were adopted and used as the basis for the development of the DCLASS code:

1. A part may be best characterized by its basic shape, usually its most apparent attribute.
2. Each basic shape may have several features, such as holes, slots, threads, and grooves.
3. A part can be completely characterized by basic shapes and features, size, precision, and material type, form, and condition.
4. Several short code segments can be linked to form a part classification code that is human-recognizable and adequate for human monitoring.
5. Each of these code segments can point to more detailed information.

After several years, an 8-digit hybrid part family code was developed.

The DCLASS part family code is composed of 8 digits partitioned into 5 code segments, as shown in Fig. 6.12. The first segment, composed of 3 digits, denotes the basic shape. The 1-digit form-features code is entered in the next segment; it specifies the complexity of the part, including features (such as holes and slots), heat treatments, and special surface finishes. The 1-digit size code is in the third segment of the part family code. From the value of this code, the user would know the overall size envelope of the coded part. The 1-digit

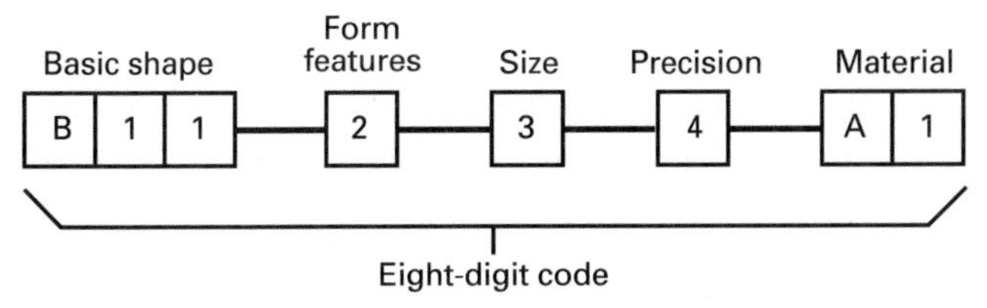

Figure 6.12 DCLASS part code segments.

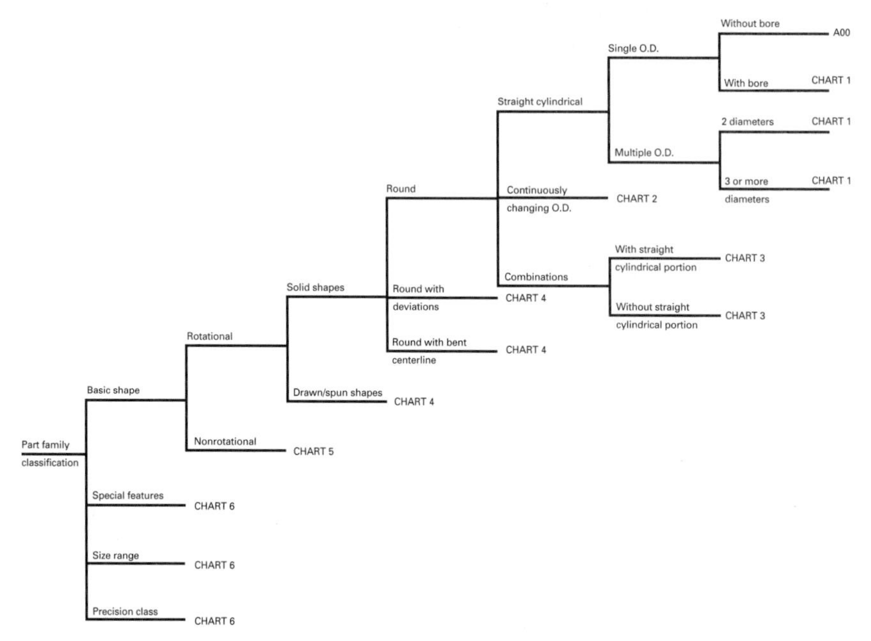

Figure 6.13 DCLASS logic tree.

fourth segment denotes precision. The final 2 digits, which compose the fifth segment of the part family code, denote the material.

The DCLASS code for the bushing in Fig. 6.7 is B0123A7. Consider the first 3 digits specifying the basic shape. Normally one would code a part by answering questions proposed by an interactive computer program. In this case, however, we will look at some figures to determine the appropriate values to assign to the code. First, look at Fig. 6.13, which depicts a DCLASS part family classification chart. (Note that this chart is structured as a logic tree. It could have been structured in some other manner, but logic trees have proven to be easy to work with. This is just one of many such charts in the DCLASS coding system.) Using this chart to code the bushing, we take a path through a logic tree that follows these branches: *basic shape, rotational, solid shapes, round, straight cylindrical, single O.D.* (outside diameter), and *with bore*. At this point we are referred to chart 1, which appears in Fig. 6.14. Looking at chart 1, we take the following branches: *round concentric bore* and *single bore diameter.* The last branch terminates with the code value B01, which corresponds to the first 3 digits of the code for the bushing.

As previously noted, the second segment of the part family code describes the complexity of the special features of the part being coded,

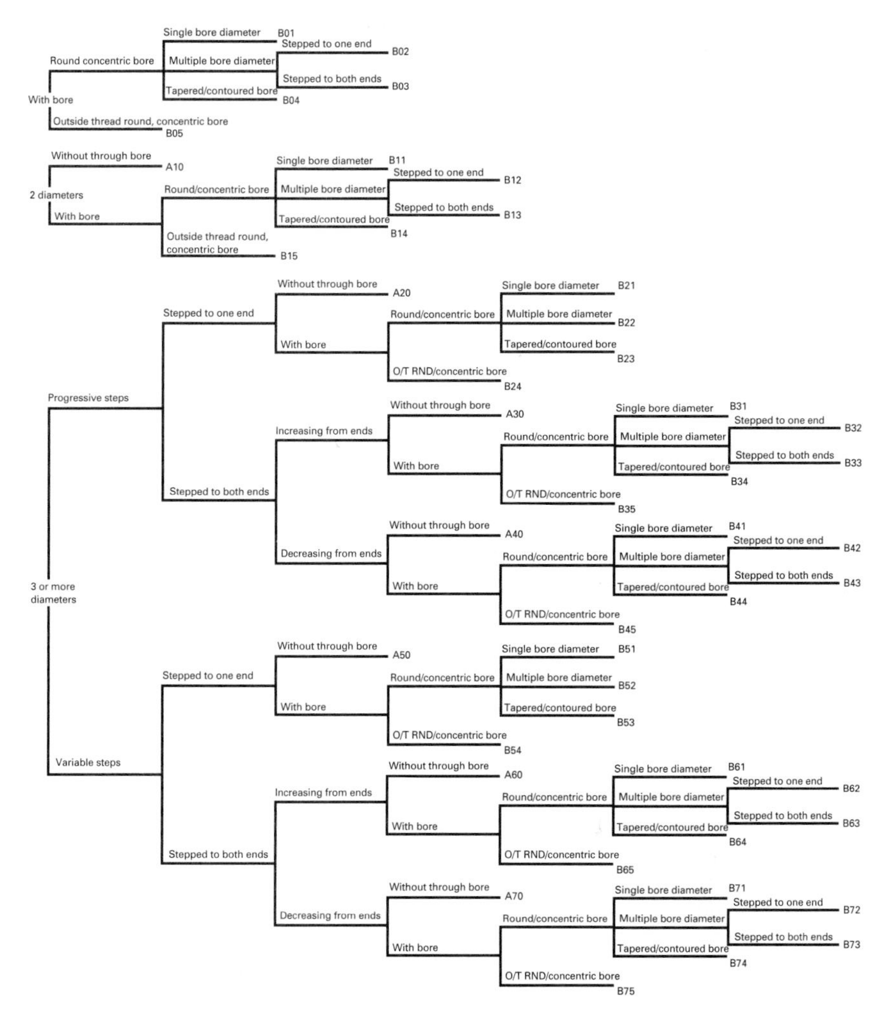

Figure 6.14 DCLASS chart 1 logic tree.

which in this case is a bushing. In the DCLASS system, special features include form features (holes, etc.), heat treatments, and surface finish treatments (plating, painting, anodizing, etc.). Table 6.3 provides the code values that can be used to code the complexity of the special features. For the bushing, a complexity code value of 1 will be assigned, since it has a hole and does not require heat treatment or surface finishing.

The third segment of the part family code refers to the size of the part. Table 6.4 is used to select the appropriate value. Using this table, we place a 2 in the code for the bushing.

TABLE 6.3 Complexity Code for Special Features

Feature complexity code	No. of special features
1	1
2	2
3	3
4	5
5	8
6	13
7	21
8	34
9	>34

SOURCE: Reprinted with permission from Ref. 4.

TABLE 6.4 DCLASS Size Code

	Maximum dimension			
Size code	English (in)	Metric (mm)	Description	Examples
1	0.5	10	Subminiature	Capsules
2	2	50	Miniature	Paperclip box
3	4	100	Small	Large matchbox
4	10	250	Medium small	Shoebox
5	20	500	Medium	Breadbox
6	40	1,000	Medium large	Washing machine
7	100	2,500	Large	Pickup truck
8	400	10,000	Extra large	Moving van
9	1,000	25,000	Giant	Railroad boxcar

SOURCE: Reprinted with permission from Ref. 4.

The fourth segment denotes the precision of the part being coded. Precision in DCLASS represents a composite of tolerance and surface finish. Table 6.5 lists the five classes of precision used. Class 1 represents tight tolerances and a precision-ground or lapped surface finish. At the other extreme, class 5 represents loose tolerances and a rough-cast or flame-cut surface. A part with a precision code of 1 requires careful processing with careful inspection. The bushing under consideration requires no special tolerances or surface finish, so a code value of 3 will be specified.

The final segment of the part family code contains the material type. Referring back to Fig. 6.7, we see that the bushing is to be made out of

TABLE 6.5 DCLASS Precision Class Code

Class code	Tolerance	Surface finish
1	≤0.0005 in	≤4 rms
2	0.0005–0.002 in	4–32 rms
3	0.002–0.010 in	32–125 rms
4	0.010–0.030 in	125–500 rms
5	>0.030 in	>500 rms

SOURCE: Reprinted with permission from Ref 4.

stainless steel. The logic tree in Fig. 6.15 is used to determine the appropriate code value for stainless steel. Looking at the logic tree, we select the following branches: *metals, ferrous metals, steels, high-alloy steels,* and *stainless steel.* The appropriate code for the material type is A7. This procedure completes the DCLASS code for the bushing.

The procedure is not difficult, and it would be easy to computerize. Dell Allen, at Brigham Young University, has developed a general-purpose computer system for processing classification and decision-making logic. This system is available for several types of computers (micros, minis, and mainframes).

6.4.7 MICLASS coding system

MICLASS stands for Metal Institute Classification System; it was developed by The Netherlands Organization for Applied Scientific Research. MICLASS is one of the more popular commercial systems available in the United States. As noted earlier, the MICLASS system consists of two major sections.

The first section is a 12-digit code (Fig. 6.16) which is used to classify the engineering and manufacturing characteristics of a part. The first 4 digits deal with the form: main shape, shape elements, and the position of these elements. The main shape of a part is the form of a final product as depicted in the drawing; it could be a rotational part, a boxlike part, a flat part, or some other nonrotational part. Shape elements are part features such as holes, slots, and grooves. The next 4 digits provide dimensional information: the main dimension, the ratio of the various dimensions, and the auxiliary dimension. The use of the auxiliary dimension varies with the main shape of the part; in general, the auxiliary dimension provides additional size information. Digits 9 and 10 contain tolerance information, and the final 2 digits provide a machinability index of the material.

The second section of the MICLASS code is optional. It can contain as many as 18 characters tailored to meet the specific needs of a com-

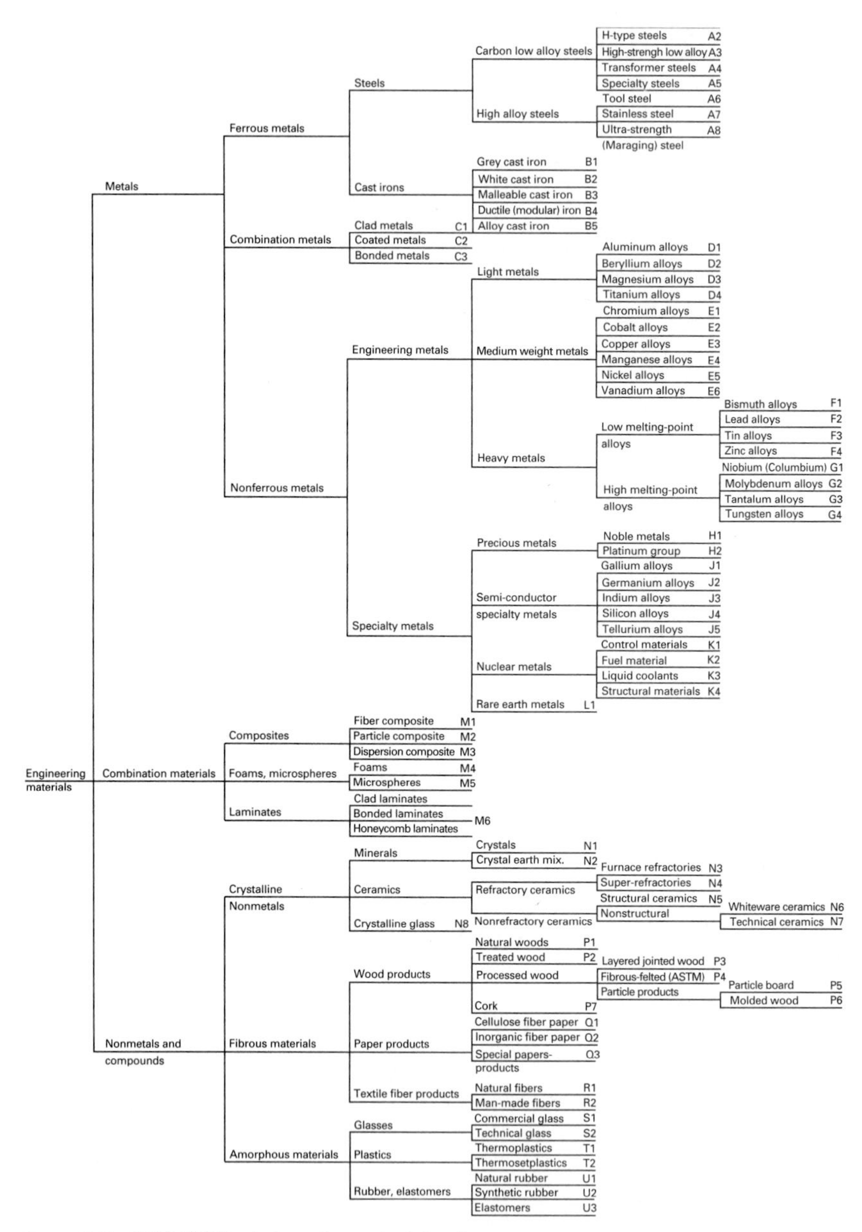

Figure 6.15 DCLASS logic tree for material code.

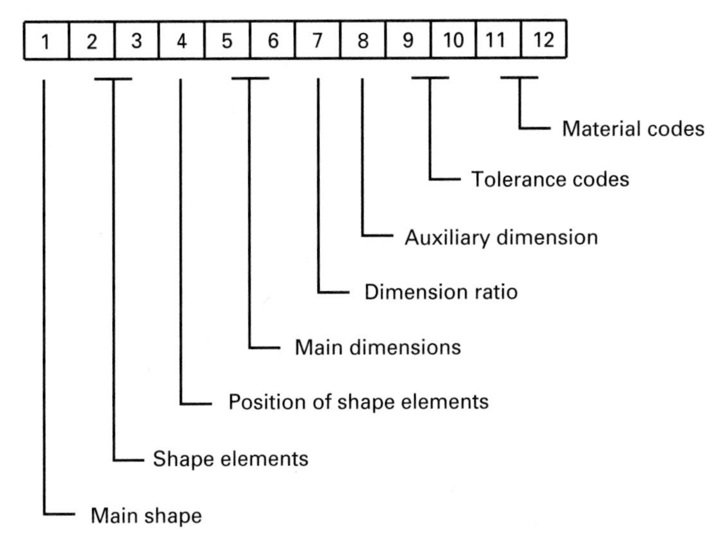

Figure 6.16 MICLASS code structure.

pany, such as vendors, lot sizes, costs, and producibility tips (similar to the expansion of CODE).

The 12 digits in the first section are universal in that the definition used for the various digits do not change from firm to firm. The advantage is that a plant or division can read the information contained in a part code received from another facility using the MICLASS code. A disadvantage arises when the code does not provide the necessary information in the first 12 digits, hence the optional extension.

Manually coding several thousand parts using the 30-digit MICLASS code would be a very tedious and time-consuming job, and errors would probably be made. Consequently, the MICLASS system provides several interactive computer programs to assist the user. For example, one of the programs prompts the user with questions about a part to be coded. Figure 6.17 represents a part to be coded by using the MICLASS interactive program, and Fig. 6.18 shows the interactive dialog between the computer and the user. Note that the questions relate to characteristics of the part such as shape, size, materials, and tolerances. Responses are yes or no or dimensions; consequently, no in-depth computer knowledge is required. As the user answers questions, the program gathers the data required to code the part. The computer then assigns a code to the part and stores the information in the MICLASS database. A code of 1271 3231 3144 is assigned to this part. Possible errors that could be made in coding a part are minimized by this procedure. Once stored in the database, this information is available for analysis.

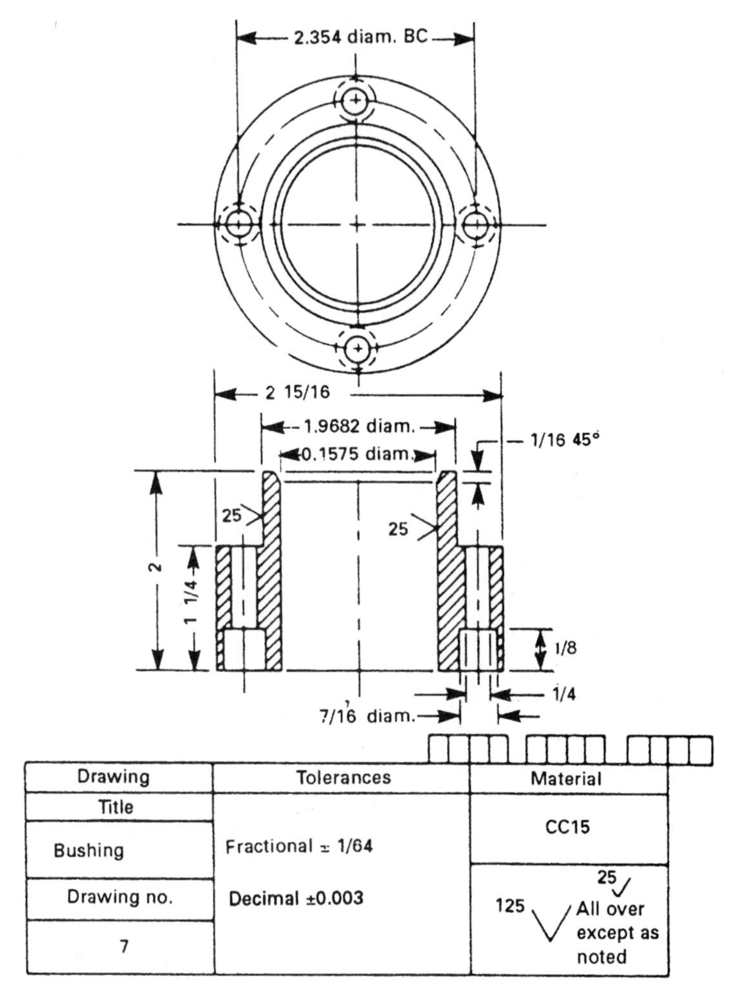

Figure 6.17 Example part to be coded by MICLASS.

As stated earlier, the MICLASS system includes several programs to assist in the analysis of coding and classification information stored in the database. These programs can be used for design classification, elimination of design duplication, manufacturing standardization, improving control and speed of material flow, optimization of machine tool purpose and use, and improving efficiency throughout design and manufacturing.

Some of the specific functions performed by the MICLASS programs are

```
VERSION-A-
3 MAIN DIMENSIONS (WHEN ROT. PART D.L AND O)? 2.9375  2  0
   DEVIATION OF ROTATION FORM? NO
   CONCENTRIC SPIRAL GROOVES? NO
TURNING ON OUTERCONTOUR (EXCEPT ENDFACES)? YES
   SPECIAL GROOVES OR CONE(S) IN OUTERCONTOUR? NO
   ALL MACH. DIAM. AND FACES VISIBLE FROM ONE END (EXC. ENDFACE + GROOVES)? YES
INTERNAL TURNING? YES
   INTERNAL SPECIAL GROOVES OR CONE(S)? NO
   ALL INT. DIAM. + FACES VISIBLE FROM 1 END (EXC. GROOVES)? YES
ALL DIAM. + FACES (EXC. ENDFACE) VISIBLE FROM ONE SIDE? YES
EXC. HOLING AND/OR FACING AND/OR SLOTTING? YES
   IN INNERFORM AND/OR FACES (INC. ENDFACES)? YES
   IN OUTERFORM? NO
ONLY KEYWAYING, ETC.? NO
MACHINED ONLY ONE SENSE? YES
   ONLY HOLES ON A BOLTCIRCLE—AT LEAST 3 HOLES? YES
FORM, OR THREADING TOLERANCE? NO
DIAM. ROUGHNESS LESS THAN 33 RU (MICRO-INCHES)? YES
   SMALLEST POSITIONING TOL. FIELD? .016
   SMALLEST LENGTH TOL. FIELD? .0313
MATERIAL NAME? CC15
CLASS.NR. = 1271 3231 3144
DRAWING NUMBER, MAX 10 CHAR? 7
NOMENCLATURE, MAX 15 CHAR? BUSHING
CONTINUE [Y/N]? N
PROGRAM STOP AT 4690
```

Figure 6.18 Interactive computer dialog for MICLASS coding example.

1. *Data management.* This provides the ability to manipulate data and files (including sorting), correct errors, and list data.
2. *Design retrieval.* Designs with the same or similar coded numbers can be retrieved. Note that similar parts as well as duplicate parts can be retrieved; the variation of the coded parts retrieved is controlled by the user.
3. *Production mix.* After thousands of parts have been coded, analysis of the information contained in the database can be very time-consuming. Programs are provided for assistance; for example, there are programs to produce graphs of the occurrence of a given classification code, or of a specific part of a code, or of combinations of a code.

One reason that the MICLASS system has been so successful is the powerful computer tools that are provided for the user.

6.5 Cost Models for Cellular Manufacturing

Several types of analytical models have been recently developed to help users understand the cost-benefit relationships of group technology in a manufacturing environment. Two such models are described here: one for production costs, the other for tooling costs.

6.5.1 Production costs

The production planning cost model assumes that the objective is to develop a manufacturing plan to minimize production costs over a planning horizon of T time periods. Normally a time period represents a month. Manufacturing costs can be separated into three types:

1. Direct labor costs
2. Direct material costs
3. Indirect burden or overhead costs

Direct labor costs consist of all types of labor costs for production workers who are engaged directly in manufacturing operations to convert raw materials into finished products. For example, it is usually possible to charge the time of a machine operator directly to each job on which the operator works. However, it is much more difficult—practically impossible, in fact—to charge tool room attendants' time directly to each job handled in the shop. Consequently, the tool room attendant's time is usually classified as an indirect cost and is allocated to all parts made in some time period, such as a month. Other names for this type of cost are burden or overhead cost. The same practice applies to material costs. The cost of the metal from which a gear is fabricated is an example of a direct material cost. The costs of the coolant, grease, and oil used by the machine tools are examples of indirect material costs.

The model described in this section can be used to minimize the production costs of group technology for a manufacturing cell while satisfying product demand. Decisions must be made regarding what value should be assigned to each decision variable. The best values for the decision variables are those that minimize the value of the objective function, Eq. (6.1):

Minimize:

$$
\begin{aligned}
Z = \sum_{t}^{T} L(I_s V_{it} P_i + W_{it} I_2) + \sum_{j} (F_{jt} + G_j + oO_t) \text{ (labor)} \\
+ \sum_{i} V_{it}(M_i + B_i) \text{ (material and burden)} \\
+ I_s h_i I_{it} \text{ (inventory holding)}
\end{aligned}
\tag{6.1}
$$

subject to:

$$V_{it} + I_{it-1} - I_{it} = d_{it} \qquad \forall i,t \tag{6.2}$$

$$\sum_{i} V_{it} P_i + \sum_{i} W_{it} I_s + \sum_{j} F_{jt} G_j - O_t + U_t = R_t \qquad \forall i,t \tag{6.3}$$

$$\sum_{j \in i} W_{it} \leq CF_{jt} \qquad \forall t,j \tag{6.4}$$

$$V_{it} \leq CW_{it} \qquad \forall i,t \qquad (6.5)$$

$$F_{jt}, W_{it} \geq 0 \qquad \forall i,j,t \qquad (6.6)$$

$$F_{jt}, W_{it} \leq 1 \qquad \forall i,j,t \qquad (6.7)$$

$$F_{jt}, W_{it} = 1 \text{ or } 0 \qquad \forall i,j,t \qquad (6.8)$$

$$O_t \leq E \qquad \forall t \qquad (6.9)$$

$$U_t \leq A \qquad \forall t \qquad (6.10)$$

$$V_{it}, U_t, O_t, I_{it} \geq 0 \qquad \forall i,t \qquad (6.11)$$

where V_{it} = production quantity of part in period t
I_{it} = inventory of part at the end of period t
U_t = undertime associated with the plan in period t
O_t = overtime associated with the plan in period t
W_{it} = number of setups for part in period p
F_{jt} = number of setups for family j in period t
t = period in the planning horizon
T = planning horizon
L = labor rate
P_i = unit processing time for part i
I_s = standard setup time for part i
G_j = standard setup time for part family j
o = percent increase in the labor rate for overtime
M_i = standard material cost per unit for part i
B_i = burden cost per unit for part i
h_i = holding cost per unit for part i
R_t = regular hour schedule for period t
d_{it} = demand for part i in period t
C = an arbitrarily large constant
E = maximum overtime permitted
A = maximum undertime permitted

The objective function measures, for a given production plan, the three types of costs noted above. Direct labor is computed by applying a single labor rate to the production time. The production time includes run and setup times, and takes into account possible differences between regular and overtime pay rates. Differences in setup times between parts in a family and between families are also permitted. If group tooling has been developed, there may be virtually no setup time required between parts in the same family; however, the setup time between families may be significant. This situation can be accommodated. Material and overhead costs are assumed to be linearly related to production volume. Inventory holding costs are applied on a per-unit, per-period basis.

Examining constraints, one can see that Eq. (6.2) ensures that the demands are satisfied for all parts in all the periods in the planning horizon. The constraints expressed in Eq. (6.3) limit the capacity of the cellular manufacturing unit and compute the amount of overtime and undertime utilized in a production plan. The next two constraints, Eqs. (6.4) and (6.5), ensure that setup time is incurred for a part and for a part family when appropriate. Note that these setups are limited to one per period and must be integer values by the constraints of Eqs. (6.6), (6.7), and (6.8). Eqs. (6.9) and (6.10) limit the amount of overtime and undertime.

Some simplifying assumptions were made in this model. For example, it represents a cellular manufacturing unit as a single resource with limited capacity and considers the machine and setup times as aggregate values. Consequently, if there is more than one machine in the cell, machine loading and balancing must be considered when the cell is designed and when parts are assigned to be processed in this cellular manufacturing unit. As a result of including part and family setups as decision variables, this model is a mixed-integer linear programming problem. Although some simplifying assumptions were made to keep the model simple, solving the model is very difficult. It belongs to a class of problems termed *NP-hard,* which means that in general no method exists for solving problems of a practical size.

The effects of group technology on production planning in cellular manufacturing was studied by Graves.[5] It is similar to a production planning model proposed by Manne in 1958 and later refined by others.[6] Faced with an NP-hard problem, Graves examined what some others had achieved to develop a more tractable formulation. He found that Hax and Meal[7] examined a similar formulation for planning production and end products. They simplified the required computations by aggregating some of the data. Parts were aggregated into families based on similarities in setup requirements. These families were aggregated into types based on similar seasonal demand patterns and production rates. With these aggregations, the modified model could be used to solve practical problems. However, the solution obtained was nonoptimal.

The fact that the solutions were nonoptimal does not indicate that they are worthless, because a good solution is usually acceptable to management in a practical environment. Graves also observed that there is a similarity in the aggregates of families and types in this model with parts and families in group technology. This led to an ap-

proach for aggregating some of the variables in the above production planning model. The resulting model is
Minimize:

$$Z = \Sigma_t^T [L(V_t P) + oO_t + V_t(M + B) + HI_t] \tag{6.12}$$

Subject to:

$$V_t + I_{t-1} - I_t = D_t \qquad \forall t \tag{6.13}$$

$$V_t P - O_t + U_t = R_t \qquad \forall t \tag{6.14}$$

$$O_t \leq E \qquad \forall t \tag{6.15}$$

$$U_t \leq A \qquad \forall t \tag{6.16}$$

$$V_t, U_t, O_t, I_t \geq 0 \qquad \forall t \tag{6.17}$$

where t = period number
T = planning horizon
V_t = aggregate production volume
P = aggregate processing time
o = overtime rate
O_t = overtime hours
M = aggregate material cost per unit
B = aggregate burden cost per unit
H = aggregate holding cost per unit
I_t = aggregate inventory
D_t = aggregate demand
R_t = regular hours available
E = maximum overtime per period
A = maximum undertime per period

Graves provides some procedures for performing the aggregation computations. Basically, weighted averages based on the planned demands are calculated for the material, holding, and burden costs. The aggregate processing time is a weighted average; however, it also includes an estimate of setup time requirements for the entire planning horizon. The production quantities and inventories are simple summations.

This formulation is identical to the fixed work-force linear cost model used with end products. Graves used this model to analyze the cost structure of a group technology in a cellular manufacturing, to perform sensitivity analysis of the model parameters, and to analyze modifications to the family and machine compositions. An interesting observation has to do with different approaches used to arrive at the same model formulation. Originally, the model was developed without consideration of group technology concepts. Hax and Meal, in the process of making their model more computationally acceptable, aggregated the parts into families and types based on setup requirements, production rates, and seasonal demand patterns. Graves demonstrated that group technology provides a logical process for doing this. Although group technology was initially developed to assist in engineering design and manufacturing engineering, following the path taken to develop the above models gives some insight into how group technology aids in simplifying the management of the manufacturing environment, particularly for affordable automation.

6.5.2 Tooling costs

Group technology clearly can provide benefits in a manufacturing environment; however, one should not assume that group tooling will cost less than conventional tooling. The cost of conventional tools required to produce a batch of parts may be expressed as

$$C_{ct} = \sum_{i=1}^{p} C_i \tag{6.18}$$

where p = number of fixtures
C_i = cost of a fixture
C_{ct} = total conventional tooling cost

The cost of the group tooling required to produce a family of parts is

$$C_{gt} = C_g + \sum_{a=1}^{q} C_a \tag{6.19}$$

where q = number of adapters
C_g = cost of group fixtures
C_a = cost of group adapters
C_{gt} = total group tooling cost

These equations can be used to calculate the unit tooling costs:

$$C_{uc} = C_{ct}/N \tag{6.20}$$

$$C_{ug} = C_{gt}/N \tag{6.21}$$

where N = number of parts produced
C_{uc} = unit tooling cost for conventional method
C_{ug} = unit tooling cost for group tooling method

TABLE 6.6 Conventional/Group Tooling Data

Items	Conventional tooling	Group tooling
Cost of fixture	$500	$480
Cost of each adapter	—	80
Number of pieces/ family to be produced	100	100

TABLE 6.7 Tooling Cost Comparison

Number of different parts in part family	C_{tc}	C_{uc}	C_{gt}	C_{ug}
1	$500	$5.00	$560	$5.60
2	1000	5.00	640	3.20
3	1500	5.00	720	2.40
4	2000	5.00	800	2.00
5	2500	5.00	880	1.76
6	3000	5.00	960	1.60
7	3500	5.00	1040	1.49
8	4000	5.00	1120	1.40
9	4500	5.00	1200	1.33
10	5000	5.00	1280	1.28

The following example will provide some additional insight into how the costs of these two methods can differ.

A family of 100 parts is to be produced. Table 6.6 shows the cost of the individual tools. Table 6.7 compares the tooling costs and the unit tooling costs for both methods as the number of parts in a family varies. The computations were made in the following way:

When $p = 1, q = 1,$

$$C_{ct} = C_1 = 500$$

$$C_{uc} - C_{ct}/N = 500/100 = 5.00$$

$$C_{gt} = C_g + C_1 = 480 + 80 = 560$$

$$C_{gt} = C_{ug}/N = 560/100 = 5.60$$

When $p = 2, q = 2,$

$$C_{ct} = pC_1 = 2 \times 500 = 1000$$

$$C_{uc} = C_{ct}/N = 1000/(2 \times 100) = 5.00$$

$$C_{gt} = C_g + qC_1 = 480 + (2 \times 80) = 640$$

$$C_{ug} = C_{gt}/2N = 640/(2 \times 100) = 3.20$$

and so forth.

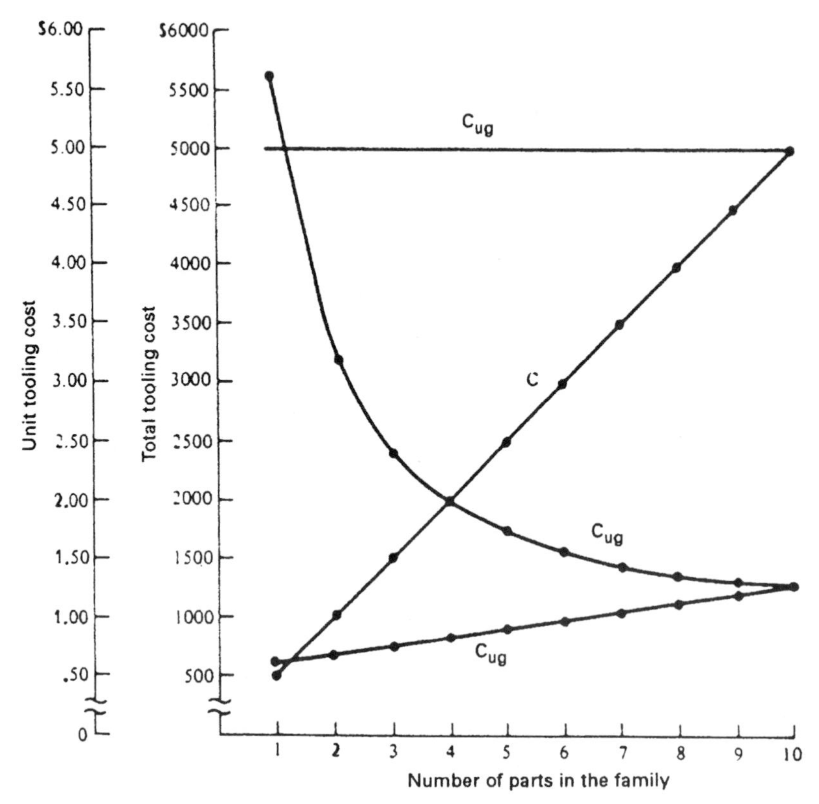

Figure 6.19 Tooling costs.

Figure 6.19 is a plot of the total tooling costs and unit tooling costs of these two methods in relation to the number of different parts in the family. The rate of increase in total cost for conventional tooling is much greater than for group tooling. Also, as the number of parts in a family increases, the unit tooling costs for group tooling become more economical.

6.6 Economics of Affordable Automation in Group Technology

As more and more companies successfully implement group technology programs, significant savings are being reported. The following savings are typical:

50% in new part design

10% in number of drawings

60% in industrial engineering time

20% in plant floor space requirements

40% in raw material stock

60% in in-process inventories

70% in setup time

70% in throughput time

In addition, a successful group technology program provides many benefits that are difficult to quantify, such as simplification of manufacturing environment, improvement of the work environment, better quality, and improved product designs.

One of the most important savings is in new part design. Whenever a new part is required, a design engineer is faced with the alternatives of finding a similar part that has been designed before and modifying it as necessary or designing a new part. If a similar part exits, it is probably filed by part number with a short descriptive title. Since most companies have many thousands of drawings on file, it is virtually impossible to retrieve the part drawing unless the engineer can remember the part number or descriptive title. Therefore the engineer usually designs a new part, which increases design costs and complicates the manufacturing environment by introducing yet another number into the system. A new design will result in a significant increase in manufacturing preparation costs. Table 6.8 lists some activities that are affected by the introduction of a new part.

In addition, manufacturing management becomes more difficult with the introduction of each new part. For example, the complexity of sequencing parts through machines increases, machine tools must be acquired and delivered, and long and frequent setups may reduce machine utilization and result in larger throughput times. This situation occurs quite often in many batch-type manufacturing organizations.

TABLE 6.8 Some Activities Affected by a New Design

Engineering activities	Manufacturing activities
Design	Equipment and facilities planning
Analysis	Process planning
Drawing	Tool design
Testing	Tools and gauges
Auditing	Time standards
Documentation	Production planning and control
	Scheduling
	Quality control
	Accounting
	Cost estimating

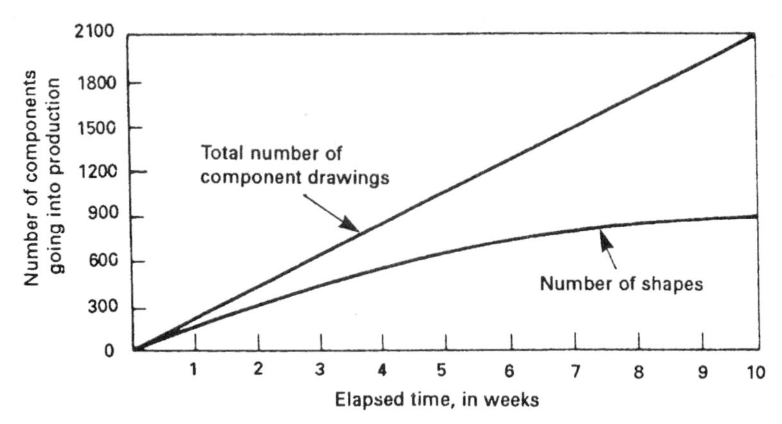

Figure 6.20 New component shapes compared to new component drawings.

In reality, when a large number of "different" parts are observed, many will be found to be similar to others. For example, Fig. 6.20 depicts data on 2100 randomly selected drawings taken from all components moving to production at one company during a 10-week period. Note that after 7 weeks the number of new shapes has nearly leveled off. Although the curve will never become level, it illustrates that a classification and coding system can save a great deal of design work by facilitating the retrieval of designs of identical or similar parts. This will reduce the number of new parts that need to be designed, so that a "new" design may only require changes to an existing design. The same type of savings can be achieved in manufacturing by reducing the number of process plans that must be prepared for new part designs. In addition, this type of environment encourages standardization of design features, simplification of the design process, and improvement of the cost-estimating system.

References

1. V.B. Soloja and S.M. Ursoevic, "Optimization of Group Technology Lines by Methods Developed in the Institute for Machine Tools and Tooling (IAMA)," Belgrade, September 1969.
2. S.P. Mitrofanov, *Scientific Principles of Group Technology* (English translation), National Library for Science and Technology, Washington, D.C., 1966.
3. J. Burbridge, *Introduction to Group Technology*, Wiley, New York, 1975.
4. "Part Family Classification and Coding," Monograph No. 3, Brigham Young University, Provo, Utah, 1979.
5. G.R. Graves, "Consideration for Group Technology Manufacturing in Production Planning," Ph.D. thesis, Oklahoma State University, Stillwater, Okla., 1995.
6. W.S. Manne, M.S. Dunn, and S.J. Pflederer, "Computerized Process Planning" (final report), Contract (DAAK 4076-C-1104), UTRC Report R77-942625, August 1977.
7. A. Hax and H. Meal, "Hierarchical Integration of Productive Planning and Scheduling," in *Studies in the Management Sciences*, vol. I, Logistics, New York, 1975.

Chapter

7

Cost-Saving Strategies

7.1 Production Loss in Automatic Assembly Machines

Multistation assembly machines may be classified into two main groups according to the method used to transfer assemblies from station to station. The first group includes those assembly machines that transfer all the work carriers simultaneously. These are known as *indexing* or *synchronous machines* (Fig. 7.1). A stoppage of any individual work cluster causes the whole machine to stop. In the other group of machines, which are known as *free-transfer* or *nonsynchronous machines* (Fig. 7.2), the work clusters are separated by buffers containing assemblies, and transfer to and from these buffers occurs when the particular work station has completed its cycle of operation. Thus, with a free-transfer machine, a fault or a stoppage of a work cluster will not necessarily prevent another work cluster from operating because a limited supply of assemblies will usually be available in the adjacent buffers.

One of the principal problems in applying automation to the assembly process is the loss in production resulting from stoppages of automatic work clusters when defective component parts are fed to the machine. With manual work stations on an assembly line, the assembly workers are able to discard defective parts quickly, with little loss of production. However, a defective part fed to an automatic work cluster can, on an indexing machine, cause a stoppage of the whole machine, and production will cease until the fault is cleared. The resulting downtime can be very high for assembly machines having several automatic work clusters. This can result in a serious loss in production and a consequent increase in the cost of assembly. The quality

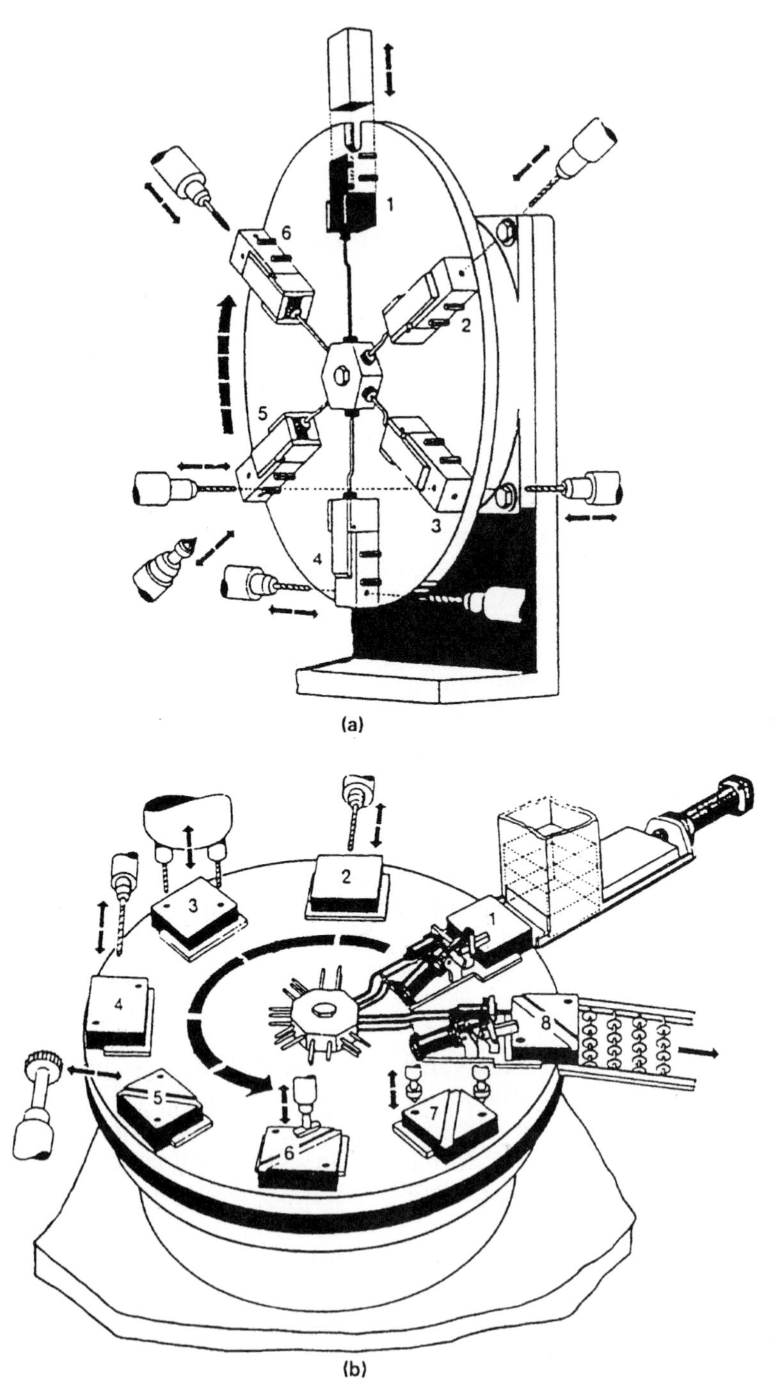

Figure 7.1 (*a*) Vertical synchronous rotary indexing system. (*b*) Horizontal synchronous rotary indexing table.

Figure 7.2 (a) Free-transfer (nonsynchronous) machine.

levels of the parts to be used in automatic assembly, therefore, must be considered when an assembly machine is designed.

7.2 Downtime in Automatic Assembly Machines

It is known that an indexing machine having n automatic work clusters and operating with cycle time of t seconds is fed with parts having, on average, a ratio of defective to acceptable parts of x.

It is also known that a proportion m of the defective parts will cause machine stoppages and, further, that it will take an operator T

Figure 7.2 (*b*) Free-transfer (nonsynchronous) machine.

seconds, on average, to locate the failure, remove the defective part, and restart the machine.

With these known factors, the total downtime due to stoppages in producing N assemblies will be given by $mxnNT$. Each time the machine indexes, all assembly tasks are completed and one assembly is delivered from the machine; hence, the machine time to assemble N assemblies is Nt seconds and, thus, the proportion of downtime D on the machine is given by

$$D = \frac{\text{downtime}}{\text{assembly time} + \text{downtime}}$$

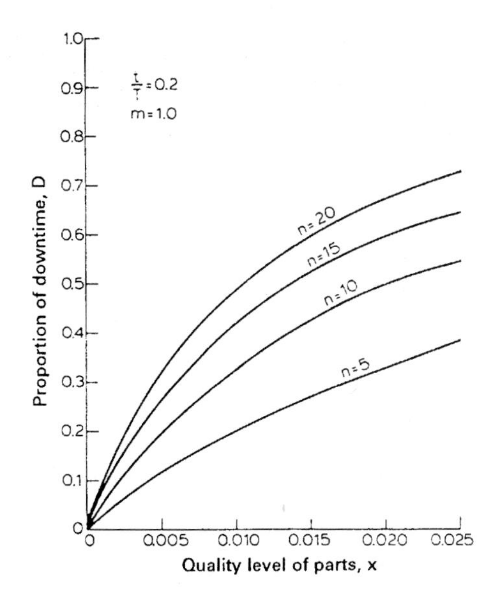

Figure 7.3 Effect of quality level on downtime.

$$= \frac{mxnNT}{mxnNT + Nt}$$

$$= \frac{mxn}{mxn + (t/T)} \tag{7.1}$$

In practice, a reasonable value of the machine cycle time t is 6 seconds. It has been observed that a typical value for the average time taken to clear a fault is 30 seconds. With these figures, the ratio t/T will be 6/30 = 0.20.

Figure 7.3 shows the effect of variations in the mean quality level of the parts on the downtime for indexing machines with 5, 10, 15, and 20 automatic work clusters. It is also observed that in this example that all defective parts will produce a stoppage of the machine and, thus, $m = 1.0$.

For standard fasteners such as screws, which are often employed in assembly processes, an average value for x might be between 0.01 and 0.02. In other words, for every 100 acceptable screws, there would be between one and two defective ones. A higher quality level is generally available, but with screws, for example, a reduction of x to 0.05 may double the price and seriously affect the cost of the final assembly. It has been documented that it is rarely economical to use an indexing machine having a large number of automatic workheads. These also illustrate the reason for allowing for downtime when considering the use of an indexing assembly machine.

7.3 Assembly Cost of Defective Parts

It was stated above that all the defective parts fed to the automatic workheads would stop the machine (that is, $m = 1.0$). In practice, however, some of these defective parts would pass through the feeding devices and automatic workheads but would not be assembled correctly and would result in the production of an unacceptable assembly. In this case, the effect of the defective part would be to cause downtime on the machine equal to one machine cycle. The time taken to produce N assemblies, whether these are acceptable or not, is given by $(mxnNT + Nt)$, where $m < 1.0$. Only about $[N - (1 - m)xnN]$ of the assemblies produced will be acceptable. The average production time t_{av} of acceptable assemblies is therefore given by

$$t_{av} = \frac{mxnNT + Nt}{N - (1 - m)xnN}$$

$$= \frac{(t + mxnT)}{1 - (1 - m)xn} \tag{7.2}$$

Typical values are

$$x = 0.01$$

$$t = 6 \text{ seconds}$$

$$T = 30 \text{ seconds}$$

$$n = 10$$

Accordingly,

$$t_{av} = \frac{30(2 + m)}{(9 + m)} \tag{7.3}$$

Equation (7.3) is plotted in Fig. 7.4 to illustrate the effect of m on t_{av}, and it can be seen that, for a maximum production rate of acceptable assemblies, m should be as small as possible.

Although it is theoretically feasible to allow a defective part to pass through the feeder and workhead and spoil an assembly rather than allow it to stop the machine, this practice is totally unacceptable for error-free environment strategy. It is highly recommended that, once a defective part is detected, the entire affected subassembly be shuttled away from the synchronous indexing machine by either of two methods:

1. *Lock out downstream work clusters.* This method allows the subassembly containing the defective part to stay in the indexing ma-

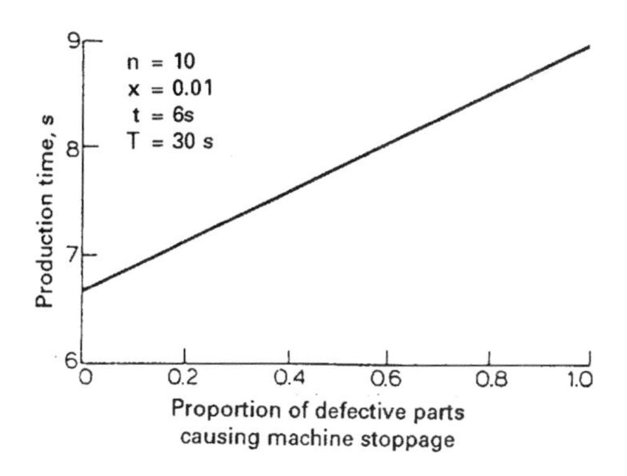

Figure 7.4 Effect of defective parts on production time.

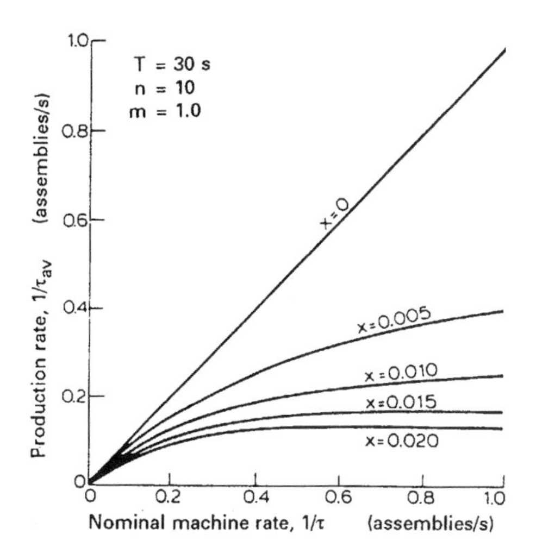

Figure 7.5 Effect of machine rate on production rate.

chine. However, it disables all sequential work clusters when the subassembly passes through, preventing delivery of further parts.

2. *Rejecting the subassembly immediately.* This method rejects the subassembly instantaneously once a defective part is detected to maintain error-free production security—particularly important in pharmaceutical and food industries.

For the case where the defective parts always stop the machine, m is equal to 1, and Eq. (7.2) becomes

$$t_{av} = t + xnT \tag{7.4}$$

Figure 7.5 illustrates how the production rate ($1/t_{av}$) is affected by changes in nominal machine rate ($1/t$) for various values of x and for typical values of $T = 30$ seconds and $n = 10$. It can be seen that, when x is small, the production rate approaches the machine speed. However, in general, x will lie within the range 0.005 to 0.02 and, under these circumstances, for high machine speeds (short cycle times), the production rate tends to become constant. Alternatively, it may be stated that, as the cycle time is reduced, only a relatively small increase in the production rate results. This explains why it is rarely practicable to have indexing assembly machines working at very high speeds.

In practical circumstances the cost of dealing with the unacceptable assemblies produced by the machine must be taken into account.

7.4 Cost Savings in Assembly

The total cost S_t of each acceptable assembly produced on an assembly machine where each work cluster assembles one part is given by the sum of the cost of the individual parts ($S_1 + S_2 + S_3 + \ldots + S_n$) plus the cost of operating the machine for the average time taken to produce one acceptable assembly. Thus,

$$S_t = O_t t_{av} + S_1 + S_2 + S_3 + \ldots + S_n \tag{7.5}$$

where O_t is the total cost of operating the machine per unit time (including operators' wages, overhead charges, actual operating cost, machine depreciation, and the cost of dealing with the unacceptable assemblies produced) and t_{av} is the average production time of acceptable assemblies [from Eq. (7.2)].

In estimating O_t, we will assume that the machine stoppage caused by a defective part will be cleared by one of the operators employed on the machine and that no extra cost will be entailed other than that due to machine downtime. Further, we will assume that, if a defective part passes through the work cluster and spoils an assembly, it will take an extra assembly worker t_{extra} seconds to dismantle the assembly and replace the nondefective parts back in the appropriate feeding devices.

Thus, the total preparation cost O_t is given by

$$O_t = O + N_{un} t_{extra} W_{rate} \tag{7.6}$$

where O is the total cost of operating the machine per unit time if only acceptable assemblies are produced and W_{rate} is the assembly workers' rate, including overhead. The number of unacceptable assemblies produced per unit time is

$$N_{un} = \frac{(1 - m)xn}{t + mxn} \tag{7.7}$$

Substituting Eq. (7.6) in Eq. (7.7) gives

$$O_t = \frac{O + (1 - m)xnt_{\text{extra}}W_{\text{rate}}}{t + mxnT} \tag{7.8}$$

In estimating the cost S_i of an individual component part, we will assume that it can be broken down into

- The basic cost α_i of the part, irrespective of quality level
- A cost ß that is inversely proportional to x and will therefore increase for better-quality parts

Thus, the cost of each part may be expressed as

$$S_i = \frac{\alpha_i + ß}{x} \tag{7.9}$$

ß, for the purposes of the present analysis, will be assumed constant, regardless of the basic cost α_i of the parts.

If Eqs. (7.2), (7.8), and (7.9) are now substituted into Eq. (7.5), the total cost of each acceptable assembly becomes, after rearrangement,

$$S_t = \frac{O(t + mxn) + (1 - m)xnt_{\text{extra}}W_{\text{rate}}}{1 - (1 - m)xn} + \sum_i^n \alpha_i + \frac{nß}{x} \tag{7.10}$$

Equation (7.10) shows that the total cost of the assembly can be broken into the following categories:

1. A cost that decreases as x is reduced
2. A cost that is constant
3. A cost that increases as x is reduced

It follows that, for a given situation, an optimum value of x exists that gives a minimum cost of assembly. The optimum value of x will be considered for the case in which $m = 1$, that is, where all defective parts cause a stoppage of the machine.

With $m = 1$, Eq. (7.10) becomes

$$S_t = O_t + Oxnt + (nß/x) + \sum_i^n \alpha_i \tag{7.11}$$

where O_t = cost of assembly operations
$Oxnt$ = cost of downtime
$nß/x$ = cost of parts quality
$\sum_i^n \alpha_i$ = basic cost of parts

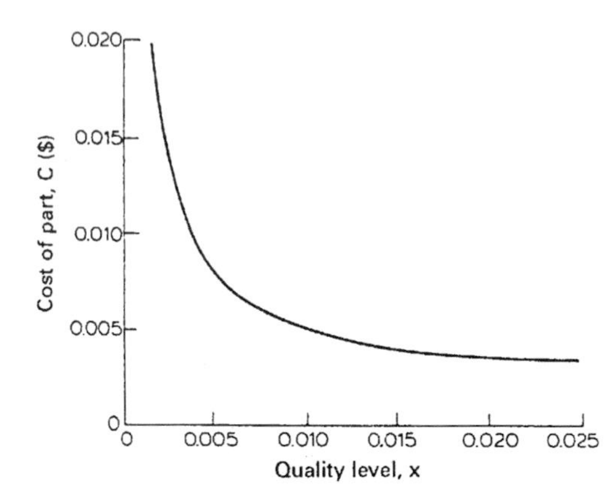

Figure 7.6 Cost per part as a function of quality level.

When Eq. (7.11) is differentiated with respect to x and set equal to zero, it yields the following expression for the optimum value of x (the value giving the minimum cost of assembly):

$$x_{opt} = [\beta/OT]^{1/2} \tag{7.12}$$

It is interesting to note that, for a given assembly machine, where O_t and ß are constants, the optimum quality level of the parts depends only on the time taken to clear a defective part from a work cluster.

Figure 7.6 shows how the cost of a part might increase as the quality level is improved. In this case, α_i = \$0.002 and ß = \$0.00003. If typical values of O_t = \$0.01/s (\$36/h), and $T = 30$ s are substituted in Eq. (7.12), the corresponding optimum value of x is 0.01 (1%). Equation (7.12) may be substituted in Eq. (7.11) to give an expression for the minimum cost of assembly:

$$S_t\,(\min) = O_t + 2n(O\beta T) + \sum_i^n \alpha_i \tag{7.13}$$

With $t = 6$ s and $m = 10$, the cost of assembling each complete set of parts [the first two terms in Eq. (7.13)] would be \$0.12. Half of this cost would be attributed to the assembly operation itself, and the other half to the increased cost due to part quality and the corresponding cost of machine downtime. Figure 7.7, where the first three terms in Eq. (7.11) are plotted, shows how these individual costs would vary as the quality level x varies, using the numerical values quoted in the example above. In this case, if parts having 0.020 defective were to be used instead of the optimum value of 0.010, the cost of assembly would increase by approximately 12 percent. This is a variation of \$0.015 per assembly and, with an average production time of

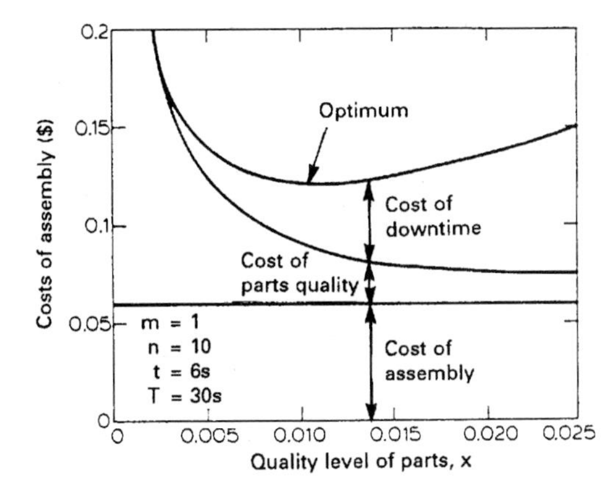

Figure 7.7 Variation of costs with part quality.

9 s [calculated from Eq. (7.2)], it represents an extra expense of approximately \$12,000 per year per shift.

In the analysis above, it was assumed that all defective parts would stop the machine. If, instead, it is possible to allow these parts to pass through the automatic devices and spoil the assemblies, the cost of assembly can be obtained by substituting $O_t = 0$ into Eq. (7.10). Thus,

$$S_t = \frac{O_t + xnt_{\text{extra}} W_{\text{rate}}}{1 - xn} + \sum_i^n \alpha_i + \frac{n\beta}{x} \tag{7.14}$$

Again, an optimum value of x arises and is found by differentiation of Eq. (7.14):

$$x_{\text{opt}} = \{n + [(t_{\text{extra}} W_{\text{rate}}/\beta) + (O_t/\beta)]^{1/2}]\}^{-1} \tag{7.15}$$

Taking $W_{\text{rate}} = \$0.002/\text{s}$, $t_{\text{extra}} = 60$ s, and the remaining values as previously noted, we find that x_{opt} is approximately 0.011. From Eq. (7.14), the minimum cost of assembly is \$0.109, which represents a savings of 9 percent in cost of assembly compared to the preceding example.

It is possible to draw two main conclusions from this example:

1. For the situation analyzed, it would be preferable to allow a defective part to pass through the work cluster and spoil the assembly rather than to allow it to stop the indexing machine, provided that the product does not infringe on public safety as in the case of pharmaceutical products or automotive brake or fuel subassemblies. This would not only increase the production rate of acceptable assemblies but would also reduce the cost of assembly.

2. An optimum quality level of parts always exists that will give minimum cost of assembly.

7.5 Asynchronous Assembly Systems

Asynchronous assembly machines often provide a higher production rate than the equivalent indexing machines. This is because the buffers of assemblies between work clusters will, for a limited time, allow the continued operation of the remaining work clusters when one has stopped. Provided that the buffers are sufficiently large, the stopped work cluster can be restarted before the other work clusters are affected and the downtime on the machine approaches the downtime on the work cluster that has the most stoppages. The following analysis will show that, even with relatively small buffers, the production rate of an asynchronous machine can be considerably higher than that of an equivalent indexing machine.

It will be assumed in the analysis that all the work heads on the asynchronous machine are working at the same cycle time of t seconds. Each work cluster is fed with parts having the same quality level of x (where x is the ratio of defective to acceptable parts), and between each pair of work stations is a buffer that is large enough to store b assemblies. A work cluster on an asynchronous assembly machine will be forced to stop under three different circumstances:

1. A defective part is fed to the work cluster and prevents the completion of its cycle of operations. In this case it will be assumed that an interval of T seconds elapses before the fault is cleared and the work cluster restarted.
2. The adjacent work cluster up the line has stopped and the supply of assemblies in the buffer storage between is exhausted.
3. The adjacent work cluster down the line has stopped, and the buffer between is full.

These last two circumstances lead to the conclusion that empty spaces in buffers cause the same problems as assemblies in buffers. Thus, to optimize the performance of an asynchronous assembly machine, the number of the spaces in buffers should be made equal to the number of assemblies in buffers. In other words, the buffers should be half-filled with assemblies. It will also will be assumed that, at any time, the buffer between any two work clusters is half-filled. Obviously, this will rarely be the case and, in fact, for half of the time, the buffer will be empty and, for the remaining time, it will be full. In the following analysis, the assumption that any particular buffer will be half-filled with assemblies will, therefore, lead to an underestimation of the downtime caused by work cluster faults.

Thus, if two adjacent work clusters have $b/2$ assemblies in the buffer between, then a fault in the first work cluster will prevent the second from working after a time delay of $bt/2$ seconds. A fault in the second work cluster will prevent the first from working after the same time. Also, over a long period, the average downtime of all work clusters must be the same. It will also be assumed that when a fault occurs, there will always be a service person available to correct it and that no work cluster will stop while another is stopped. The errors resulting from this latter assumption will become large when the quality level of the parts is poor (large x) and with a large number of automatic work clusters (large n). However, calculations show that these errors are negligible with practical values of x and n and produce an overestimate of the machine downtime.

7.5.1 Assessment of an asynchronous machine

For a typical work cluster on an asynchronous machine (Fig. 7.8) producing N assemblies, Nx stoppages will occur if m is unity. If each fault takes T seconds to correct, the downtime on the first work cluster due to its own stoppages is given by NxT. This same average downtime will apply for stoppages on the second work cluster down the line, but the first will be prevented from working only for a period of $Nx(T - bt/2)$ seconds. Similarly, stoppages of the third work cluster will prevent the first from working only for a period of time $Nx(T - bt)$ seconds. Similar expressions can be derived for the effect on the second and third work clusters up the line. Thus, it can be seen that the total downtime d on any work cluster while the machine produces N assemblies is given by

$$d/Nx = (T - 3bt/2) + (T - bt) + (T - bt/2) + T + (T - bt/2)$$
$$+ (T - bt) + (T - 3\,bt/2) + \ldots$$

or

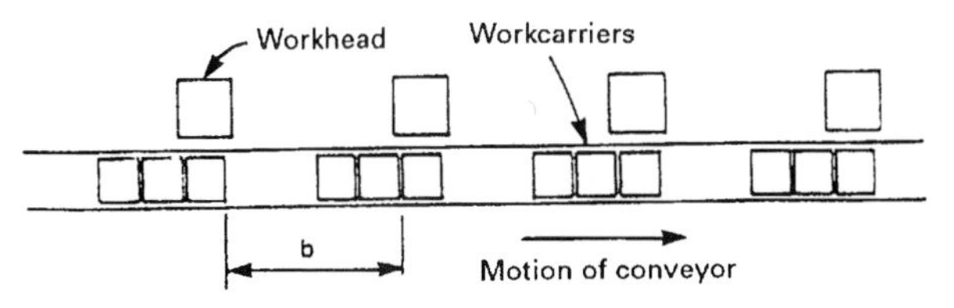

Figure 7.8 Work clusters on an asynchronous machine.

$$d/Nx = T + (2T - bt) + (2T - 2bt) + (2T - 3bt) + \ldots \qquad (7.16)$$

It should be noted that if any term in Eq. (7.16) is negative, it should be omitted. It is necessary at this stage to note the effect of the relative values of T and t. If, for example, the ratio

$$T/t = 5$$

and

$$b = 4$$

Eq. (7.16) becomes

$$d/Nx = 5T - 3bt \qquad (7.17)$$

or

$$d = 13Nxt \qquad (7.18)$$

The proportion of downtime D on the asynchronous machine may now be obtained as follows:

$$D = \frac{\text{downtime}}{\text{assembly time + downtime}} = \frac{d}{Nt + d} \qquad (7.19)$$

Substituting d from Eq. (7.18) gives

$$D = \frac{13xT}{t + 13xt} \qquad (7.20)$$

For different values of b, other terms of Eq. (7.16) will become zero or negative, and new values of d will be obtained. In general, the asynchronous machine downtime d may be expressed as

$$d = KNxt \qquad (7.21)$$

where K is a factor that depends on values of T/t and b. Table 7.1 gives the values of K for various values of b when $T/t = 5$.

The proportion of downtime D is now given by

$$D = \frac{d}{Nt + d} = \frac{Kx}{1 + Kx} \qquad (7.22)$$

For $b \geq 10$, all terms in Eq. (7.16) after the first are omitted, and the equation for the downtime on any work cluster becomes

$$d/Nx = T \qquad (7.23)$$

TABLE 7.1 Relationship between Buffer Capacity *b and the Factor* K

b	K
0	25
2	19
4	13
6	9
8	7
10	5

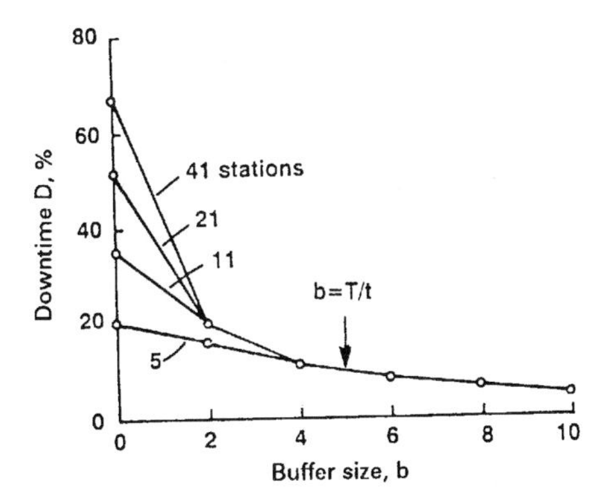

Figure 7.9 Effect of buffer size on downtime ($x = 0.01$).

The downtime on the machine equals the downtime on any work cluster and thus, for $b \geq 10$,

$$D = \frac{5x}{1 + 5x} \tag{7.24}$$

Figure 7.9 shows how the proportion of downtime is affected by the size of the buffers when $x = 0.01$ for machines with 5, 11, 23, and 41 stations. It can be seen that significant improvements in performance can be obtained with only small buffers, specially with large machines. Also, with large buffers, the machine downtime is independent of the machine size and approaches the downtime on a single station when it is unaffected by stoppages at other stations.

In theory, when the size of the buffer storage $b \geq T$, the work clusters are completely isolated one from the other, and further increases in the size of b have no effect. However, the greatest benefit occurs with the smaller buffers. Therefore, as a practical guide, it is often assumed that, for a good asynchronous assembly machine design,

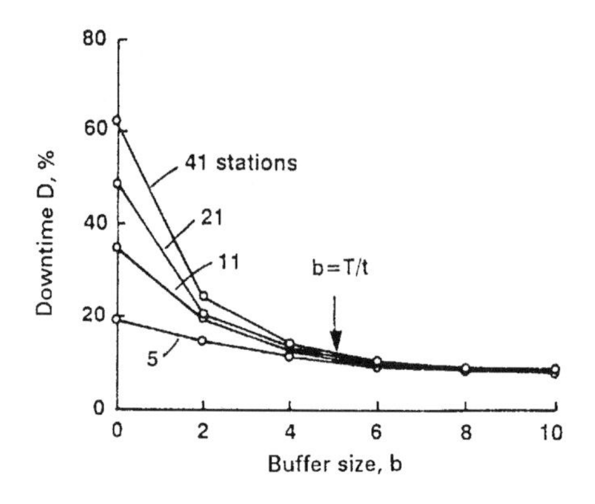

Figure 7.10 Effect of buffer size on downtime.

$$b = T/t$$

It should be pointed out that the method of analysis presented here is only approximate, although computer simulations presented in Fig. 7.10 for the same conditions as those in Fig. 7.9 show that the analysis gives a good approximation to the true performance of a machine when $b \leq T/t$.

7.5.2 Productivity of an asynchronous machine

In considering the economics of an asynchronous machine, the cost of providing the buffers has to be taken into account. Ideally, an economic analysis should use a mathematical model in which the size b of buffers between each station is a variable and then study how the assembly costs C_{pr} are affected by the magnitude of b. However, such a complex model is not necessary if certain realistic assumptions about b are made.

When the number of parts n to be assembled is small, the asynchronous machine is uneconomical compared to a synchronous indexing machine. When n is large, the reverse is true. Thus, when n is small, the optimum value of b is zero (representing an indexing machine) and, when n is large, the optimum value is close to that at which the various work clusters do not affect one another significantly. Figures 7.9 and 7.10 show the effect of buffer size on the proportion of downtime. They also indicate that, when $b \leq 10$, the proportion of downtime D on the machine approaches that of a synchronous isolated work cluster. In general, this is true if $b \geq 2T/t$, where T is the downtime due to one defective part and t is the work cluster cycle time.

However, analysis of the economics of individual machines shows that a value of $b = T/t$ results in cost savings closely approaching those for $b = 2T/t$. Therefore, to simplify analysis and to make comparisons, it will be assumed that on an asynchronous machine $b = T/t$ and that the resulting downtime for the machine approaches twice that of an individual isolated station, as illustrated in Figs. 7.9 and 7.10.

Hence, for an asynchronous machine of any size, the average production time can be estimated from Eq. (7.25)

$$t_{av} = 2xT \tag{7.25}$$

where x is the part quality level. Obviously, it is assumed in this regard that all defective parts will cause a stoppage of the work clusters. Thus Eq. (7.25) can be compared to Eq. (7.4) for a synchronous indexing machine.

7.5.3 Error management and fault recovery

In the computer simulations, the assembly machines were provided with an excess of personnel so that all faults caused by defective parts would receive immediate attention. However, the number of personnel needed in order to achieve this situation on large systems would be uneconomical, and it would be better to allow some faults to go unattended for a short time in order to increase the utilization of the fewer personnel. Clearly, for optimum working, the number of personnel required will be proportional to the number of stations of the machine.

During the production of N assemblies, each containing n parts automatically assembled, and with each part having x defectives causing faults, the total faults will be given by Nxn, and the total time spent by personnel correcting faults will be $NxnT$, where T is the average time to correct each fault. Since the total production time on an asynchronous machine is given by the Eq. (7.25), the minimum number of technicians N_{tech} required to correct faults is given by

$$\mathrm{N}_{tech} = \frac{\text{total fault correction time}}{\text{total production time}}$$

$$= \frac{Nxnt}{N(2xT + t)}$$

$$= \frac{xn}{2x + t/T} \tag{7.26}$$

Alternatively, the maximum number of stations n_{Smax} that one technician can tend is given by

$$n_{Smax} = \frac{n}{N_{tech}} = 2 + \frac{t}{xT} \tag{7.27}$$

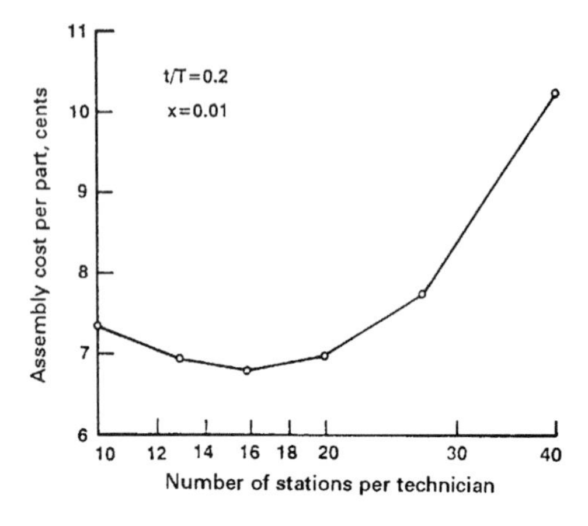

Figure 7.11 Assembly cost as a function of number of stations attended by a technician.

Using $t/T = 0.2$ and $x = 0.01$, Eq. (7.27) should show that one technician could tend no more than 22 stations on an asynchronous machine having these characteristics. However, since faults cannot occur on an asynchronous machine while an earlier fault is still being corrected, the number of stations per technician will be less than that given by Eq. (7.27). Figure 7.11 shows that the optimum number of stations is closer to 16. Thus, Eq. (7.27) could be modified to give the economical number of stations per technicians as follows:

$$n_S = 2 + \frac{2t}{3xT} \tag{7.28}$$

With a synchronous indexing machine, any fault causes the whole machine to stop and, therefore, no more than one technician, theoretically, would be needed to correct fault.

7.6 Capital Equipment Economics of Synchronous and Asynchronous Systems

In a simple analysis of the economics of automation equipment, it is advantageous to convert the capital cost of the equipment to an equivalent assembly worker rate. For this purpose, a factor Q_e can be defined as the cost of the capital equipment that can economically be used to do the work of one assembly worker on one shift. This factor, which must be determined for a particular company, is the basis for the following economic comparisons. For example, if a simple piece of automation equipment is being compared and it can do the task of one assembly worker, then the economical cost of the equipment should be Q_e for one shift working, $2Q_e$ for two shifts working, etc.

The rate for a piece of equipment initially costing C_e is then given by $C_e W_a / s_n Q_e$, where W_a = the rate for one assembly worker and S_n = the number of shifts.

In fact, many companies will determine Q_e by multiplying the annual cost of an assembly worker by a factor that usually lies between 1 and 3, representing a payback period of 1 to 3 years.

The cost of assembly C_a for a complete assembly is given by

$$C_a = t_{\text{pr}} \left(\frac{W_t + C_e W_a}{S_n Q_e} \right) \tag{7.29}$$

where t_{pr} = average time between deliveries of complete assembly for a fully utilized system

W_t = total rate of machine personnel

C_e = total capital cost for all equipment, including engineering setup and debugging cost.

For the purpose of comparing the economics of assembly systems, the cost of assembly per part will be used and will be nondimensionalized by dividing this cost per part by the rate for one assembly worker, W_a, and the average manual assembly time per part, t_a. Thus, the dimensionless assembly cost per part, C_d, is

$$C_d = \frac{C_a}{(n W_a t_a)} \tag{7.30}$$

Substituting Eq. (7.29) into Eq. (7.30) gives

$$C_d = \left(\frac{t_{\text{pr}}}{t_a} \right) \left(\frac{W_r + C_e}{n S_n Q_e} \right) \tag{7.31}$$

where

$$W_r = \frac{W_t}{(W_a n)} \tag{7.32}$$

is the ratio of the cost of personnel to the cost of one manual assembly worker, expressed per part. Thus, the dimensionless assembly cost per part for an assembly worker working without any equipment will be unity, which forms a useful basis for comparison purposes.

Finally, it should be pointed out that, for a particular assembly machine, Eq. (7.31) holds true only if the required average production time t_q for one assembly is greater than or equal to the minimum production time t_{pr} obtainable for the machine. In other words, if $t_{\text{pr}} \le t_q$, then t_q must be substituted for t_{pr} in Eq. (7.31) because the machine is not fully utilized. If $t_{pr} > t_q$, then more than one machine will be needed to meet the required production rate. For example, two machines producing N assemblies per hour will give the same assembly costs per assembly as one machine producing $N/2$ assemblies per hour. Hence, it will be assumed that, as the average required production time for one assembly t_q is reduced below the production time t_{pr} obtainable from one machine, the assembly costs become constant and are given by Eq. (7.31). To obtain t_q, it is necessary to know the required annual production volume per shift V_S and the plant efficiency P_e, the latter being defined as time actually worked in the plant divided by the time available.

Thus,

$$t_q = 0.072\, P_e / V_S \tag{7.33}$$

In this formula, t_q is in seconds, V_S is in millions, P_e is expressed as percentage, and the factor 0.072 arises from the assumption that 7.2 million seconds is the maximum time available in one shift-year.

In analyzing the economics of special-purpose assembly machines, the following nomenclature and numerical values will be used.

Term	Definition	Value
C_t	Cost of transfer device per station for an indexing machine	\$10,000
C_B	Cost of transfer device per space (work station or buffer space) for asynchronous machine	\$ 5,000
C_C	Cost of work carrier	\$ 1,000
C_F	Cost of automatic feeding device and delivery track	\$ 7,000
C_W	Cost of work cluster	\$10,000
W_{tech}	Rate for one technician engaged in correcting faults on the machine	\$0.012/s
W_a	Rate for one assembly worker	\$0.008/s
Q_e	Equivalent cost of one assembly worker in terms of capital investment	\$90,000
t_a	Average annual assembly time per part	8 s

It should be noted that equipment costs include the purchase cost, the assembly machine design costs, engineering and debugging costs, and the cost of controls. It has been established that purchase cost often forms only 40 percent of the total cost of the equipment.

7.6.1 Synchronous indexing machine

For an indexing machine, the total equipment cost C_e is given by

$$C_e = n(C_T + C_W + C_F + C_C) \tag{7.34}$$

Assuming two assemblies to load and unload assemblies, the total rate for the personnel will be

$$W_t = 2W_a + W_{\text{tech}} \tag{7.35}$$

Substituting for C_e from Eq. (7.34), for W_t from Eq. (7.35) and for t_{pr} from Eq. (7.4) into Eq. (7.31) gives as the dimensionless cost of assembly per part

$$C_d = \frac{1}{nt_a}(t + xnT)\left[2 + \frac{W_{\text{tech}}}{W_a} + \frac{n(C_T + C_W + C_F + C_C)}{S_nQ_e}\right] \tag{7.36}$$

7.6.1 Asynchronous indexing machine

For an asynchronous machine, where the number of work carriers is chosen so that, on average, the machine will be half full, the total equipment cost C_e is

$$C_e = n\{C_W + C_F + (b + 1)\,[(C_B + C_C/2)]\} \tag{7.37}$$

Again assuming two assembly workers to load and unload assemblies and using Eq. (7.28), we find the total rate for the personnel will be

$$W_t = 2W_a + [2 + (2t/3xT)]W_{\text{tech}} \tag{7.38}$$

Substituting $b = T/t$, t_{pr} from Eq. (7.25), C_e from Eq. (7.37), and W_t from Eq. (7.38) into Eq. (7.31) gives

$$C_d = nt_a(t + 2xT)2 + \left(\frac{nW_{tech}}{2 + 2t/3xT}\right)\left(\frac{1}{W_a}\right) + n\left(\frac{1}{S_nQ_e}\right)\left[C_w + C_F + \left(1 + \frac{T}{t}\right)\left(C_B + \frac{C_C}{2}\right)\right] \tag{7.39}$$

provided the number of technicians is not less than one.

For a cycle time t of 6 seconds, a downtime T of 30 seconds, a quality level x of 0.01, two shifts working ($S_n = 2$), and the numerical values of equipment and operator cost described earlier, Fig. 7.12 shows, for the two machines, how the dimensionless cost of assembly per part varies with the size of a machine. It can be seen that for a small value of n (small machine), the indexing machine is the more economical of the two whereas, for large value of n, the asynchronous machine is the more economical. The rise in cost for the indexing machine is attributable entirely to the increasing downtime as the

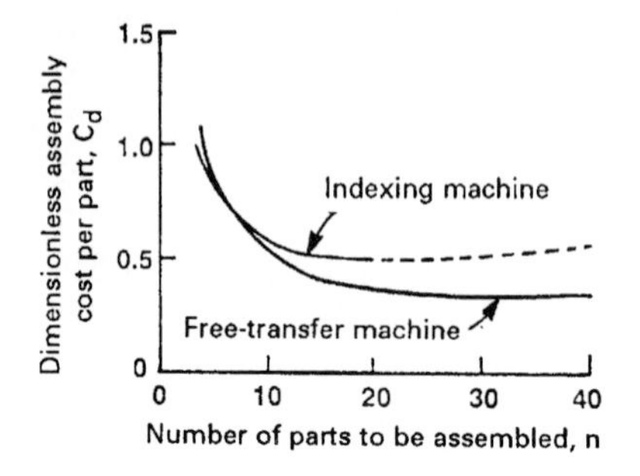

Figure 7.12 Assembly cost versus number of parts for two types of machine.

machine becomes larger. In fact, the region of the curve shown dashed represents downtime proportions of greater than 50 percent.

7.6.2 Volume sensitivity of synchronous and asynchronous machines

A special-purpose assembly machine is designed to work at a particular rate. This means that, under normal circumstances, the output can be obtained from a knowledge of t_{pr}, the mean production time obtainable. Thus, the obtainable annual production volume per shift V_S in millions is given by

$$V_S = 0.072\, P_e / t_{pr} \tag{7.40}$$

In this formula, P_e is the plant efficiency percentage and is defined as the time actually worked in the plant divided by the time available, and it is assumed that 7.2 million seconds are available in one shift-year. Thus, on a graph of assembly cost versus annual production volume such that shown in Fig. 7.13, a special purpose assembly ma-

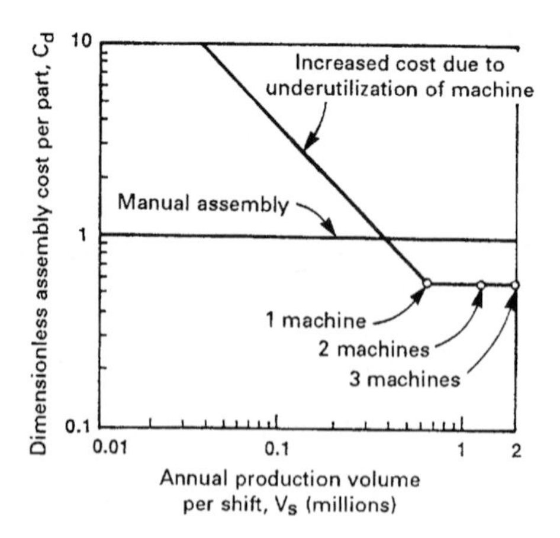

Figure 7.13 Assembly cost versus annual production volume.

chine is represented by a single point corresponding to the production volume obtainable from the machine working at maximum capacity. However, for the purpose of comparison, it is desirable to be able to show the relationship between assembly cost and the required annual production volume per shift.

If the volume required is less than that obtainable from the machine, then, as explained previously, the assembly cost will be a function of the average production time t_q, given by Eq. (7.33). In this case, when the assembly costs given by Eq. (7.31) and the annual production volume per shift are plotted on logarithmic scale, a linear relationship results, as shown in Fig. 7.13. This relationship simply shows how costs increase because the machinery is being fully utilized.

If the required production volume is greater than that obtainable from the machine, then it is assumed that backup will be provided in the form of manual assembly stations. Then, when the required volume approaches twice that obtainable, a second machine will be employed, and so on. Since the assembly cost is dependent on the number of assembly machines, the relationship between dimensionless assembly cost and volume will be a horizontal line (Fig. 7.13), because of the smoothing provided by the use of manual assembly where required.

The multistation assembly machines considered above were both assumed to be dedicated to one product. However, it should be realized that the output from one machine is on the order of 1 to 5 million assemblies per year and that application of these systems is limited to mass production situations in which the product design is stable for a few years. It should also be realized that mass production constitutes only about 15 to 20% of all production in the United States, and, for this reason, much attention is being given to automating batch assembly processes. It is clear that, for these applications, it will be necessary to employ programmable or flexible automation such as assembly robots.

7.7 Robots as Flexible Assembly Systems

One of the problems in making general economic comparisons between assembly systems is that, typically, some of the preparations required for a particular product cannot be automated; one of the key characteristics of a system is its ability to handle this situation. Various studies have been conducted to determine the types of assemblies that are generally considered candidates for robot assembly. It was found that these assemblies have similar characteristics and that a typical profile could be deduced for a candidate assembly. Using the typical candidate assembly, the economics of some configurations of robot assembly systems can be considered.

The following discussion will be restricted to those situations in which general-purpose assembly robots can be applied. Three basic systems will be modeled:

1. A single-station machine with one fixture and one robot arm
2. A single-station machine with one fixture and two robot arms
3. A multistation asynchronous machine with a single robot arm at various stations where appropriate

When the suitability of a product for robot assembly is determined, it is important to consider the available methods of part presentation. These range from bin picking using the most sophisticated vision systems to the manual loading of parts into pallets or trays.

Much research has been conducted on the subject of programmable feeders, and the impression has been given that these can be used for systems that can assemble a variety of products. However, investigations have shown that most, if not all, of the "programmable" assembly systems presently being developed in the United States will be devoted to assembly of one product or one family of products. Thus, these systems will be flexible in the sense that they can be adjusted quickly to accommodate different members of the product family (different styles) or to accommodate product design changes. Nevertheless, the equipment used on these systems will be dedicated to the product family, regardless of whether it is "special-purpose" equipment or "programmable."

When the economical choice of a part presentation method is to be determined, there will usually be only two basic types of method to consider:

1. Those involving manual handling and loading of individual parts into part trays, pallets, or magazines
2. Those involving automatic feeding and orienting of parts from bulk

In some cases, it is possible to obtain parts premagazined. Since these magazines have usually been filled by hand or automatic from bulk, then the cost of the magazining is born by the supplier and is included in the cost of parts. Thus, the economics of using premagazined parts will have to be considered on a case-by-case basis with knowledge that part cost will increase.

For present purposes, no distinction will be made between different types of automatic equipment because the cost of feeding each part automatically, C_f, will be given by multiplying the rate for the equipment (cents per second) by the time interval in which parts are required. Hence,

$$C_f = (C_F W_a / S_n Q_e) t_{at} \tag{7.41}$$

Where C_f is the cost of feeding one part (cents); C_F is the cost of a feeder for one part type, including tooling, engineering, and debugging (cents per second); and t_{at} is the average station cycle time (seconds).

The cost C_{mm} of using a manually loaded magazine to present one part is given by

$$C_{mm} + (C_M W_a / S_n Q_e) t_{at} + W_a t_m \tag{7.42}$$

where C_M is the cost of the magazine, t_m the manual handling and insertion time for loading one part into the magazine, and W_a the assembly worker rate (cents per second).

For a simple part, the value of t_m is approximately 2 seconds. A typical value of W_a is 0.8 cents per second.

It will be assumed that a vibratory-bowl feeder, when supplied with simple tooling, engineered, debugged, and fitted with the necessary delivery track, costs \$7000, and the cost of special-purpose magazines or feed tracks is \$1000. If, as illustrated previously, $S_n = 2$ and $Q_e = 90$, the comparison of part presentation costs shown in Fig. 7.14 is obtained.

It can be seen that the special-purpose feeder will be more economical than the manual loading of magazines for a station cycle time less than about 60 seconds. Also, for longer cycle time, the use of feeder equipment rapidly becomes exorbitant.

7.8 Criteria for Robotic Assembly

Analysis of many robotic assemblies produced the following criteria, where M is the total number of different parts and n is the total number of parts involved in each assembly:

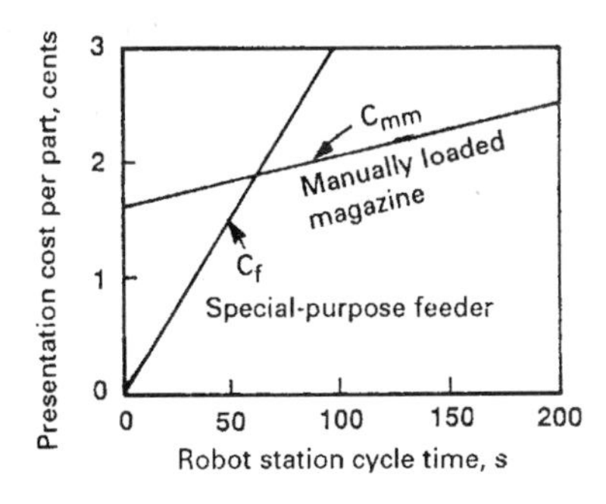

Figure 7.14 Comparison of part presentation costs.

1. $M = 0.78n$: This approximates the case in which an assembly contains 10 parts, three of which are identical. Information such as this is important because, for example, in a single-station machine using one arm, these three parts could all be presented by the same feeder.
2. $M = 0.22n$: These parts can be assembled automatically but not fed automatically. Such parts would be manually loaded into pallets, magazines, or feed tracks.
3. $M = 0.71n$: These could be fed and oriented by $0.5n$ special-purpose feeders.
4. $M = 0.12n$: These require manual handling and insertion and some manual manipulation. On a transfer machine, these would require a separate manual station. On a single-station system, the partially completed assembly would be delivered to a manual station external to the system and then returned to the system after completion of the manual operation. Thus, two pick-and-place operations involving a robot arm would be required in addition to the manual operation.
5. $M = 0.18n$: This requires special grippers or tools in addition to the basic grippers.
6. $M = 0.15n$: This requires transfer of the assembly to a special-purpose work cluster or tool to complete a required operation.

The same analysis showed that the average assembly time per part for a single-station assembly system with one fixture and two robot arms was 3.29 seconds; for a system with one arm, it was 5.4 seconds. If the product were to be manually assembled, the average assembly time per part would be 8.15 seconds.

Using these figures, it is possible to estimate the cost of assembly using different configurations of robot assembly systems and to compare the results with the cost of manual assembly.

7.8.1 Single-station robotic system

For a single-station robot with one work fixture and one arm, the costs of the assembly equipment for an assembly containing small n parts are listed in Table 7.2. Summing these costs gives the total system cost:

$$C_{e1} = 110 + 5.72n \tag{7.43}$$

Similarly, the cost of a single-station system with one work fixture and two robot arms is found to be

TABLE 7.2 Costs for Single-Station Robotic System

Item	Number required	Cost/unit	Total cost, thousands of dollars
Robot arm, sensor controls, etc.	1	100	100
Work fixture	1	5	5
Standard grippers	1	5	5
Special grippers or tools	$0.18n$	5	$0.9n$
Magazines, pallets, etc.	$0.22n$	1	$0.22n$
Special-purpose feeders	$0.5n$	5	$2.5n$
Manual station and associated feed track	$0.12n$	5	$0.6n$
Special-purpose work cluster	$0.15n$	5	$1.5n$

$$C_{e2} = 165 + 6.62n \tag{7.44}$$

The first term has increased because of the additional robot arm and gripper. The second term has increased because, with a two-arm system, in order to keep the cycle time low, it was found best to share repeated operations between the two arms and to duplicate the part feeders.

One of the major problems in modeling automatic assembly systems is how to estimate the cost associated with the technician responsible for attending the machine and clearing stoppages due to faulty parts. On a multistation high-speed assembly machine, faults can occur so frequently that at least one full-time technician, who can also maintain adequate supplies for parts in the various automatic feeders, will be required. Under these circumstances, assemblies are produced at high rates, and the cost of the technician forms a relatively small portion of the total cost of the assembly. However, with a low-speed single-station machine using robot arms, the cost of a full-time technician to attend the machine can constitute more than 50 percent of the cost of assembly.

For the purpose of this analysis, estimates will be made of the actual manual time involved in producing each assembly. Where parts are loaded to magazines or pallets or must be manually inserted, the total time required will be estimated. It will be assumed that if this total manual time is less than the system cycle time, the assembly worker can be engaged in other tasks unrelated to the particular assembly machine being modeled. Similarly, the time necessary for tending the machine will be estimated, and it will be assumed that the technician can be fully occupied by tending several such machines.

It should be emphasized that these assumptions will give the most optimistic estimates of assembly costs, especially while robot assembly is in the development stage. It can be argued, however, that, unless machines are eventually designed that can run with a minimum of attention, they will simply not provide an economical alternative to manual assembly. In fact, if a single-station general-purpose assembly machine were to require one full-time technician, then, given assistance in the form of fixtures etc., that same individual could probably perform the assembly task at a rate similar to that of the robot.

If the average time taken to clear a fault and restart the system is T, the number of parts in the assembly is n, and the average proportion of defective parts that will cause a fault is x, then the average production time peer assembly t_{pr} will given by

$$t_{\mathrm{pr}} + t_{\mathrm{at}} + xnT \tag{7.45}$$

where t_{at} is the station cycle time (neglecting a downtime due to defective parts).

Using the basic equation for the dimensionless assembly cost per part, Eq. (7.31), gives

$$C_d = \left(\frac{t_{\mathrm{pr}}}{nt_a}\right)\left(\frac{W_t}{W_a} + \frac{C_e}{S_n Q_e}\right) \tag{7.46}$$

where C_e = total cost equipment in thousands of dollars, including debugging etc., as given by Eqs. (7.43) and (7.44), and W_t = total personnel rate, given by $1.363nW_a + xnTW_{\mathrm{tech}}$.

Substituting all these values and the equipment costs determined earlier into Eq. (7.46) gives the following equations. For a one arm system,

$$C_{\mathrm{dt}} = 0.662 + 0.22n \tag{7.47}$$

and for a two-arm system,

$$C_{\mathrm{dt}} = 0.647 + 0.016n \tag{7.48}$$

These equations are plotted in Fig. 7.15, where it can be seen that the economics of one- and two-arm single-station systems are very similar. Therefore, the decision as to which type of system to use must be made on an individual basis. However, it can be seen that single-station automatic systems used for assemblies containing more than

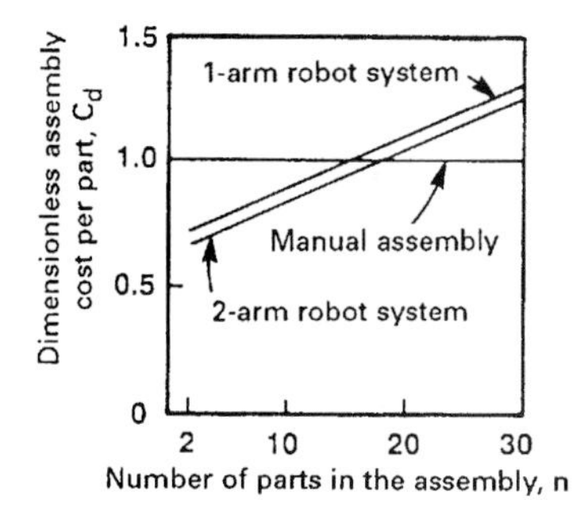

Figure 7.15 Assembly cost per part versus number of parts for robotic and manual systems.

about 18 parts are unlikely to be economical for the conditions assumed. An average station cycle time for such an assembly would be approximately 97 seconds for a single-arm station and 59 seconds for a two-arm station.

7.8.2 Multistation robotic system

When the demand for a product is greater than can be assembled on a single-station assembly machine, duplicates of such a machine can be installed or, alternatively, a multistation asynchronous machine can be employed. The typical assembly profile can be again used to study the economics involved.

On an asynchronous assembly machine, it is usually not possible to perform manual operations at any station where automatic operations are carried out. Similarly, those operations involving special-purpose work clusters must be carried out at separate stations. From the information on the profile of the candidate assembly (Table 7.2), the number of manual stations will be $0.12n$, and the number of stations with a special-purpose work cluster will be $0.15n$. The remaining parts ($0.7n$) will be assembled by robots, and the number of stations will depend on the required assembly rate. From Eq. (7.33), if the plant efficiency P_e is 80 percent, the station cycle time (neglecting downtime) is $5.76/V_S$, where V_S is the annual volume per shift.

The figure for the average assembly time per part for a one-arm flexible system used earlier was 5.14 seconds, so that the maximum number of parts assembled at one robot station, n_r, is given by

$$n_r = 5.76/5.14V_S = 1.12/V_S \tag{7.49}$$

and the number of robot stations n_t on the asynchronous machine would then be

$$n_t = 0.7n/n_r = 0.625nV_S \tag{7.50}$$

TABLE 7.3 Breakdown of Equipment Costs for a Multistation Asynchronous Machine

Item	Number required	Cost/unit, thousands of dollars	Total cost, thousands of dollars
Robot arm, sensors, controls, etc.	$0.72nV_S$	75	$54nV_S$
Grippers	$0.72nV_S$	5	$3.6n$
Transfer device, 3 work carriers	$0.72nV_S$	29	$20.9nV_S$
Magazines, pallets, etc., for manual assembly	$0.22n$	1	$0.22n$
Special-purpose feeders	$0.65n$	5	$3.25n$
Special-purpose work clusters	$0.15n$	10	$1.5n$

Table 7.3 gives a breakdown of the equipment costs for a multistation asynchronous machine. Summing these equipment costs gives the total cost C_{et} for an asynchronous system,

$$C_{et} = (12.77 + 78.5V_S)n \tag{7.51}$$

It should be noted that this equation is valid only for assemblies containing at least 6 parts and with station cycle time of at least 8.15 seconds to allow for a typical manual assembly operation. This figure corresponds to an annual production volume per shift V_S of 0.6 million.

As with the analysis of a synchronous machine, it will be assumed that assembly workers are required only for direct assembly and fault-correction procedures and can be engaged in tasks unrelated to the particular assembly for the remaining time. The downtime on a properly designed asynchronous machine approaches twice that for one individual station. Thus, the number of parts n in Eq. (7.45) should be regarded as twice the number of parts n assembled at one robot station and given by Eq. (7.49).

For an asynchronous machine, therefore,

$$t_{\text{pr}} = 6.43/V_S \tag{7.52}$$

Substituting this value for t_{pr} in Eq. (7.46) gives

$$C_{\text{dt}} = 0.567 + 0.056/V_S \tag{7.53}$$

This result is plotted in Fig. 7.16 to illustrate the effects of annual production volume on the cost of assembly. For comparison purposes,

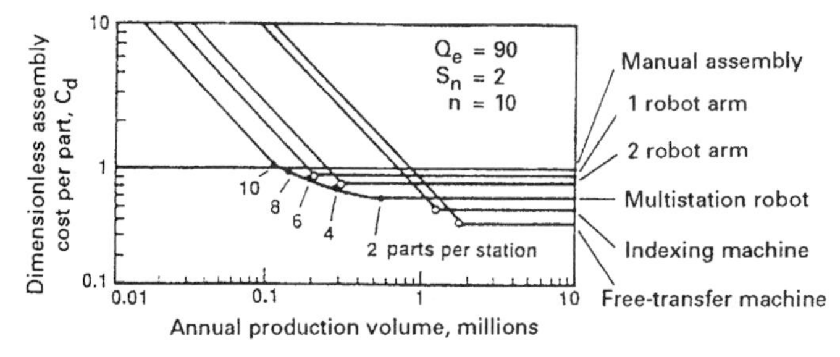

Figure 7.16 Effect of annual production volume on assembly cost.

the results for single-station robot machines assembling 10 parts and for special purpose-machines assembling 10 parts are also illustrated in Fig. 7.16.

It can be seen that the single-station robot machines and special-purpose machines are each represented by one point on the graph where, if these machines are underutilized, the cost of assembly increases rapidly. If volumes greater than those produced by one machine are needed, further identical machines could be used, giving the same assembly cost. However, with the multistation robot machine, a degree of flexibility is available. By selecting the number of parts assembled at each station, the machine can be made to match, within limits, the required production volume.

From Fig. 7.16 it can be seen that, for the conditions represented, special-purpose machines would be recommended for production volumes greater than 1.5 million per year. For volumes lower than 200,000, innovative mechanization together with manual bench assembly is preferable. Innovative mechanization utilizing certain assembly systems, together with robots, can be economical.

Chapter

8

Affordable Control Systems in Manufacturing

8.1 Today's Business

Rising costs. Shorter lead time. Complex customer specifications. Competition from across the street—and around the world.

Today's businesses face an ever-increasing number of challenges. The manufacturers that meet these challenges will be the ones that develop more effective and efficient forms of production, development, and marketing.

Advanced sensor and control technology can make a fundamental contribution to manufacturing based on simple and affordable integration. With sensor and control technology, one can integrate manufacturing processes, react to rapidly changing production conditions, help people make complex qualitative decisions, lower the costs, and improve product quality throughout the manufacturing enterprise.

The first step in achieving such flexibility is establishing an information system that can be reshaped whenever necessary. This will enable it to respond to the changing requirements of the enterprise—and the environment. This reshaping must be accomplished with minimal cost and disruption to the operation.

Undoubtedly, sensor and control technology will play a key role in economical information systems. However, sensor and control technology alone can not shorten lead time, reduce inventories, and minimize excess capacity to the extent required by today's manufacturing operation. The various sensors must be integrated with appropriate control means throughout the manufacturing operation according to a computer-integrated manufacturing (CIM) strategy. The individual manufacturing processes will then be able to flow, communicate, and respond together as a unified cell, well-structured for its functions.

In order to develop a sensory and control information system that will achieve these objectives, the enterprise must start with a long-range architectural strategy that provides a foundation for today's needs as well for tomorrow's. These needs include supporting new manufacturing processes, incorporating new data functions, and establishing new databases and distribution channels. The tools for this control and integration are available today.

Advanced sensor and control technology is more than an implementation of new technologies. It is a long-range strategy that allows the entire manufacturing operation to work together to achieve the business's qualitative and quantitative goals. It must have top management's commitment. It may entail changing the mind-set of people in the organization and managing the change. The success of CIM strategy is largely credited to the success of implementing affordable advanced technology of mechanisms and sensor and control systems.

Integrating affordable flexible systems for assembly using sensors and control systems is the way of the future. In times of disaster, even the most isolated outposts can be linked directly into the public telephone network through portable versions of satellite earth stations called *very small aperture terminals* (VSATs). This plays a vital role in relief efforts for disasters like the eruption of Mount Pinatubo in the Philippines; the massive oil spill in Valdez, Alaska; the 90,000-acre fire in the Idaho forest; and Hurricane Andrew in south Florida and coastal Louisiana. These VSATs are unique types of sensor and control systems. They can be shipped and assembled quickly to facilitate communication by powerful antennas that are much smaller than conventional satellite dishes. This type of sensor and control system provides an excellent alternative to complicated conventional communication systems, which in times of disaster often are damaged or overloaded.

Multispectral sensors and control systems will play an expanding role to help offset the increasing congestion on America's roads, by creating "smart highways." At a moment's notice, data could be gathered to help police, tow trucks, and ambulances respond to emergencies. Understanding flow patterns and traffic composition would also help traffic engineers map out future traffic control strategies. The result of less congestion will be billions of vehicle-hours saved each year.

The spacecraft Magellan is close to completing its third cycle of mapping the surface of the planet Venus. The key to gathering data is a synthetic aperture radar as a sensor and information-gathering control system, the sole scientific instrument aboard Magellan. Even before the first cycle ended, in mid-1991, Magellan had mapped 84 percent of Venus's surface, returning more digital data than all previous U.S. planetary missions combined, with resolutions 10 times better than those provided by earlier missions. To optimize radar performance, a unique and simple computer software program was developed, capable of handling the nearly 950 commands per cycle. Each cycle takes one Venusian day, the equivalent of 243 Earth days.

8.2 Intense Competition in the United States

Manufacturing organizations in the United States are under intense competitive pressure. Major changes are occurring with respect to resources, markets, manufacturing processes, and product strategies. As a result of international competition, only the most productive and cost-effective industries will survive.

Today's innovative affordable automation, together with sensors and control systems, has explosively expanded beyond its traditional production base into far-ranging commercial ventures. It will play an important role in the survival of innovative industries. Its role in information assimilation and control of operations to maintain an error-free production environment will help enterprises to stay on the competitive course.

8.3 Establishing an Affordable Automation Program

Manufacturers and vendors have learned the laborious way that *technology alone does not solve problems.* A prime example is the gap between the information and the control worlds. Because of the gap, production planners have set their goals according to dubious assumptions concerning plant-floor activities, while plant supervisors have not been able to isolate production problems until well after their occurrence.

The challenge of effective affordable automation for an error-free production environment has drawn forth a long list of solutions. Some of these solutions are as old as the concept of computer-integrated manufacturing itself. However, in many cases, the problem has turned out to be not inadequate technology, but the inability to integrate equipment, information, and people.

8.3.1 Belt tightening

Recent economic belt tightening has forced industry to justify every capital expense, and CIM has drawn fire from budget-bound businesspeople in all fields. Too often, the implementation of flexible automation within CIM has become a compatibility nightmare in today's multivendor factory-floor environments. Too many end users have been forced to discard previous automation investments and/or spend huge sums on new equipment, hardware, software, and networks in order to effectively link together data from distinctly dissimilar sources. The expense of compatible equipment and the associated labor cost for elaborate networking are often prohibitive.

The claims made for CIM open systems are often misleading. This is largely because of proprietary concerns, limited-access databases, and operating system compatibility restrictions. The implementations fail to provide the transparent integration of process data and plant business information that allows CIM to work.

In order to solve this problem, it is necessary to establish a clear automation program. A common approach is to limit the problem description to a workable scope by eliminating the features that are not amenable to analysis. The problem can then be replaced by a simpler, workable model. Action can then be taken on the basis of model predictions.

The dangers associated with this strategy are obvious. If the simplified model is not a good approximation of the actual problem, the solution effort will be inappropriate and may even worsen the problem.

8.3.2 Robust automation

Robust automation programs can be a valuable asset in solving complex automation problems. Advances in flexible automated systems and sensor technology have provided the means to make rapid, large-scale improvements and have contributed in essential ways to the manufacturing technology of today.

The infrastructure of an automation program must be closely linked with the implementation of sensors and control systems, within the framework of the organization. The situation becomes more difficult whenever problem solving is extended to include the organizational setting. Organization theory is based on a fragmented and partially developed body of knowledge, and can provide only limited guidance in the formation of problem models. Managers commonly use their experience and instinct in dealing with complex production problems that include organizational aspects. As a result, the creation of a competitive manufacturing enterprise—one that involves advanced automation technology, based on sensors and control systems, and organizational aspects—is a task that requires an understanding of both the techniques of establishing an automation program and the methods of integrating it within a dynamic organization.

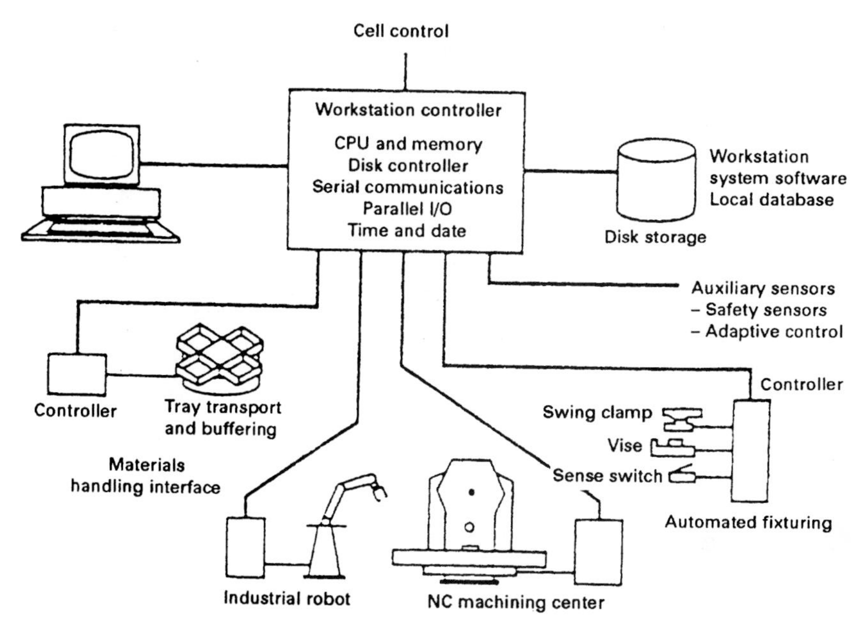

Figure 8.1 Computer-controlled manufacturing system.

The automated manufacturing system has to be built from compatible and intelligent subsystems. Ideally, each manufacturing system has to be computer-controlled and has to communicate with controllers and material handling systems at high levels of the hierarchy (Fig. 8.1).

8.4 Understanding Workstations, Work Cells, and Work Centers

Workstations, work cells, and work centers represent a coordinated cluster in a production system. A production machine with several processes is considered a workstation. A machine tool is also considered a work station. Integrated workstations form a work cell. Several complementary workstations may be grouped together to construct a work cell. Similarly, integrated work cells may form a work center. This structure is the basic concept in modeling a flexible manufacturing system. The flexible manufacturing system is also the cornerstone of the computer-integrated manufacturing strategy (Fig. 8.2).

The goal is to provide the management and project development team with an overview of major tasks to be solved during the planning, design, implementation, and operation phases of computer-integrated machining, inspection, and assembly systems. Financial and

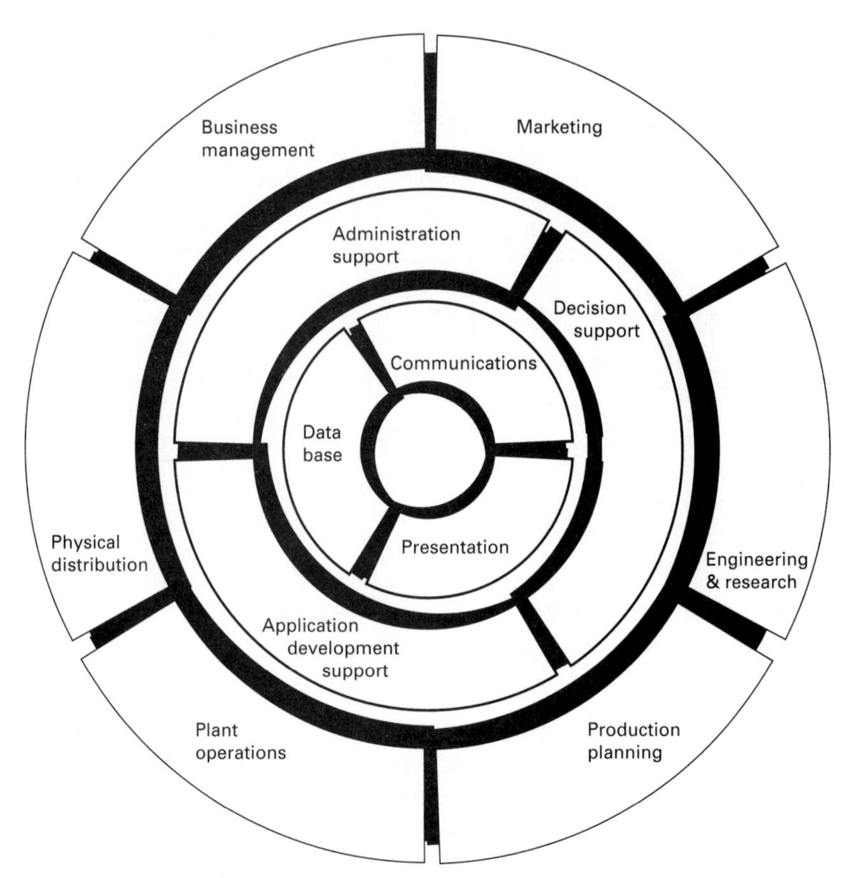

Figure 8.2 Major functional areas of manufacturing.

technical disasters can be avoided if a clear understanding of the role of sensors and control systems in the computer-integrated manufacturing strategy is asserted.

Affordable automation, sensors, and control systems provide the means of integrating different inputs to create the expected outputs. Inputs may be raw material and/or data which have to be processed by various auxiliary components of a system, such as tools, fixtures, and clamping devices. Sensors will provide the feedback data to describe the status of each input. The outputs may also be data and/or materials which can be processed on further cells of the manufacturing system. The flexible manufacturing system, which contains workstations, work cells, and work centers—all equipped with appropriate sensors and control systems—constitutes a distributed management information system, linking together subsystems of machining, packaging,

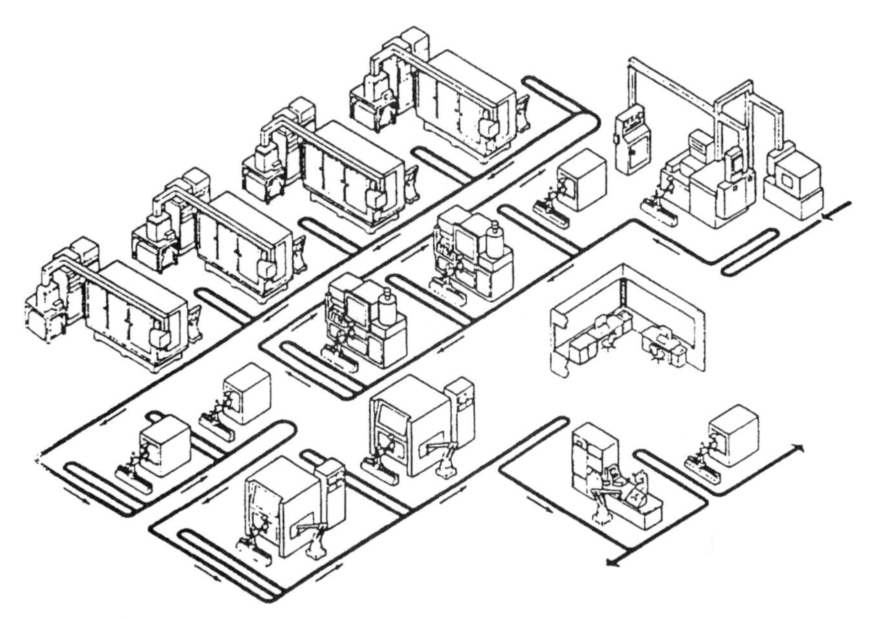

Figure 8.3 Workstation, work cell, and work center.

welding, painting, flame cutting, sheet-metal manufacturing, inspection, and assembly with material handling and storage processes.

In design of a flexible manufacturing system according to the computer-integrated manufacturing strategy, the basic strategy is to create a variety of sensors interconnecting different material handling systems, such as robots, automated guided vehicle systems, conveyers, and pallet loading and unloading carts, so that they communicate with data processing networks for successful integration with the system.

Figure 8.3 shows a cell consisting of several automated workstations with their inputs and outputs, with facilities for storage of workpieces, linking to materials handling systems of other cells, and providing data communication to the control system.

The data processing links enable communication with the databases (containing part programs, inspection programs, robot programs, packaging program, and machining data) and sensors sending data for real-time control. The data processing links also carry feedback data to the upper level of control hierarchy. Accordingly, the entire work cell facility is provided with data for immediate analysis and for real-time fault recovery.

A cluster of manufacturing cells grouped together for particular production operations is called a *work center.* Various work centers can be linked together via satellite communication links irrespective of the location of each center. Manufacturing centers therefore can be located several hundred feet apart or several thousand miles apart. Adequate sensors and control systems, together with communication links, will provide practical real-time data for analysis.

The output of the cell is not only a finished or semifinished part, but also data in a computer-readable format. The information will provide instructions to the next cell so that it can achieve its output requirements, through the distributed communication networks. If, for example, a part is required to be surfaced to a specific datum in a particular cell, sensors will be adjusted to read the required datum during the surfacing process. Once successfully completed, the part must be transferred to another cell for further machining or inspection processes, accompanied by the pertinent data, in a computer-readable format. The next cell is not necessarily physically adjacent.

8.4.1 Integration of automated assembly and fabrication systems

Sensor and control technology can achieve impressive results if effectively integrated with corporate manufacturing strategy. The following benefits can be achieved:

1. *Productivity.* Greater output and lower unit cost.
2. *Quality.* Product is more uniform and consistent.
3. *Production reliability.* The intelligent, self-correcting sensory and feedback system increases the overall reliability of production.
4. *Shorter lead time.* Parts can be randomly produced in batches of only one or in reasonably high numbers and the lead time can be reduced by 50 to 75 percent.
5. *Lower expenses.* Expenses are lower because the overall capital expenses are 5 to 10 percent lower. The cost of integrating sensors and feedback control systems is less than the cost of stand-alone sensors and feedback systems.

8.4.2 Integration of sensory and control systems with flexible manufacturing systems

This vital integration is the only available technology to date that yields utilization of a machine tool as high as 85 percent, as Table 8.1 indicates. The time spent in useful work in stand-alone machines without integrated sensory and control systems is as little 1.5 percent, as Table 8.2 shows.

To achieve the impressive results indicated in Table 8.1, the integrated manufacturing system must maintain a high degree of flexibil-

TABLE 8.1 Time Utilization of an Integrated Manufacturing Center with Sensor and Control System

	Active	Idle
Tool positioning and tool changing		25%
Machining process	5%	
Loading and inspection	15%	
Maintenance	20%	
Setup	15%	
Idle time		15%
Total	85%	15%

TABLE 8.2 Time Utilization of a Stand-alone Manufacturing Center without Sensory and Control Systems

	Active	Idle
Machine tool in wait mode		35%
Labor control		35%
Support services		15%
Machining process	15%	
Total	15%	85%

ity. If any cell breaks down for any reason, the production planning and control system can reroute and reschedule production; in other words, it can reassign the system environment. This can be achieved only if not just the processes but also the routing of parts is programmable. The sensory and control systems will provide an instantaneous description of the status of the part to the production and planning system.

If different processes are rigidly integrated into a special-purpose, highly productive system such as a transfer line for large batch production, then neither modular development nor flexible operation is possible. However, if the cells and their communication links to the outside world are programmable, much useful feedback data may be gained. Data on tool life, dimensions of machined surfaces measured by in-process gauging, production control, and fault recovery can enable the manufacturing system to increase its own productivity and to learn its own limits and inform the part programmers of them. The data may also be very useful to the flexible manufacturing system designers for further analysis. In non-real-time systems, such data cannot usually be collected, except by manual methods which are time-consuming and unreliable.

8.5 Classification of Control Processes

An engineering integrated system can be defined as a machine responsible for a certain production output, a controller to execute certain commands, and sensors to convey the status of the production processes. The machine may be a computer numerical control (CNC) machine tool, a packaging machine, or a high-speed press, for example. The controller contains certain commands arranged in a specific sequence designed for particular operations. The controller sends its commands in the form of signals, usually electric pulses. The machine is equipped with various devices, such as solenoid valves and step motors, designed to receive the signals and respond according to their functions. The sensors provide a clear description of the status of the machine performance. They give detailed accounts of every process in the production operation. Figure 8.4 shows the role of sensors and control logic in directing the entire manufacturing operation in a machine cell.

Once each process is executed successfully, according to a specific sequence of operation, the controller can begin sending additional commands to execute further processes until all processes are executed. This completes one cycle. At the end of each cycle a command is sent to begin a new loop until the production demand is met.

In an automatic process a distinction is made among three components: the *machine,* the *controller,* and the *sensors.* These three components interact with one another to exchange information. Mainly, there is interaction between the controller and the rest of the system through either an *open-loop control system* or a *closed-loop control system.*

An open-loop control system (Fig. 8.5) can be defined as a system in which there is no feedback. Motor motion is expected to faithfully follow the input command. Stepping motors are an example of open-loop control.

A closed-loop control system (Fig. 8.6) can be defined as a system in which the output is compared to the command, with the result being used to force the output to follow the command. Servo systems are an example of closed-loop control.

8.5.1 Open- and closed-loop control systems

In an open-loop control system, the actual value (mass flow in Fig. 8.5) differs from the reference value (input variable). In a closed-loop system, the actual value is constantly compared to the reference value (Fig. 8.6).

In the mass flow example in Fig. 8.7, the amount of matter flowing through a pipeline in a specific unit of time must be regulated. The current flow rate can be recorded by a measuring device, and a correcting device such as a valve actuator can be used to set a specific

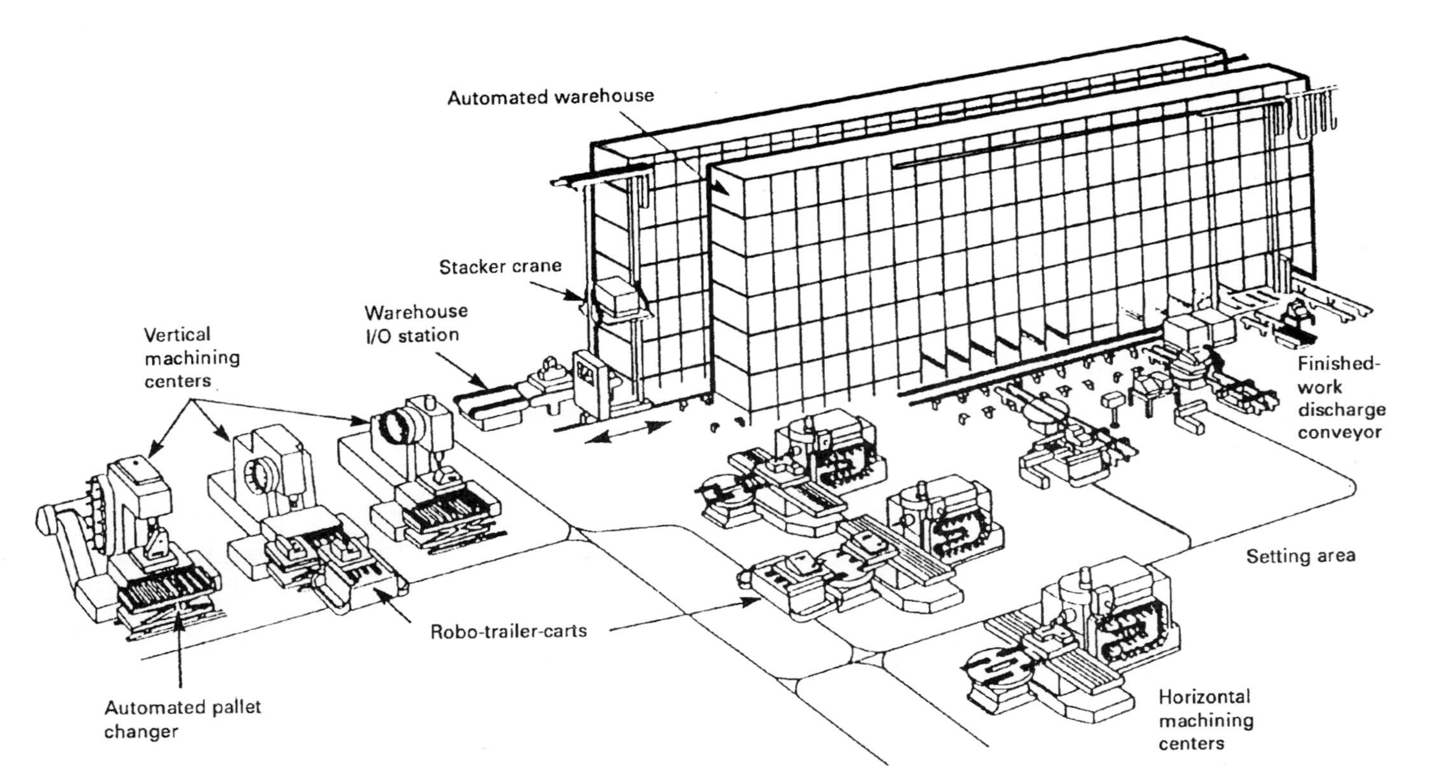

Figure 8.4 The importance of sensors and control logic in manufacturing cells.

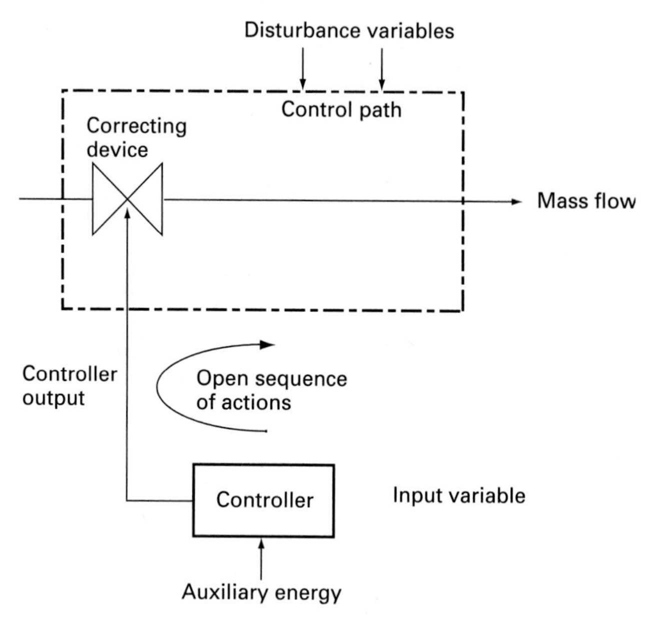

Figure 8.5 An open-loop control system.

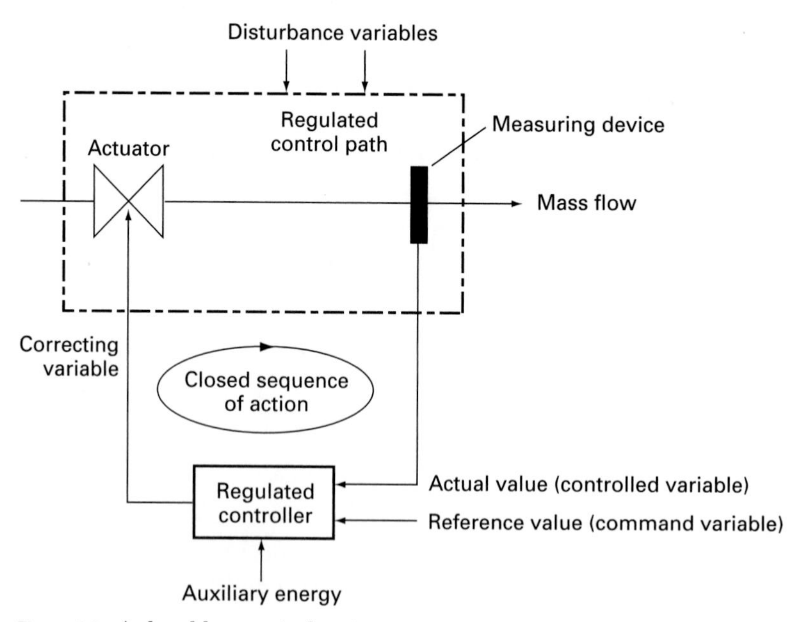

Figure 8.6 A closed-loop control system.

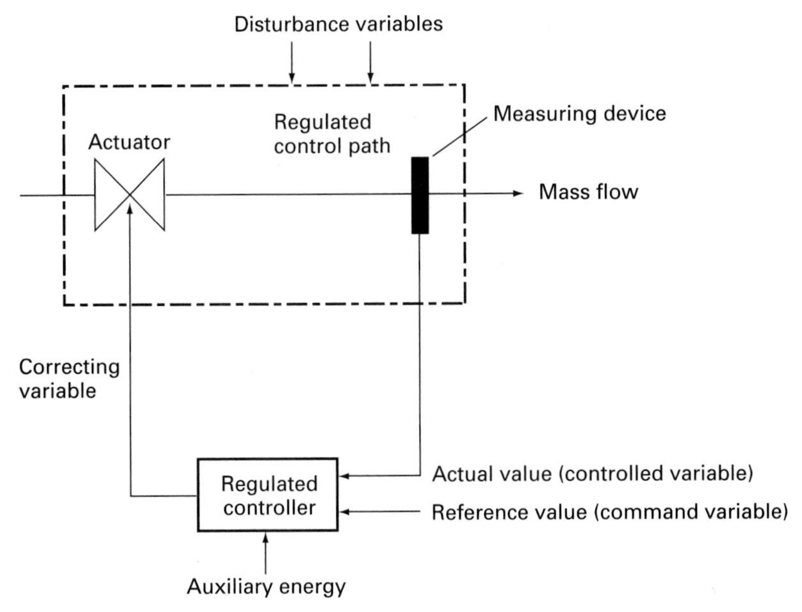

Figure 8.7 Regulation of mass flow.

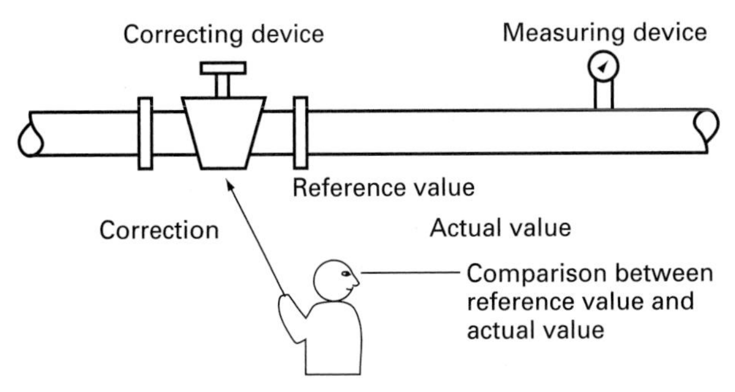

Figure 8.8 Model of closed-loop correction.

flow. The system, if left on its own, may suffer fluctuations and disturbances which will change the flow rate. The reading of the current flow rate is the *actual value.* The *reference value* is the desired value of the flow rate. If the reference value differs from the actual value and remains unaltered, the system is considered an open-loop system. If the flow rate falls below the reference value due to a drop in pressure, as illustrated in Fig. 8.8, the valve correcting device must be opened further to maintain the desired actual value. Where disturbances occur, the actual value must be continuously observed. Therefore, adjustment is made to continuously regulate the actual value. The loop of measurement, comparison, adjustment, and reaction within the process is called a *closed loop.*

8.6 Sensors and Their Environment

Understanding the environment in which sensors must operate is important to effective implementation of error-free production. Awareness of the characteristics of photoelectric controls, for example, and the different ways in which they can be used will establish a strong foundation for an error-free manufacturing environment. This understanding also will allow the user to picture the condition of each manufacturing process in the production environment.

Table 8.3 highlights key questions the user must consider.

8.7 Automated Cartesian Workstation

The unique design concept of the automated cartesian workstation (Fig. 8.9) provides outstanding productivity, speed, and economy for a variety of automation applications such as pick-and-place operations and probing. The overhead gantry form offers efficient use of *X-Y-Z* space and high total system performance in terms of accuracy, rigidity, speed, and low settling time. The automated cartesian workstation robotic positioner is functionally expandable through multitasking and contouring software. As a result, it can be effectively applied to other tasks as well, such as automated screw insertion, precision dispensing, and vision inspection applications.

The automated cartesian workstation is a three-axis robotic positioning system integrated with a controller-motor. The positioning ac-

TABLE 8.3 Key Characteristics of Sensors

Key point	Consideration
1. Range	How far is the distance to the object to be detected?
2. Environment	How dirty or dark is the environment?
3. Accessibility	What accessibility is there to both sides of the object to be detected?
4. Wiring	Is wiring possible to one or both sides of the object?
5. Size	How big or small is the object?
6. Consistency	Is the object consistent in size, shape, and reflectivity?
7. Requirements	What are the mechanical and electrical requirements?
8. Output signal	What kind of output is needed?
9. Logic functions	Are logic functions needed at the sensing point?
10. Integration	Is the system required to be integrated?

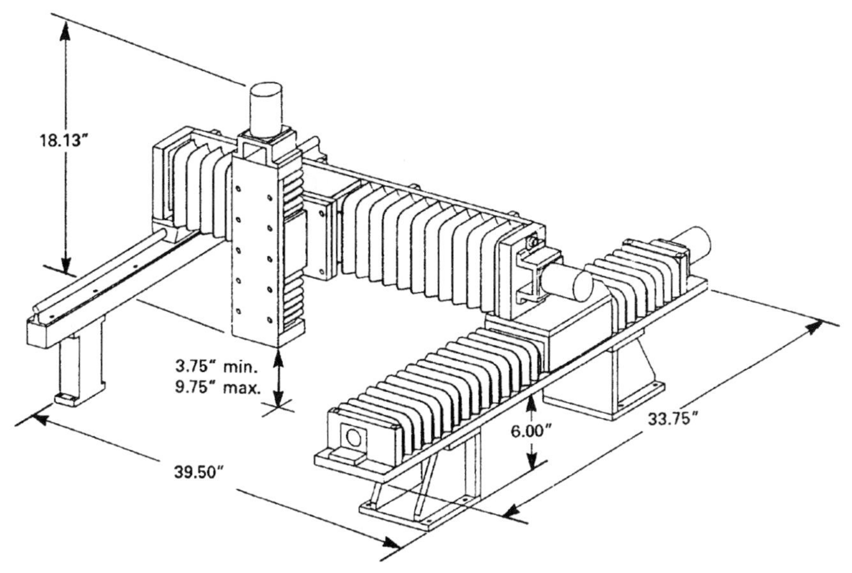

Figure 8.9 Automated cartesian workstation.

tuators are lead-screw-driven and use recirculating bearings to move a shuttle along a track. Bellows covers protect the lead screw and ways from contamination.

The primary machine-communication interface is an advanced position controller that features programming simplicity, a powerful three-axis microcomputer-based control, integrated motor drives, and power supplies. The memory can store anywhere from 2 to 2000 unique *X-Y* positions. Subprogram calls allow contour routines to be created and nested in main programs to save program space. Uncommitted input/output (I/O) allows motion commands to be synchronized with external events.

The automated cartesian workstation controller has adjustable ramping, selectable within a program, to optimize move times and acceleration/deceleration characteristics. A parabolic ramping algorithm adjusts step pulse timing to optimize acceleration, reduce move time, and maximize the utilization of available motor torque.

The system accommodates expansion cards that allow for motion controller I/O expansion up to 32 cards. Each card accepts up to 24 points. Any point can be used as an input or an output.

Continuous-path contouring software offers a constant vector speed on any two axes in the system. The software is useful for complex parts that contain straight lines and radii and where travel at constant speed is required in applications such as dispensing, welding, and laser precision cutting.

TABLE 8.4 Cartesian Robot Resolution

Accuracy grade	Motor/drive	Resolution, in
Commercial	Stepping, open loop	0.001, 0.0005
Commercial	Microstepping, open loop	0.001, 0.0005 0.0001, 0.00005
Commercial	Microstepping, closed loop	0.0001
Precision	Microstepping, open loop	0.001, 0.0005 0.0001, 0.00005
Precision	Microstepping, closed loop	0.0001
Precision	DC servo	0.0001

TABLE 8.5 Cartesian Workstation Characteristics

Work envelope	24×18×6 in	24×18×6 in
Rated payload	20 lb	10 lb
Velocity range:		
X-Y	0.20–12 in/s	0.02–15 in/s
Z	0.20–6 in/s	0.02–6 in/s
Position accuracy		
X-Y (TPR)*	0.005 in	0.0015 in
Z	± 0.0005 in/in	± 0.0002 in/in
Repeatability	± 0.001 in	± 0.0002 in
Controller	3-axis programmable with operator front panel containing thumbwheels, position displays, and push buttons	

*True position radius.

Multitasking software provides the capability to run multiple programs simultaneously and turn inputs/outputs on or off on the fly, based on position, velocity, or time.

Table 8.4 lists the characteristics of the various types of motor/drives available. Table 8.5 lists the characteristics of the cartesian work station.

8.7.1 Positioning system for cartesian workstation

The programmable linear positioning system for cartesian robots offers exceptional productivity and economy for the automation of a variety of applications such as pick and place, probing, scanning, fastening, and inspecting (Fig. 8.10).

The single-axis positioning system provides programmable electromechanical motion with I/O capability for travel to 24 in at speeds up

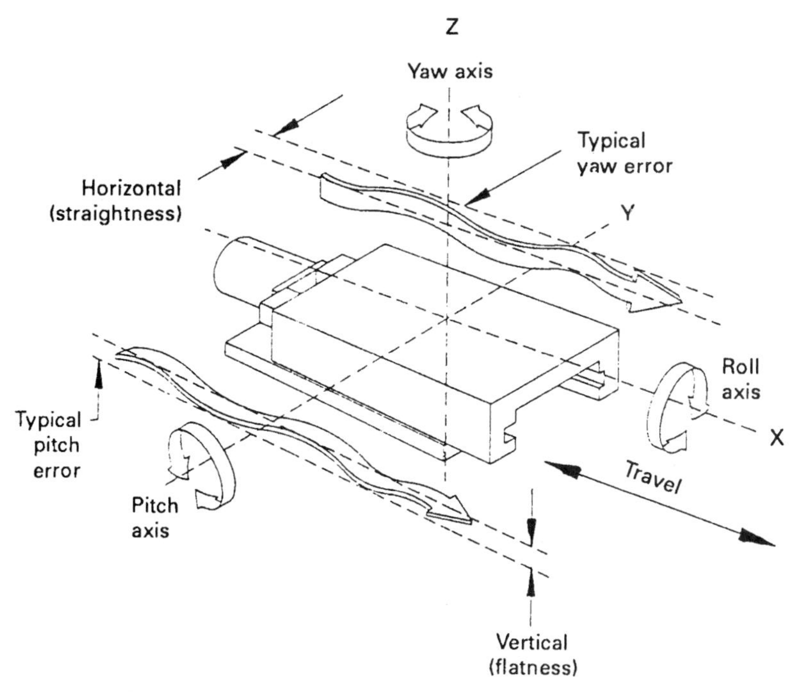

Figure 8.10 Programmable linear positioning system.

to 20 in/s. The automated slide is ideally suited for applications requiring repetitive, precision, and high-speed positioning at precisely controlled feed rates. The intelligent linear positioning system ia a superior alternative to systems based on less flexible pneumatic, hydraulic, cam, or electric-limit-switch linear actuators.

The positioning slide system includes a precision linear positioning table, a microprocessor-based motion controller, and a high-performance motor drive. The system offers a variety of selectable positions to suit a wide range of travel applications, resolution, speed, repeatability, and open- or closed-loop control requirements (Fig. 8.11).

The controller features battery-protected memory for stand-alone operation and an RS-232C communication port for real-time operation from a host computer or programmable controller. An integral programmable logic interface can be used to operate external solenoid valves, relays, switches, and indicators. Inputs can be used to trigger moves or to branch to different programs using conditional jump program commands.

The Motion Expert™ software provides solutions to motion equations to calculate velocity, acceleration, and move time for specified loads, saving valuable engineering time.

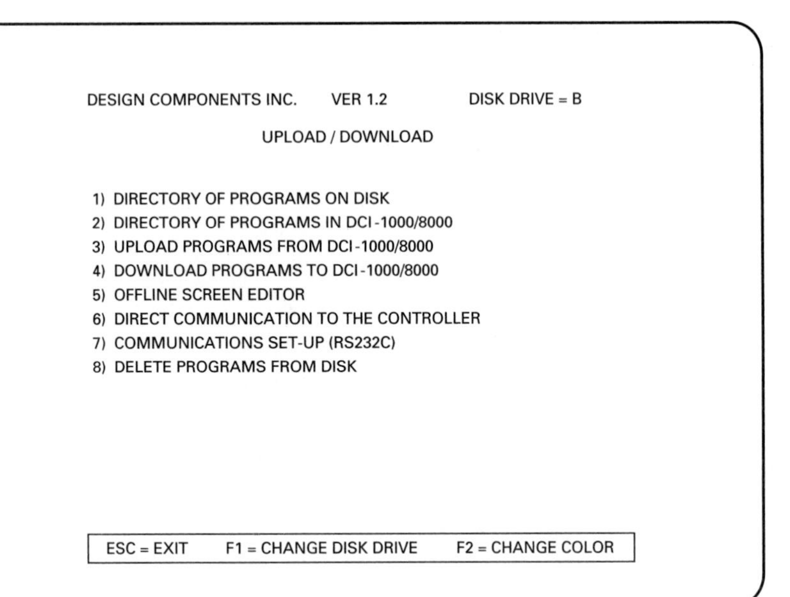

(a)

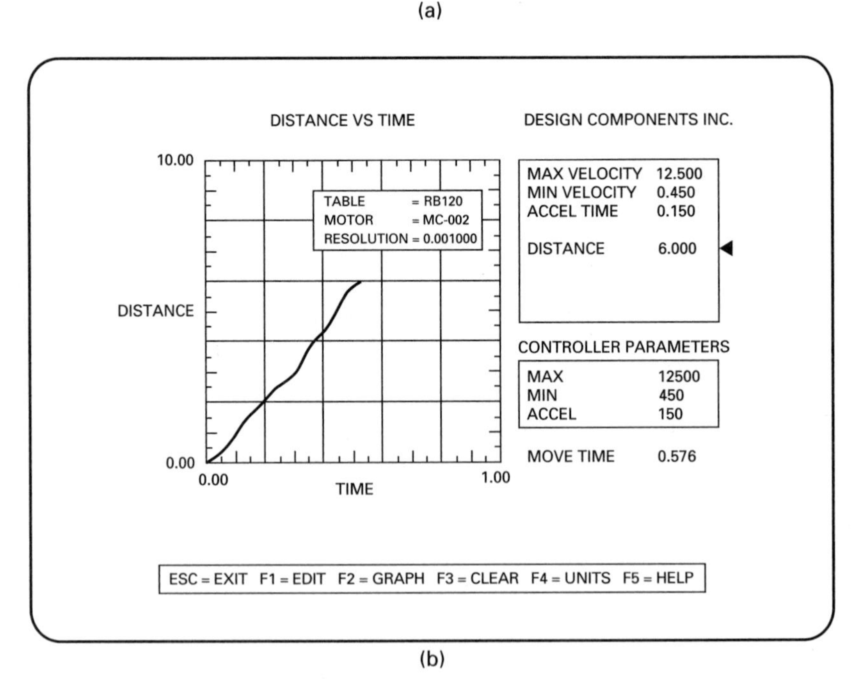

(b)

Figure 8.11 (*a*) Positioning table program menu, (*b*) distance traveled versus time of positioning table.

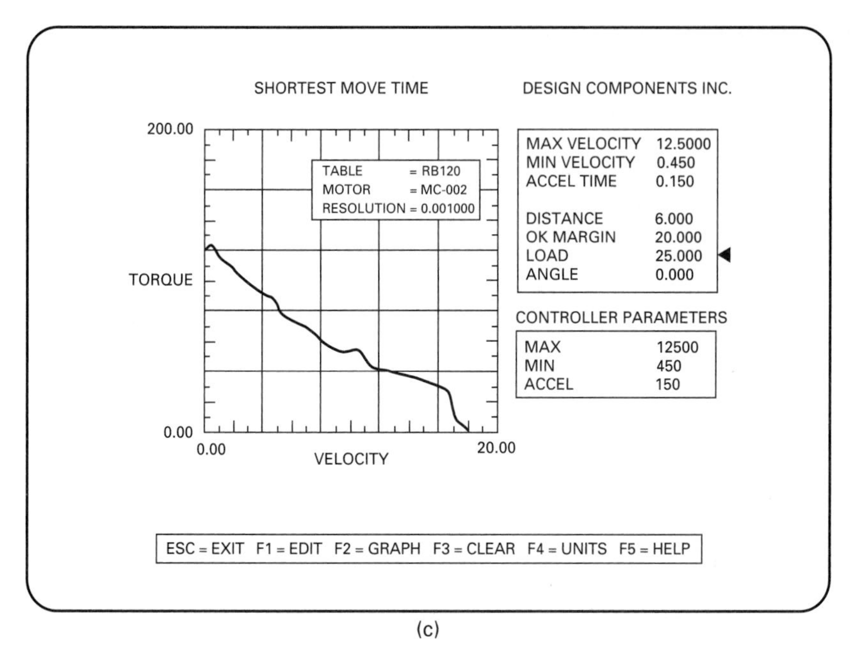

(c)

Figure 8.11 *(Continued)* (*c*) Torque versus velocity of positioning table.

The user may select an appropriate model to satisfy the length-of-travel requirements. The software requires a PC-compatible computer with 256K memory, 5¼-in disk drive, and an IBM-compatible CGA, EGA, or VGA graphics card.

The software can be uploaded/downloaded to enable the creation and editing of motion control programs off-line and setting up communication with the motion controller. The standard software allows the personal computer to store motion control programs in the computer's memory and transfer them to the controller's battery-backed random-access memory (RAM) and vice versa. A thumbwheel switch can be provided to allow simple entry of distance, feed rate, program calls, and/or other application variables when a program is running. Optional I/O cards offer uncommitted I/O expansion, up to 32 points.

The positioning system has the following features:

1. Payload is 200 lb maximum
2. Moment loading is 25 ft · lb
3. Solid-state overtravel limit provided by home switches
4. Optically isolated I/O
5. Ball screw drives

6. Protective bellows
7. Storage capacity for 100 programs
8. Automatic homing
9. RS-232C communications
10. 10-ft motor limit-switch cable
11. 30 Kbyte memory
12. Teach mode or keyboard programmability

8.7.2 Linear accuracy of cartesian robot

Linear motion. A positioning table in motion has six degrees of freedom along or about the three perpendicular coordinate axes. To produce the highest accuracy in any given axis of linear motion, the error in the remaining five degrees of freedom needs to be controlled and minimized.

Linear error. The linear translation error is represented by a horizontal or vertical deviation from the theoretical, true straight line of travel (Fig. 8.12).

Pitch, yaw, and roll. A rotational or angular error perpendicular to the axis of travel (pitch and yaw) and about the axis of travel (roll) is also present in all positioning tables (Fig. 8.12).

Straight-line accuracy. Straight-line accuracy is defined as the total deviation from the theoretical perfect line travel measured as displacement in the horizontal and vertical planes.

Horizontal displacement or straightness is the sum of the linear translation, and angular yaw and roll error. *Vertical displacement* or flatness is the sum of the linear translation, and angular pitch and roll error.

Orthogonality. Orthogonality is the perpendicularity of travel of one positioning table with respect to another in either the *X-Y* or *X-Z* planes. Error is measured as angular displacement from a perfect right angle.

Precision perpendicularity is achieved through the alignment process and the utilization of precision measuring tools and procedures.

Position accuracy. This is defined as the maximum error between the specified or commanded position along the axis of travel and the actual position (Fig. 8.13).

Repeatability. This is the extent to which a move to a defined, specific position will repeat itself, from one direction (unidirectional) or two directions (bidirectional). Positioning accuracy and repeatability are

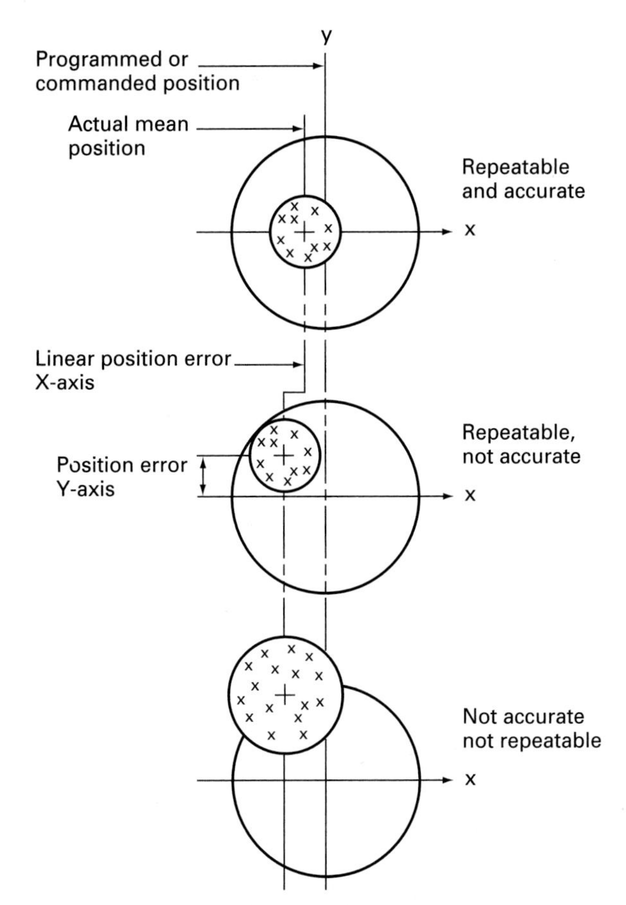

Figure 8.12 Linear and rotational error.

most affected by the pitch variation and backlash of the drive train (lead screw, gear train, or pulley). Other factors such as thrust bearing quality, rigidity, and assembly methods affect structural compliance or lost motion and can decrease accuracy (Fig. 8.13).

Absolute positioning accuracy. This is required when the specified or commanded distance is the basis of a point-to-point measurement. To achieve absolute accuracy of positioning, a precision external linear encoder or interferometer feedback is required. Abbé error offset must be addressed when linear encoders are used and the point of measurement is offset from the mounting position of the encoder. High-precision tables are recommended to reduce the position error due to imperfect straight-line accuracy and orthogonality.

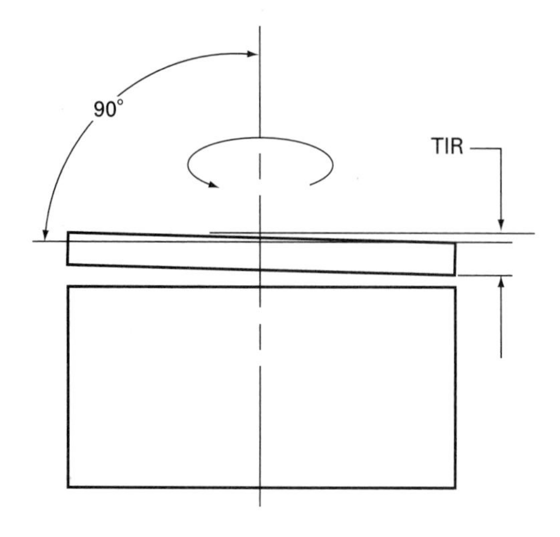

Figure 8.13 Position accuracy.

For most automation applications, the ability to repeat a programmed move within a defined error tolerance is usually most important and sufficient to achieve design results. The cost to achieve absolute position accuracy can be significant and should be avoided unless necessary. The cost to achieve a high degree of repeatability is relatively less expensive.

Resolution. This is the smallest increment of motion or adjustment of position that is attainable. The positioning table or actuator resolution is defined by the drive train. System resolution is a function of a motor/encoder and linear actuator resolution when no linear encoder is employed.

It must be noted that fine microstepping requires very low levels of incremental torque. If the frictional resistance of the actuator exceeds the available incremental torque of the motor, motor resolution will not be translated to system resolution. Under this condition position accuracy and repeatability will be adversely affected.

8.7.3 Rotary accuracy of cartesian robot

The accuracy grade of a rotary motion table is based on the degree of error due to the eccentricity of the center of rotation and the angular deviation of the plane of the work face.

Rotary motion precision is a function of the quality and stiffness of the bearings, gear accuracy, and preciseness of manufacturing and assembly of the components and mating parts.

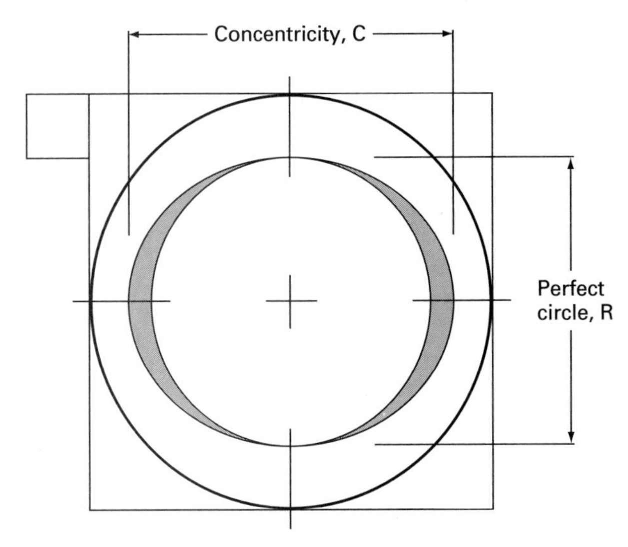

Figure 8.14 Eccentricity.

Concentricity. This is the maximum deviation between a perfect surface and the actual path of a point on a table's work surface as it travels a full 360° (Fig. 8.14).

Position accuracy. This is defined as the maximum error between the specified or commanded position and the actual position, measured as an angular deviation.

Repeatability. This is the extent to which a move to a defined, specific position will repeat itself, from one direction (unidirectional) or two directions (bidirectional).

Runout. This is the maximum distance a point on the table work surface will deviate from a perfect flat plane which is perpendicular to the axis of rotation as it rotates through 360° (Fig. 8.15).

Loading. Load capacity is the maximum load a table can carry without permanent deformation of the top and base sections of the table or the bearing elements. Load capacity values assume a unit load, central to the work surface, with the center of gravity a maximum of 2 in above the table top.

Moment loads. Cantilever, or moment, loads increase bearing element stress and deflection of supporting members. The increased stresses and misalignment of the slide/drive train assembly will affect operating life. Maximum moment loads for tables are based on the

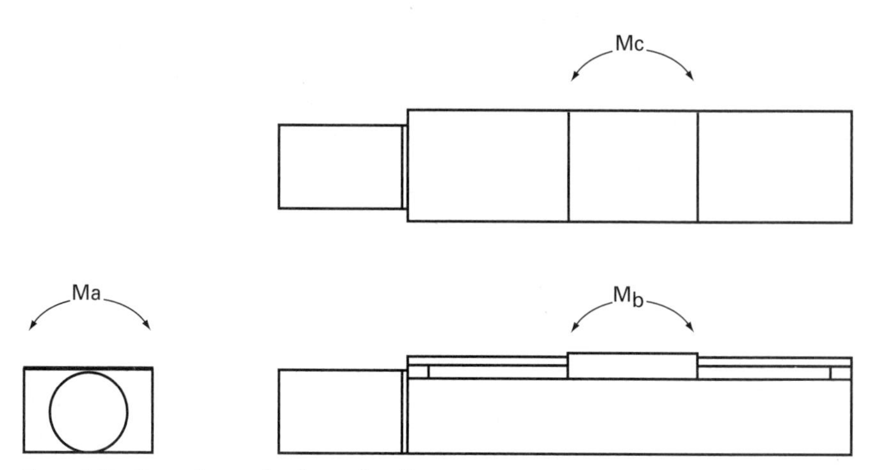

Figure 8.15 Runout error in plane of motion.

linear bearing rating and a conservative allowable deflection of supporting members (Fig. 8.16).

Life. The expected life of a linear bearing is predictable in terms of dynamic load capacity. That capacity is the load at which 90 percent of a sample bearing will endure 1,000,000 in of travel without loss of geometric integrity (i.e., surface wear or fretting).

Crossed-roller-bearing tables have higher load capacities than ball-bearing tables because of the larger contact area between bearing elements. They are also less susceptible to damage from shock and, for a given load, offer longer life.

If the load on the table is other than the rated dynamic capacity, the expected life of the bearing ways will vary for ball or rollers assemblies as follows.

For ball-bearing ways:

$$L_i = 10^6 \, (\mathrm{SDC}/L_o)^3 \tag{8.1}$$

For crossed-roller ways:

$$L_i = 10^6 \, (\mathrm{SDC}/L_o)^4 \tag{8.2}$$

where L_i = expected life
SDC = safe dynamic capacity
L_o = applied load

As seen in Eqs. (8.1) and (8.2), the life of crossed roller bearings

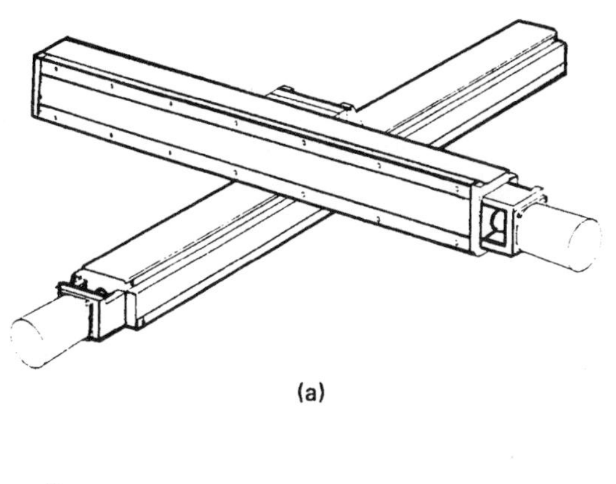

(a)

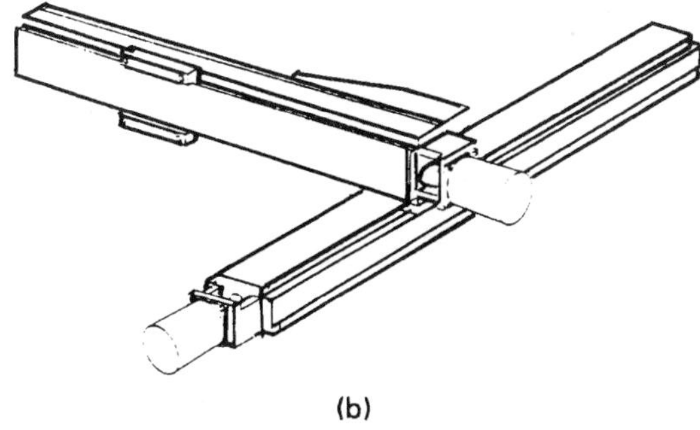

(b)

Figure 8.16 Moment loads: (*a*) Optional top-to-top upright (moving actuator), (*b*) standard size mounting.

varies as $(1/L_o)^4$ whereas ball bearings vary as $(1/L_o)^3$. This makes rollers more suited for use near their stated capacity or for very long duty cycles.

Under any load, the life of a slide mechanism will vary, depending on the operating conditions it encounters. In calculating expected life of a linear bearing table an adjustment factor f_s for imperfect operating conditions applies:

$$\text{Safe dynamic capacity} = \text{dynamic capacity}/f_s$$

Representative values of f_s are listed below.

Operating condition	f_s
Smooth, shock-free	1
Gradually varying loads	2
Shock, vibration	4

8.7.4 Robotic actuators

Robotic actuators are designed to provide accurate, reliable position repeatability in a lightweight, stiff design. The stiff slider design is ideal for cartesian robotic extended-arm configurations for overconveyers and automated assembly applications including pick and place, dispensing, fastening, and testing. They have the following features:

1. Precise, smooth motion
2. Light weight, rigid construction
3. Broad range of resolution and speed
4. Repeatability to 0.0002 in
5. Payload to 100 lb

The actuators are available in standard side mounting, top-to-top mounting, and other mounting configurations (Fig. 8.17).

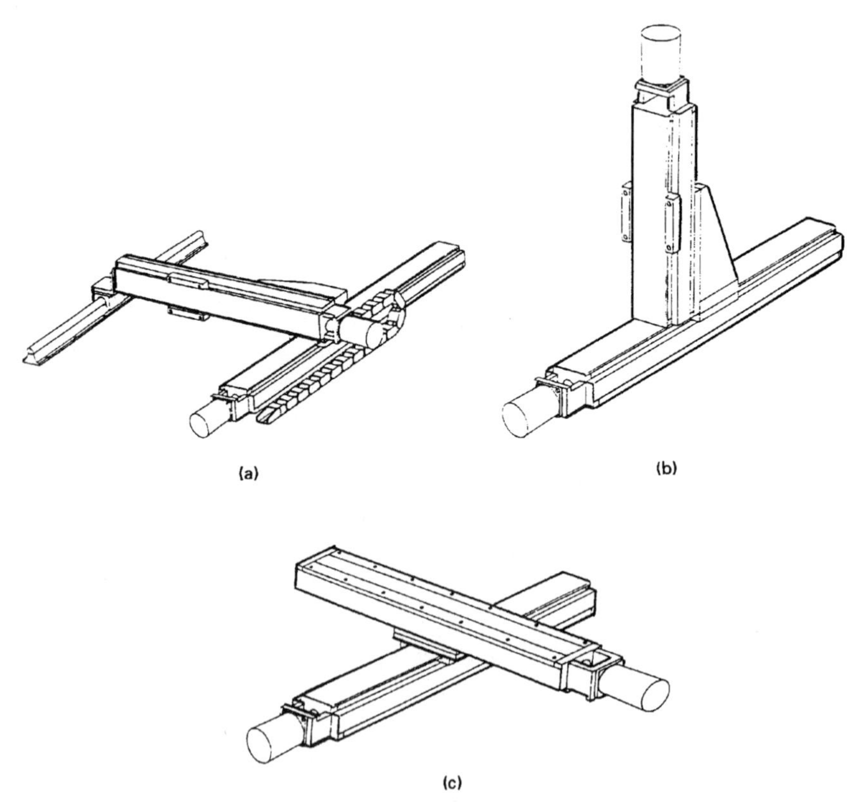

Figure 8.17 Mounting arrangements for a linear actuator: (*a*) Standard side mount with optional cable track and outboard support, (*b*) optional x–z mounting, (*c*) standard top-to-top (moving actuator).

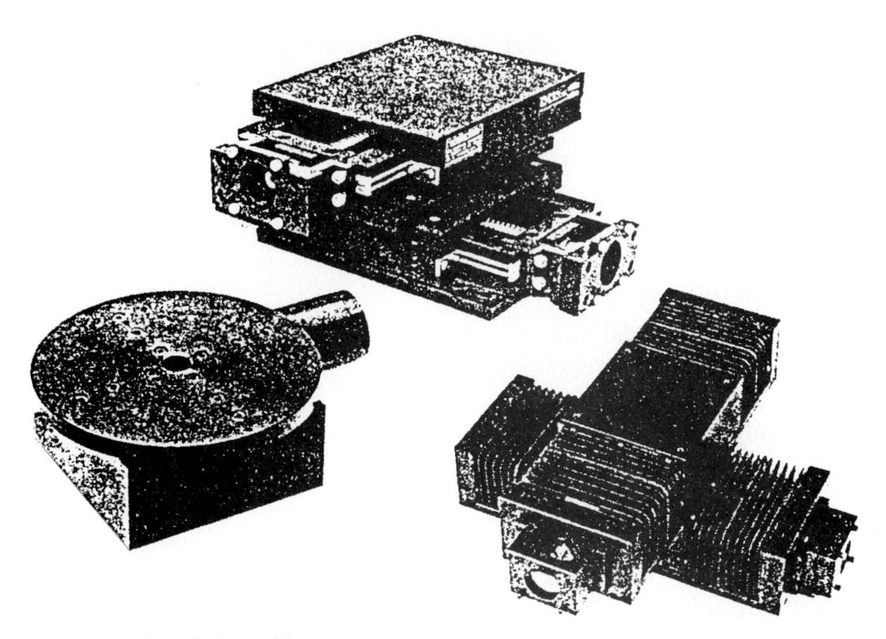

Figure 8.18 Positioning tables.

8.7.5 Positioning tables

Positioning tables (Fig. 8.18) are available in basic sizes and shapes to accommodate a range of applications in a single axis, *X-Y, X-Y-Z, X-Y-θ*, and *X-Y-Z-θ*. They feature:

1. Linear nonrecirculating ball or crossed roller bearings
2. Linear recirculating ball bearings with bellows
3. Rotary angular contact ball bearings
4. Lead screw or ball screw drives in a range of resolutions

Nonrecirculating ball or crossed roller bearings offer the smoothest and most economical precision positioning for applications with:

- Light to medium load, central and uniform
- Short travel (0 to 12 in)
- Smooth, low-vibration motion
- Linear speed of 0.15 in/s
- Less than ⅓ *g* acceleration

Recirculating bearings allow full support of the load over the entire travel and the addition of bellows covers. These tables are recommended for applications requiring:

- Precision and accuracy
- Medium to heavy loads
- Moment loading
- Long travel
- Linear speeds to 30 in/s
- Acceleration to 1g

Preloaded, rotary angular-contact ball bearings with fine-pitch worm gear drive provide precise angular positioning for:

- Commercial accuracy
- Central loads to 200 lb
- Low initial loads

8.8 Conversion of Motion for Affordable Automation

The motion of an object may be rotary or linear in nature. The components used to implement motion usually consist of an actuator (e.g., motor) coupled to the physical object by a mechanical transmission. This transmission is used to direct the actuator motion to the object in order to change rotational direction, change axis, multiply or reduce torque, or multiply or reduce speed. The transmission may be used to convert rotary motion to linear motion or to provide a match between the actuator and load so that the maximum energy is transferred to the load. Some devices produce linear relationships between input and output motions and thus are well-suited for use in closed-loop control. However, some of the more traditional mechanical schemes have highly linear input/output relationship, which can create severe control problems.

The word *transmission* has been used to define all the components from the actuator to the physical object. These components may be shaft couplers, multiple sets of gears, or lead screws, to name a few. Although it is possible to implement a motion by direct drive (i.e., actuator and jointed object are one component with no transmission), it may not be practical for cost or size reasons. One commercially available system using a direct-drive approach is a robot developed by ADEPT Corporation.[1]

[1]ADEPT Inc., San Jose, Calif., U.S.A.

TABLE 8.6 Mechanical and Electrical Analogies

Translational	Rotational	Electrical
Force F	Torque T	Current i
Velocity V	Angular velocity ω	Voltage v
Mass M	Moment of inertia J	Capacitance C
Linear spring constant K	Torsion spring constant K	Reciprocal inductance $1/L$
Linear viscous B	Rotational viscous B	Conductance $1/R = G$
Friction constant B	Friction constant B	

Models of mechanical systems express mathematically the relationships between quantities such as input torque to a system and the physical position of various components. The combination of elementary mechanical components such as lumped inertias or springs results in a *mechanical network* which may be analyzed by differential equations, Laplace transformations, or simulation techniques. In addition, electrical analogs of the variables and parameters of the mechanical network can be formed. These analogous circuits can provide a very simple and cost-effective method of physically modeling a mechanical system without the need to fabricate mechanical components or directly solve the differential equations. If a test signal (such as a sine wave or step function) is applied to the analogous electric circuit, the response of the mechanical network may be predicted; also, if the values of the electrical components are varied, the effects of parametric changes in the mechanical system may be observed.

It has been established that translational and rotational mechanical quantities and electrical quantities are all analogous, as indicated in Table 8.6.

8.8.1 Rotary-to-rotary motion generation

Rotary-to-rotary generation of motion is associated with such components as gear trains, belts and pulleys, and sprockets and chains. These components will take the torque and speed provided by a rotary actuator connected to the input shaft and provide rotary motion on the output shaft. Depending on the type of component and its characteristics, the motion on the output shaft may be reversed, have an increased torque (with a reduced speed), or have an increased speed (with a reduced torque).

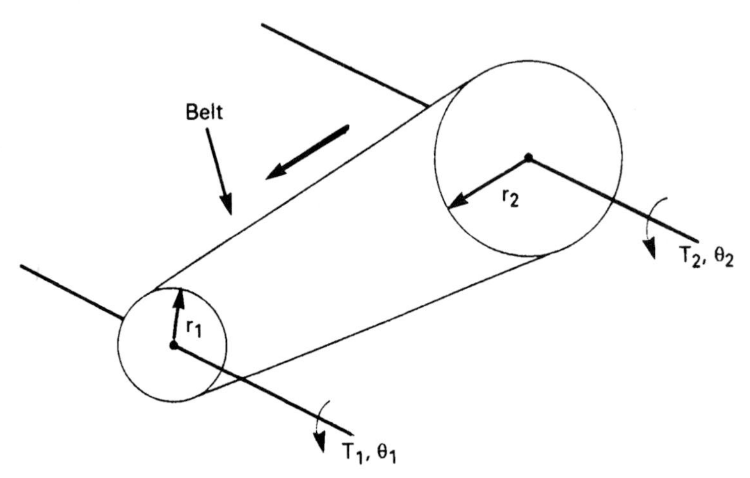

Figure 8.19 Model of belt-and-pulley system.

8.8.2 Belt-and-pulley system

Figure 8.19 can be used to model an ideal belt-and-pulley or chain-and-sprocket system. The analysis of the system is similar to that of a gear train. Note that in this case the relative motions of both shafts are in the same direction. Timing belts and chains perform essentially the same function as gear trains. However, they allow the transmission of the rotational motion and torque over longer distances without the need for multiple gears. Figure 8.20 shows a timing belt and toothed pulleys. Unlike a belt and pulley, this method will not allow slippage between the belt and the sprocket because it uses teeth; however, it will exhibit a spring constant if the output is locked and a torque applied to the input.

8.8.3 Rotary-to-linear motion generation

Rotary-to-linear motion generation takes the rotational motion and torque from an actuator and produces a linear motion and force on the output. Components whose ideal model produces a linear relationship between the input shaft rotation and the output linear displacement are the lead screw, rack and pinion, and belt and pulley. Nonlinear relationships are available with slider-crank assemblies and cams. This type of motion generation is important in manufacturing applications.

Lead screws. Figure 8.21 shows a lead screw driving a payload along a single axis. In this configuration, the screw is fixed with its ends free to rotate. As the screw is turned, the nut moves along a shaft

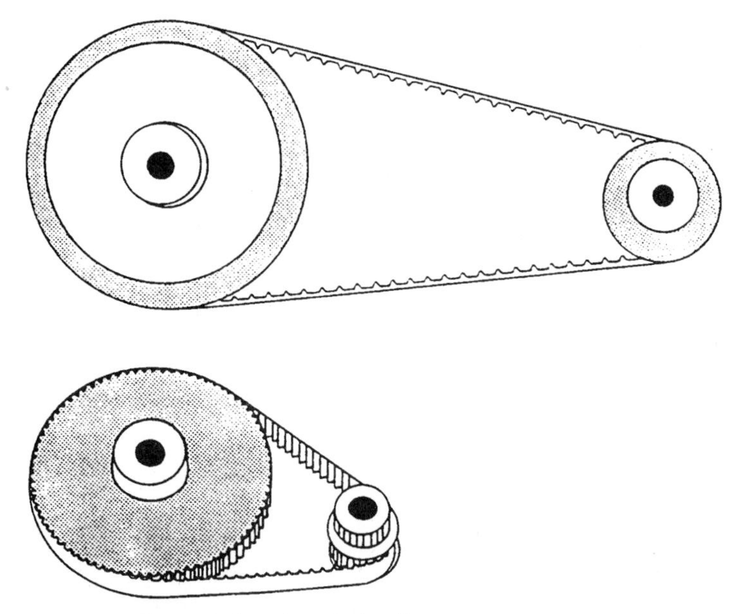

Figure 8.20 Timing belt and toothed pulleys.

with the payload attached. Typically, the payload is supported and moves along a surface with very low friction. A rotary displacement of the input shaft, q_1, causes a linear motion of the payload, χ. The coupling ratio is defined as the input shaft rotation per unit of linear motion. The typical rating of a lead screw is turns per millimeter. This is also referred to as the pitch *P.* An alternative coupling ratio is given by the reciprocal of the pitch, which is called the lead *L.* This factor is defined as the axial distance a screw or nut travels in one revolution and may be given in such units as millimeters per revolution or millimeters per turn. A lead screw having a pitch of 10 turns/mm has a lead of 0.1 mm/turn. Therefore, relationships can be expressed as

$$q = P \tag{8.3}$$

$$\chi = Lq \tag{8.4}$$

These equations may be differentiated three times to obtain the linear velocity, acceleration, and jerk and rotational velocity, acceleration, and jerk, respectively.

Just as in the case of rotary motion generation, we are interested in how a load on the output is seen by the input. That is, is there an equivalent torque-inertia system for Fig. 8.21? The derivation of the inertia reflected back to the input shaft is based on a balanced kinetic

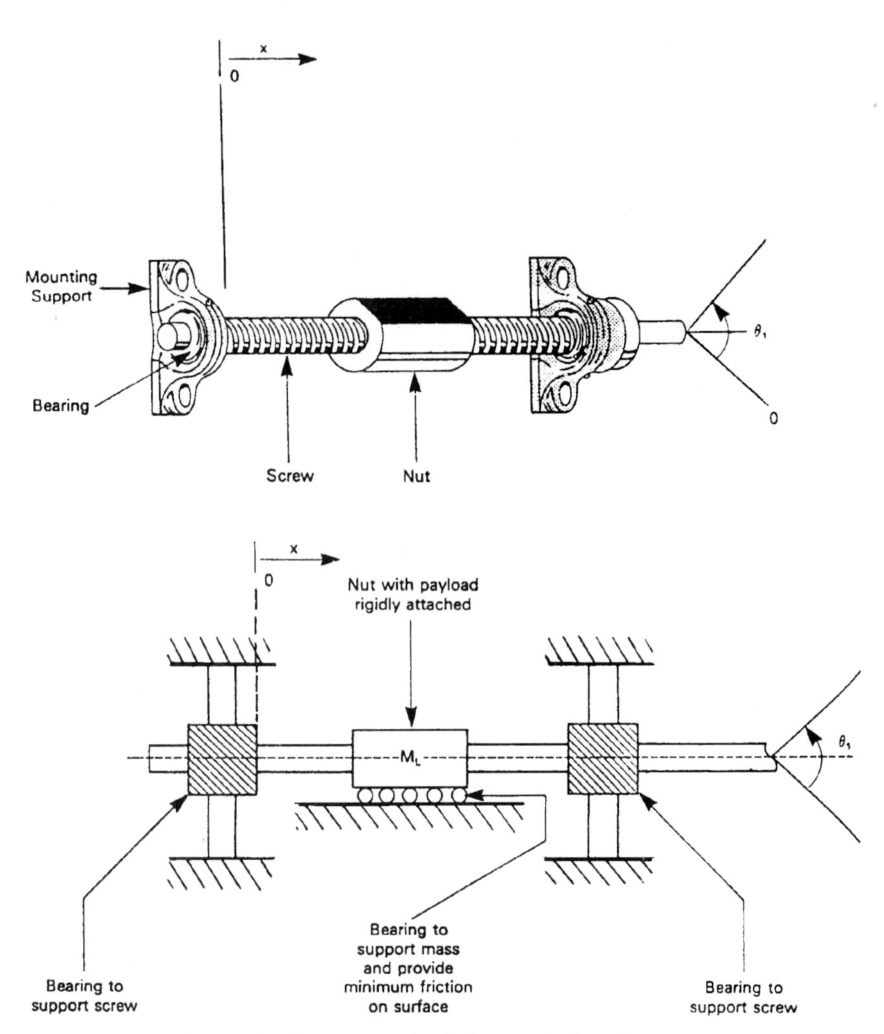

Figure 8.21 Lead screw for driving a payload along a single axis.

energy. For the linear motion of the payload's mass M_L, kinetic energy can be expressed as

$$E_k = \tfrac{1}{2} M_L v^2 \tag{8.5}$$

The corresponding kinetic energy of a torque-inertia system is defined as

$$E_k = \tfrac{1}{2} J_{eq} w^2 \tag{8.6}$$

Equating the kinetic energies of Eqs. (8.5) and (8.6) and solving for the

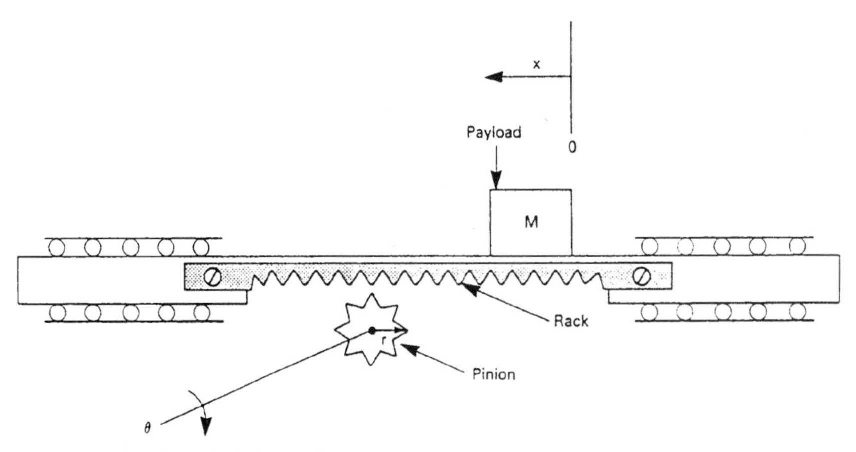

Figure 8.22 Rack-and-pinion train.

inertia J_{eq}, after relating rotary and linear velocities by a pitch, yields:

$$J_{eq} = \frac{M}{(2pP)^2} = \frac{W}{g/(2pP)^2} \tag{8.7}$$

Equation (8.7) shows that the reflected inertia of a load given by a load screw can be reduced by choosing a lead screw with a greater pitch.

Rack-and-pinion system. Figure 8.22 shows a representation of a rack-and-pinion train. The pinion is the small gear attached to the actuator, and the rack is a linear member with gear teeth on one side. The transfer relationship of such a mechanical system is

$$\_ = 2prq \tag{8.8}$$

As defined by Eq. (8.8), the linear distance traveled is proportional to the input shaft rotation (in revolutions), with the constant of proportionality equal to the circumference of the pinion. That is, the linear distance traveled is equal to the distance traveled by the pinion. The reflected inertia, as seen by the input shaft, can be shown to be

$$J = Mr^2 \tag{8.9}$$

Belt-and-pulley system. Another method of generating linear motion from rotary motion is shown Fig. 8.23. Note that both pulleys have the same radius. The pulley connected to the input is called the *driving pulley* and the second pulley is called the *idler pulley.* The distance traveled by the load is equal to the distance traveled by the drive pulley. Equation (8.8) describes the transfer relationship of the

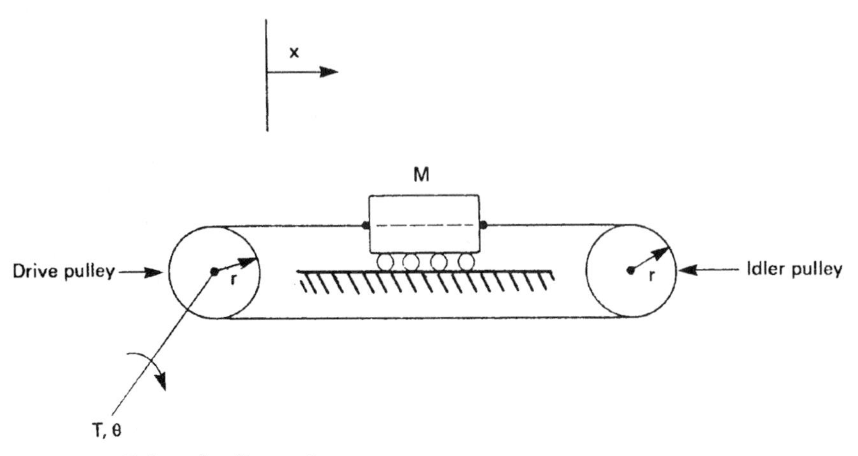

Figure 8.23 Belt-and pulley system.

belt and pulley system, while Eq. (8.9) is the reflected inertia as seen by the input shaft.

In practical manufacturing implementations of the lead screw, rack and pinion, and linear drive via belt and pulley, there are physical limitations on the linear motion. For example, if the linear motion exceeds its designated range, it typically hits an end stop. This mechanical contact forces the rotary shaft to lock. In the event that a torque continues to be applied to the shaft in this locked position, it is possible for extreme forces to be generated, causing the devices to fail (e.g., the threads of the screw or nut may strip, or gear teeth may break). The practical solution to this problem is to place an electrical limit sensor some distance before the physical limit. The signal from this sensor is used to stop the actuator before a damaging force can be applied at the extreme limit of motion.

Slider-crank system. The slider-crank mechanism is an extremely cost-effective means of generating linear motion from rotary motion. Figure 8.24 represents a crank driving a linear stage (or load). In this implementation, the crank portion is the wheel that rotates about its center and has a rod of fixed length mounted to a point on its circumference. The other end of the rod is attached to a linear stage which is constrained to move in only one dimension on a relatively frictionless surface. At its connections to the wheel and the linear stage, the rod is free to rotate. Thus, the angle formed with the horizontal will change as a function of the wheel's position.

As the disk travels from 0° to 180° in the counterclockwise direction, the linear stage moves a distance equal to *r*. If the disk continues to travel from 180° back to 0° (still in the counterclockwise direction),

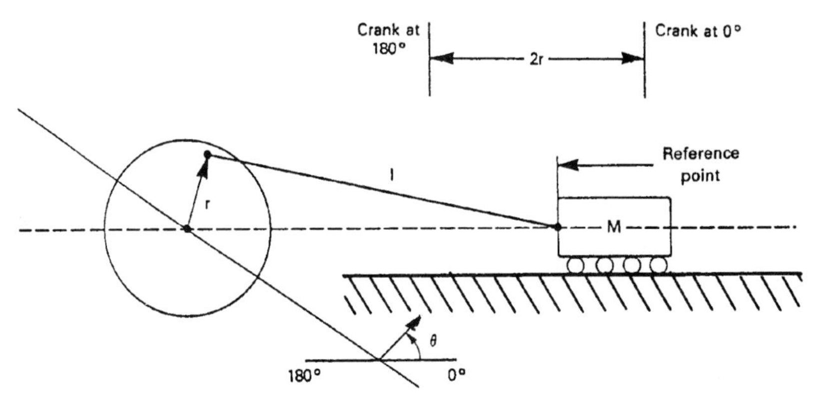

Figure 8.24 Slider-crank mechanism.

the load will move in the opposite direction over exactly the same linear distance. Note that there are two distinct angular positions of the input shaft that correspond to a single linear position of the load.

The angular positions that correspond to full extension and retraction of the linear stage are called *dead center positions* or *dead points*. As in a gasoline reciprocating engine, a flywheel with sufficient inertia must be connected to the crank's axis of rotation. The reason for this is that, when the connecting rod and crank are in line at a dead point, the direction of rotation may go either way. Thus, the inertia of the flywheel is necessary to bias the mechanism for proper operation.

If the input shaft is rotated continuously, the motion of linear stage is reciprocating. As can be observed by a careful examination of Fig. 8.24, the torque seen by the input and the relationship between the input shaft's position and linear stage's position are nonlinear in nature.

Cam system. Cams can also provide a nonlinear transformation of rotary motion to linear motion. The cam is an irregularly shaped body which is driven by some type of motor. Touching the surface of the cam is a follower bearing to which the cam surface imparts motion. Figure 8.25 shows a cam with a follower while Fig. 8.26 shows the relevant displacement diagram. For the system in Fig. 8.25, as the cam rotates, the follower touches its surface. Thus the rotary profile of the cam is imparted to the follower, which in turn causes a linear motion of the arm. The actual velocity, acceleration, and jerk of the linear motion can be defined by the shape of the cam.

It should be noted that cams are very common in automatic assembly and manufacturing machines because they provide a simple method of providing almost any desired follower motion. By using a simple limit sensor and motor, a cam may be driven one revolution

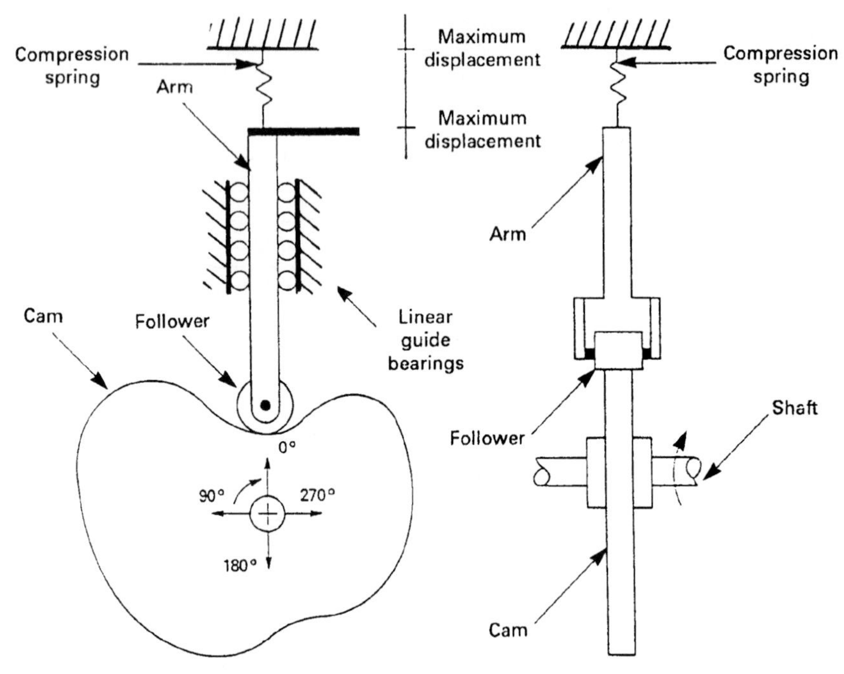

Figure 8.25 Cam and follower system.

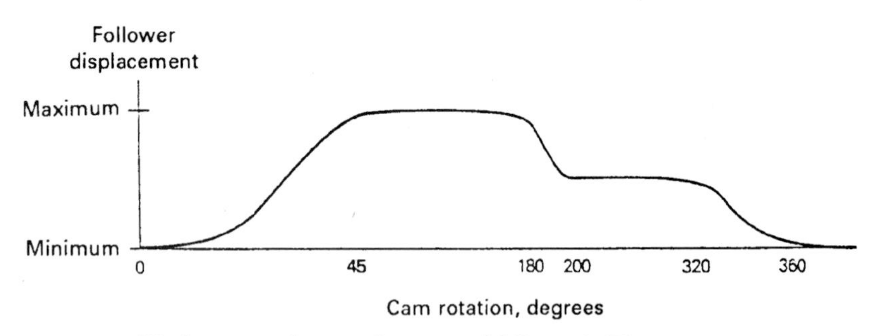

Figure 8.26 Displacement diagram for cam and follower in Fig. 8.25.

and thus produce some desired motion of its follower. As can be seen from the displacement diagram of Fig. 8.26, the follower may extend rapidly, dwell, retract partially, dwell, and then return to its original position as the limit stop is again reached.

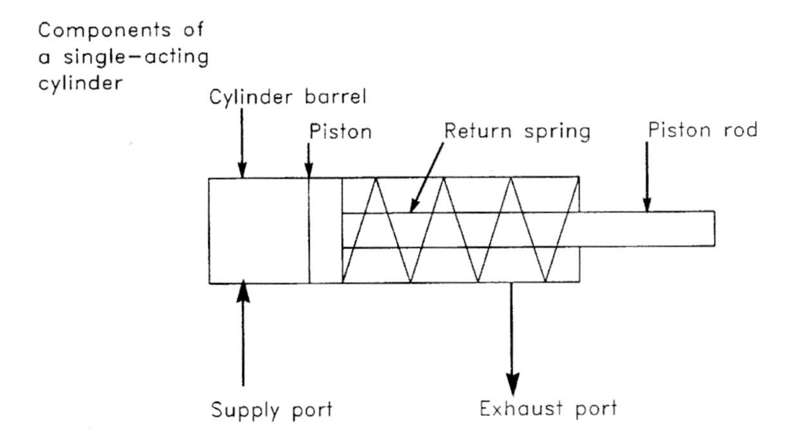

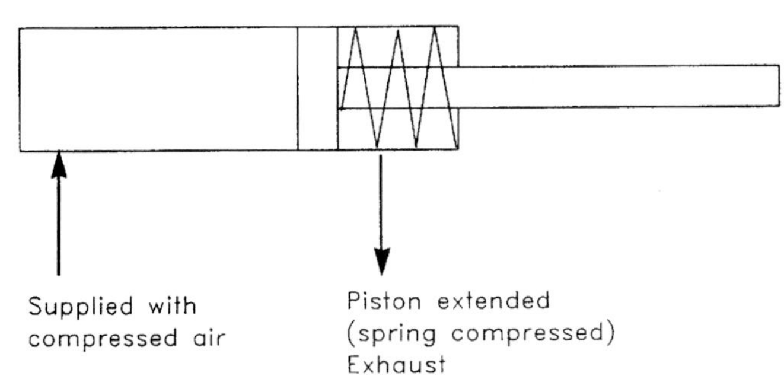

Figure 8.27 Single-acting pneumatic cylinder.

8.9 Pneumatic Actuators

8.9.1 Single-acting pneumatic actuator

The pneumatic cylinder is a proven actuator for the generation of linear movements. There is a wide variety of pneumatic cylinder designs. A single-acting pneumatic cylinder is constructed from a cylinder barrel, a piston with a piston rod, a return spring, and supply and exhaust ports (Fig. 8.27).

If the cylinder chamber is supplied with compressed air, a pressure acts on the piston surface. The force F generated by the pressure P on a piston of surface area A is

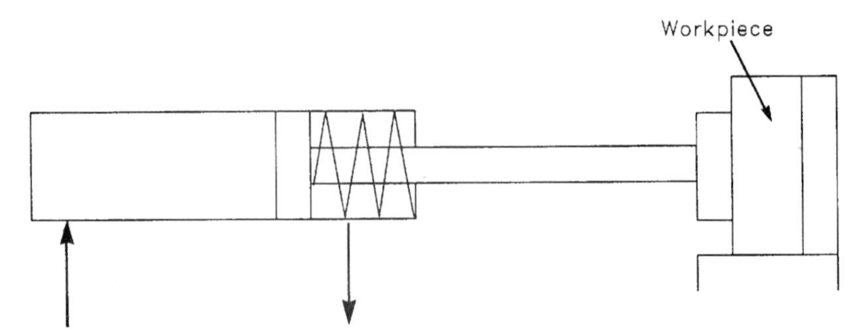

Figure 8.28 Application for a single-acting pneumatic cylinder.

$$F = PA$$

If F is greater than the counterforce of the spring, the piston moves with the piston rod to the right. This movement is called the *forward stroke.*

The compressed air in the cylinder chamber escapes through the exhaust port. If the compressed air supply is interrupted, the spring force moves the piston to the left. This movement is called the *return stroke.*

The designation *single-acting cylinder* points to the fact that mechanical work can be carried out only during the forward stroke movement. The force which is exerted by a piston and piston rod increases in accordance with the piston area and the air pressure in the cylinder chamber. The stroke range of this type of cylinder rarely exceeds 100 mm. Figure 8.28, shows an application for a single-acting pneumatic piston actuator.

In addition to the piston cylinder, the pneumatic diaphragm actuator is also a single-acting actuator. In this type, a stiff diaphragm is clamped between two cylinder halves. The diagram is pressurized, causing it to deflect. The degree of deflection determines the stroke of the piston rod, which is directly connected to the diagram. A distinction is made between the simple diaphragm cylinder and the rolling diaphragm cylinder (Fig. 8.29). The stroke range for the simple diaphragm cylinder is between 1 and 5 mm, and that of the rolling diaphragm cylinder is between 50 and 80 mm.

8.9.2 Control of single-acting pneumatic actuator

The piston of a single-acting actuator (cylinder) executes a forward stroke when it is pressurized; the return stroke takes place when the cylinder is exhausted. The compressed air supply and exhaust can be regulated by a 3/2-way valve. Figure 8.30 illustrates the interaction of the two components. The 3/2-way valve has three ports and two switching positions; the method by which it is switched from one posi-

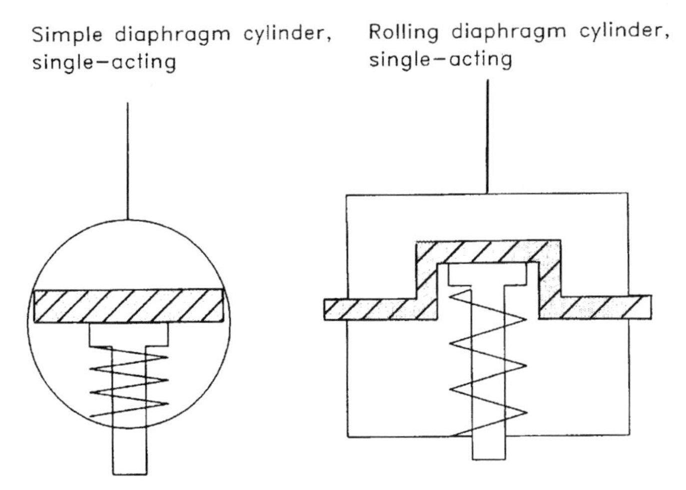

Figure 8.29 Diaphragm cylinders.

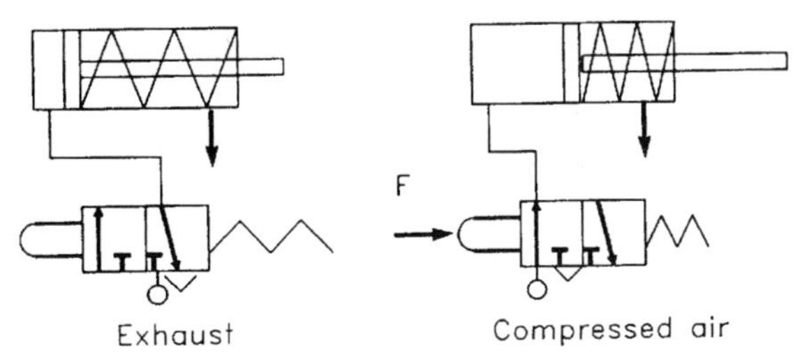

Figure 8.30 Control of compressed air supply and exhaust by a 3/2-way valve.

tion to another may vary. The example uses a valve which is held in a normal position by a spring. In this position, the compressed air is shut off from the valve. This type of valve is in the normal position.

The rounded-bar symbol to the left of the valve rectangle indicates that valve is moved to the working position by mechanical actuation (e.g., by a stem actuator). In the left part of the diagram, the valve is shown in the normal position; there is a connection from the cylinder port to the exhaust via the valve. The cylinder chamber is exhausted and the piston is retracted. Arrow *F* in the right part of the diagram indicates that the valve is activated. Accordingly, there is a connection between the valve supply port and the cylinder port so that the cylinder chamber is pressurized and the piston extends. The 3/2-way valve is an example of how a specific component within a control system is able to assume a variety of functions.

In the piston, for example, the 3/2-way valve can take over the function of limit switch sensor. Because of this, the valve is also referred to as a *final control element.* This element is itself influenced by the control signal (control value). In the example of Fig. 8.30, this is the mechanical actuating force *F.*

8.9.3 Double-acting pneumatic actuator

The construction of a double-acting pneumatic actuator is similar to that of the single-acting actuator. However, there is no return spring and the ports may be used alternatively as supply or exhaust ports (Fig. 8.31). If cylinder chamber A is supplied with air and chamber B exhausted, the forward stroke is carried out. Similarly, the return stroke is effective when B is supplied with air and A is exhausted. This reciprocating action has the advantage that the cylinder is able to carry out work in both directions of motion. The force transferred by the piston rod is somewhat greater in the forward stroke than in the return stroke because the effective surface is reduced on the piston rod side by the cross-sectional area of the piston rod.

The double-acting cylinder can be controlled by the valve shown in Fig. 8.32. The valve has four ports and two switching positions. Thus, it is referred to as a 4/2-way valve. As shown in the figure, one cylinder chamber is supplied with air and the other exhausted in each valve piston.

The double-acting piston is the basis for various cylinder configurations for special applications (Fig. 8.33), several of which are described below.

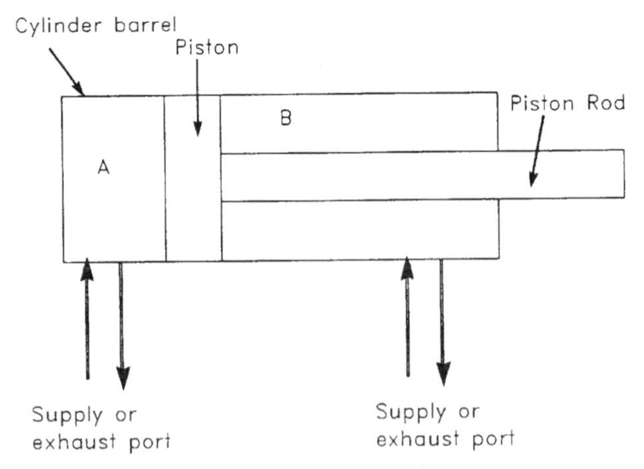

Figure 8.31 Double-acting pneumatic actuator.

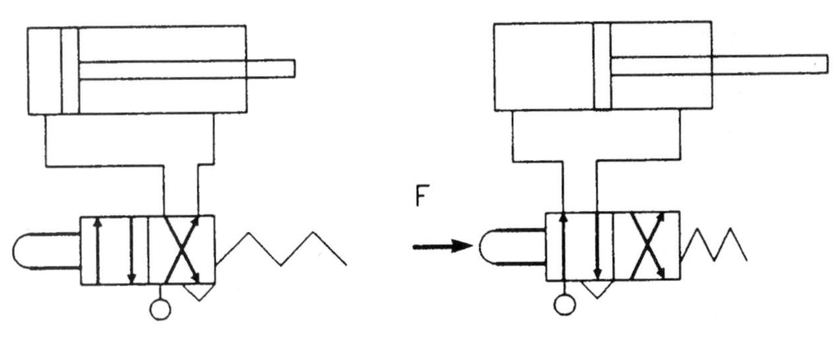

Figure 8.32 Control of a double-acting pneumatic actuator.

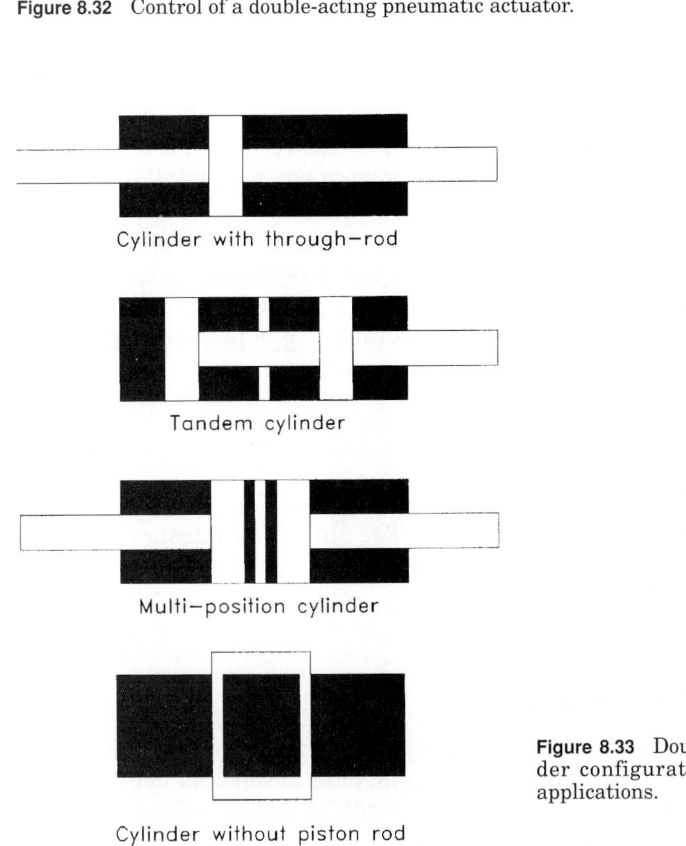

Figure 8.33 Double-acting cylinder configurations for special applications.

Through-rod pneumatic actuator. This is mainly used where power is necessary on each side of the cylinder alternately. It is also used where a power function is to be carried out on one side of the piston, while the other side of the piston rod is used for triggering a signal.

Tandem actuator. In a tandem cylinder, two separate double-acting cylinders are joined to form a single unit with an intermediate wall. They are located one behind the other so that the two forces arising when pressure is applied can be combined. For this reason, this cylinder is used where greater forces are required but the cylinder space is restricted.

Multipiston actuator. A cylinder of this type can be constructed from a minimum of two separate cylinders which are assembled base to base. In the multiposition cylinder, three or four fixed switching positions are possible. Four positions can also be achieved with two cylinders having different stroke lengths.

Pneumatic linear actuator. An actuator of this type is a rodless, double-acting cylinder. The piston in the cylinder is freely movable according to pneumatic actuation, with no positive external connection. The piston is fitted with an annular permanent magnet. The slide is located externally on the cylinder barrel and is also fitted with an annular permanent magnet. Thus, a magnetic coupling is produced between slide and piston. As soon as the piston is moved by compressed air, the slide moves synchronously with it. The components to be actuated are mounted on the slide. This cylinder type is specifically used for extremely long stroke lengths of up to 10,000 mm.

8.10 Hydraulic Actuators

The cylinder is the most important actuator for linear movements in hydraulics as well as in pneumatics. The same basic design exists in hydraulics as in pneumatics. Obviously, hydraulic cylinder construction is different because of the sealing problems created by the high pressures and the medium (oil).

8.10.1 Pneumatic-hydraulic feed unit

These devices provide constant-feed movements even with the most varied load. The unit is made up of a pneumatic working cylinder and a hydraulic cushioning cylinder (Fig. 8.34). If the pneumatic cylinder is pressurized, the piston in the hydraulic cushioning cylinder is drawn along with it. This piston drives the oil to the reverse side of the piston via a one-way flow-control valve. Feed speed can be set with the one-way flow-control valve. In the return stroke, the oil can pass quickly to the other piston side via the nonreturn valve. Therefore, the return stroke is considerably faster than the forward stroke.

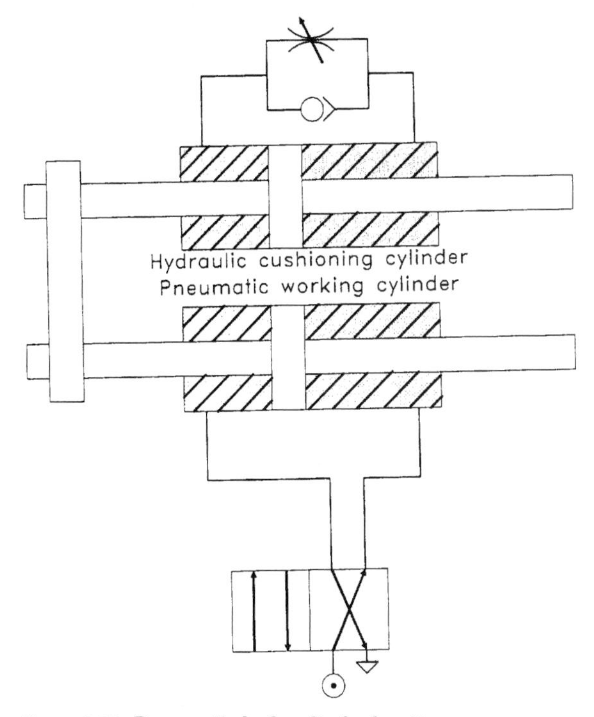

Figure 8.34 Pneumatic-hydraulic feed unit.

8.11 Electrical Actuators

There are two fundamental types of electrical linear actuators:

- An electromechanical drive converts rotary motion of a conventional motor into linear motion by means of suitable mechanical equipment (e.g., worm, worm wheel, and spindle, Fig. 8.35)
- A linear motor, which does not require conversion of motion to provide linear movement (Fig. 8.36)

There is a part in the linear motor that corresponds to the stator of a three-phase rotary motor. This part is a comb-shaped laminated inductor with an ac winding inserted in the grooves. In general, two inductors are located opposite one another.

The part of the linear motor corresponding to the squirrel cage rotor is referred to as the *armature.* It is located between the two inductors and consists of a solid conductor of aluminum, for example. (An armature of a magnetic material such as steel makes one of the

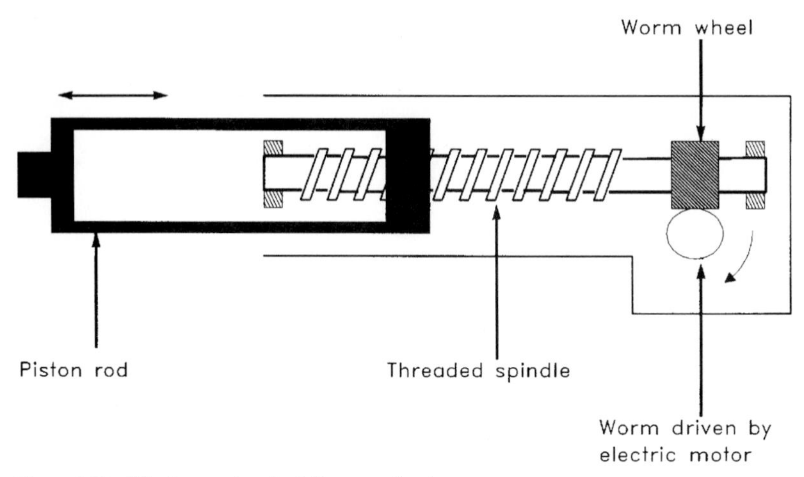

Figure 8.35 Electromechanical linear actuator.

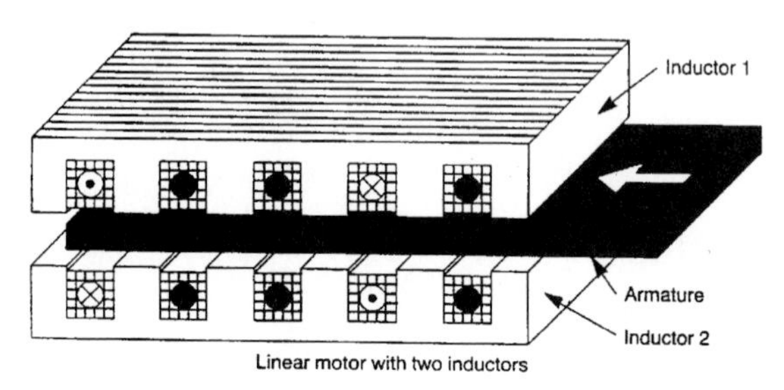

Figure 8.36 Linear-motor-based linear actuator.

two inductors unnecessary because the magnetic lines of flux are guided through the steel to the next pole. The steel armature may also be coated with a conductor material such as aluminum.)

The alternating current in the inductor winding generates a magnetic traveling wave. This induces powerful eddy currents in the armature, creating a force in the direction of the traveling wave.

If the inductor is fixed and the armature is movable, e.g. when transporting material, the armature moves with the traveling wave. If, on the other hand, the inductor is movable and the armature is fixed, e.g., when driving a trolley on a hoist, the inductor moves in the opposite direction to its traveling wave.

Linear motors are used as drives for transportation of materials, for conveyer belts, and for shunting tracks, in addition to drives for gates and large panes of glass. Their use in high-speed railways is being tested.

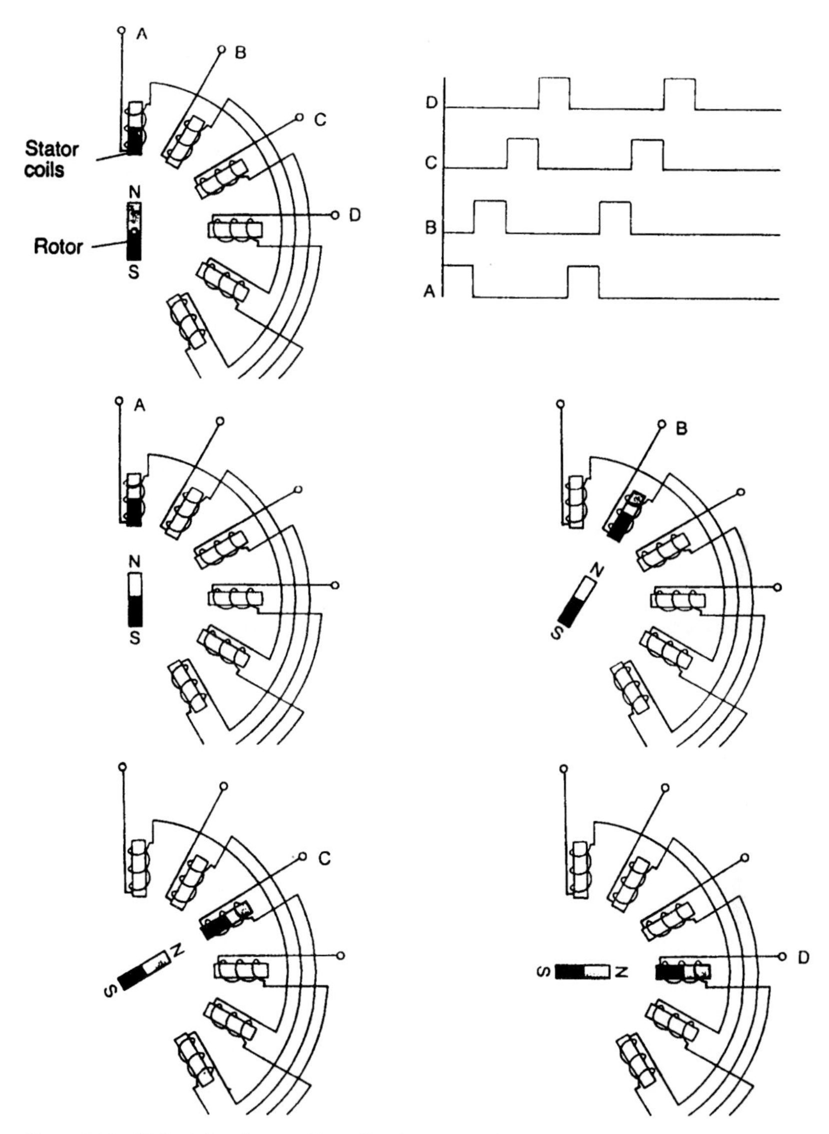

Figure 8.37 Principle of operation of a stepper motor.

8.11.1 Stepper motor with linear motion

This type of stepper motor provides linear actuation and directly carries out a linear movement. Figure 8.37 shows the principle of operation of one type of stepper motor. The rotor is constructed from a permanent magnet with north and south poles. The stator coils also generate a magnetic field when an electric current passes through

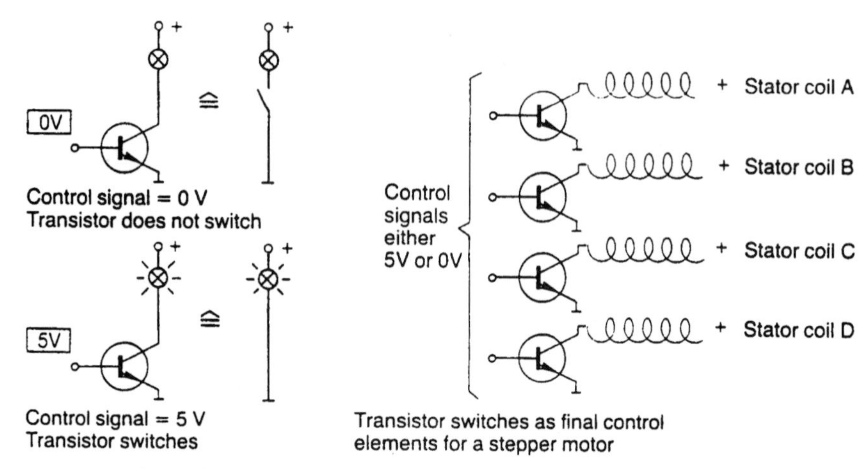

Figure 8.38 Control of stator coil changeover by transistor switches.

them. The opposite poles attract, so the rotor north pole moves toward the stator south pole. There is always only one stator south pole, located on the stator coil through which current is flowing at that particular time.

If the various coils are switched on and off in a specified order, the rotor north pole follows the migratory stator south pole step by step. Every time a coil changeover takes place, a step is carried out, corresponding to a fixed angle of rotation. The complete cycle of changeovers corresponds to the complete rotation angle.

Changeover of the stator coils is effected by electronic switches (transistor switches). Here, the transistors take over the function of the final control elements. The transistors receive the changeover commands by means of simple binary control signals (Fig. 8.38).

8.11.2 Application of the stepper motor as linear actuator

In the linear actuator illustrated in Fig. 8.39, the rotor is equipped with an internal thread into which a screw spindle is inserted. Each rotation (rotary step) of the stepper motor pushes the screw spindle forward or backward a specified distance. There is a simple correlation here between the angle of rotation a and the linear movement s:

$$s = (h/360°)a \tag{8.10}$$

where h is the diameter of the spindle head.

The linear actuator enables relatively accurate positioning with high resolution. Values of resolution—the shortest distance which can be executed per step—of 0.05 mm can be achieved. Since a given

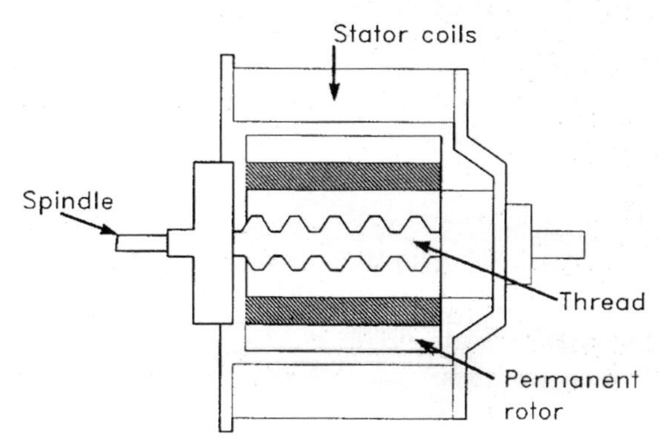

Figure 8.39 Application of a stepper motor as a linear actuator.

number of steps always covers the same distance, no feedback signal is required.

Because digital information can be directly transformed into linear motion, the linear actuator is particularly suitable for use in digital systems.

The linear actuator is used to drive metering pumps for chemical analysis. It is also used in medical technology, industrial electronics, and wherever valves to regulate flow of gases and liquids need to be driven accurately.

8.12 Rotary Actuation

8.12.1 DC motors

One of the most important actuators for the generation of rotary movements is the electric dc motor. The operating principle of a dc motor is shown in Fig. 8.40. The rotary actuator works on the principle that a force F acts on a live conductor as soon as it is moved into a magnetic field. If a conductor loop is brought into the magnetic field, a moment which turns the conduction loop around the axis of symmetry is developed as a result of the different current directions relative to the magnetic field. The current supply must be commutated to prevent the conduction loop from moving back in the opposite direction after turning by 180°. This is achieved by attaching the connecting contacts mechanically to the rotary axis and supplying the current via brushes. An arrangement of this type is called a *commutator* (reversing switch).

In order to achieve maximum turning moment, the motor is generally made up of many conduction loops. The speed of the dc motor is dependent on the voltage supply and, thus, on the current flowing;

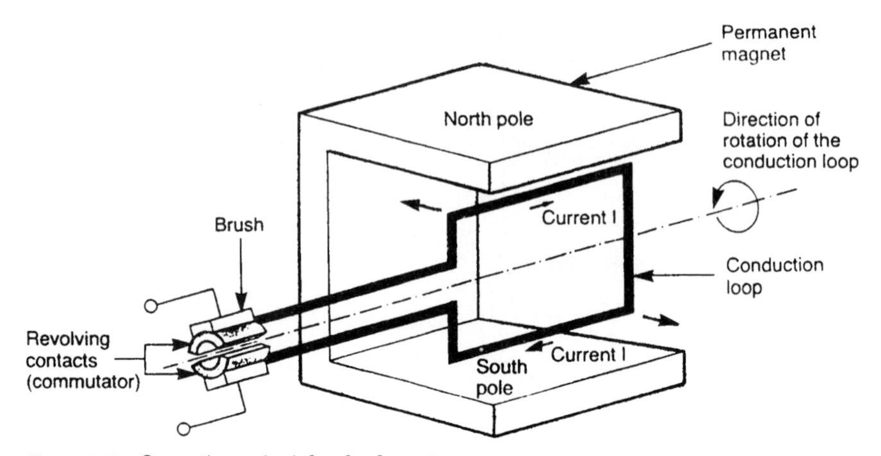

Figure 8.40 Operating principle of a dc motor.

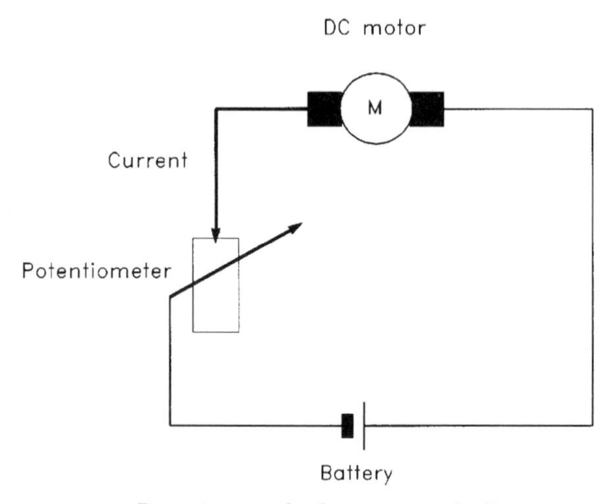

Figure 8.41 Potentiometer for dc motor speed adjustment.

therefore, a change in speed can easily be achieved by changing the terminal voltage.

The direction of rotation is changed by commutation of the terminal voltage. Just as final control elements are required by a pneumatic cylinder to extend and retract a piston rod, suitable final control elements are necessary for the dc motor to change the direction of rotation.

The speed adjustment can be achieved by a variable-resistance coil (potentiometer) (Fig. 8.41). Both current and speed are influenced by

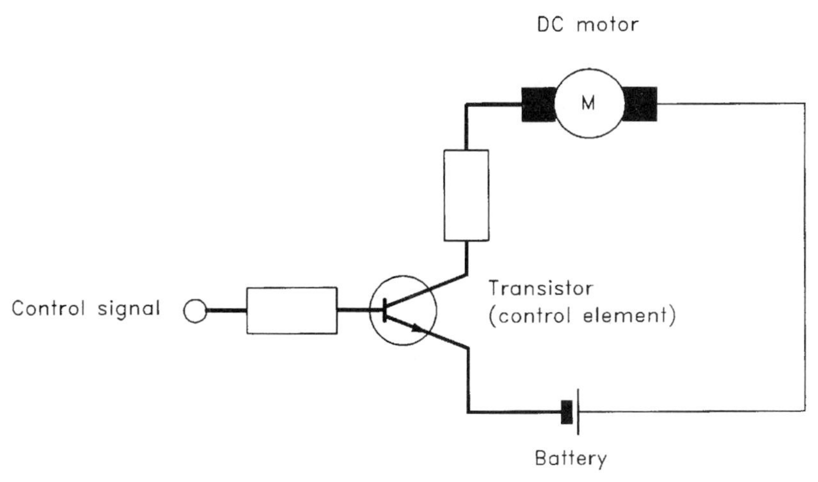

Figure 8.42 Transistor for dc motor speed adjustment.

the potentiometer. The smaller the resistor setting, the greater the current flow through the motor and, thus, the greater the speed. The potentiometer operates as a final control element. The correcting variable corresponds to the force acting on the potentiometer which changes the resistance (e.g., by turning of the adjusting knob).

For purely electrical or electronic systems, speed adjustment by means of a transistor is preferred (Fig. 8.42). An analog control signal is connected to the base of the transistor. It affects the collector current and, thus, current conduction through the motor.

Reversal of direction with a solenoid. In a solenoid relay (Fig. 8.43), the magnetic effect of an electric current is exploited. As soon as current flows to the coil (terminals S1, S2), it behaves like a magnet and attracts the armature which in turn closes the contact via the lever mechanism.

The type of contact illustrated in the figure closes when current is flowing in the coil. It is termed a *normally open* contact. The type of contact which becomes open in this situation is called a *normally closed* contact. In practice, relays exist with one or a number of these contacts or combinations of them.

Another type of contact is the *changeover* contact, which switches from one contact to the other. Figure 8.44 shows how the direction of rotation is changed with a relay fitted with two changeover contacts:

1. *Clockwise rotation.* The control signal is set to 0 V, releasing the relay and allowing the current to flow in the direction shown by the arrows.

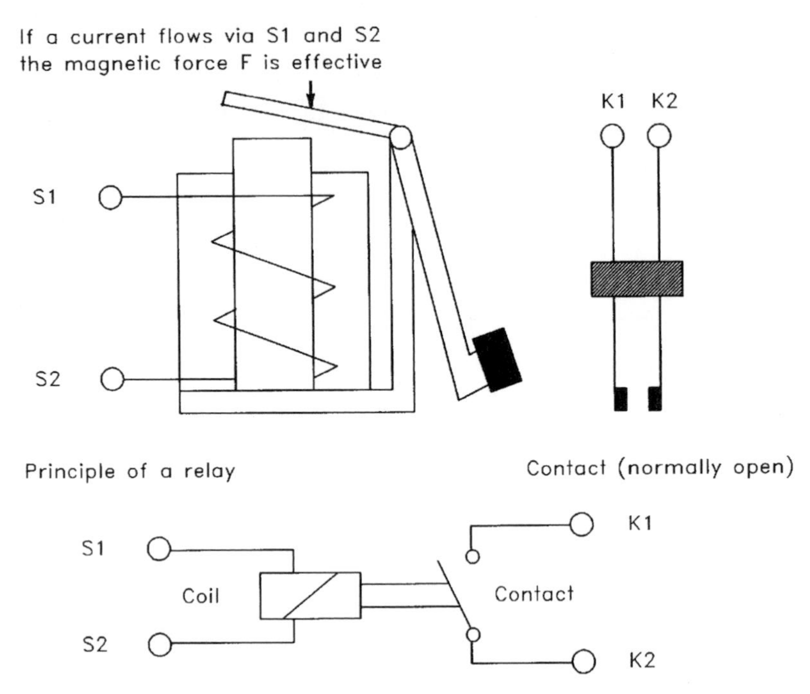

Figure 8.43 Solenoid relay operation.

2. *Counterclockwise rotation.* The control signal is set to 5 V, attracting the relay and permitting the current to flow in the reverse direction.

Reversal of direction with transistors. As shown in Fig. 8.45, the transistors functioning as electronic switches are triggered in pairs. The electronic switch control function generally determines both the speed and direction of rotation, as follows:

1. *Clockwise rotation.* Control signal Y1 is set to 0 V and Y2 is set to 5 V; thus, the transistors T2 and T3 are on and the motor current flows in the direction shown by the arrows in the figure. The magnitude of the current determines the speed.
2. *Counterclockwise rotation.* Control signal Y1 is set to 5 V and Y2 is set to 0 V, turning transistors T1 and T4 on. The motor current now flows in the opposite direction. Again the magnitude of the current determines the speed.

8.12.2 AC synchronous motors

With the synchronous motor, the alternating current generates an alternating magnetic field around the motor stator. At the frequency of

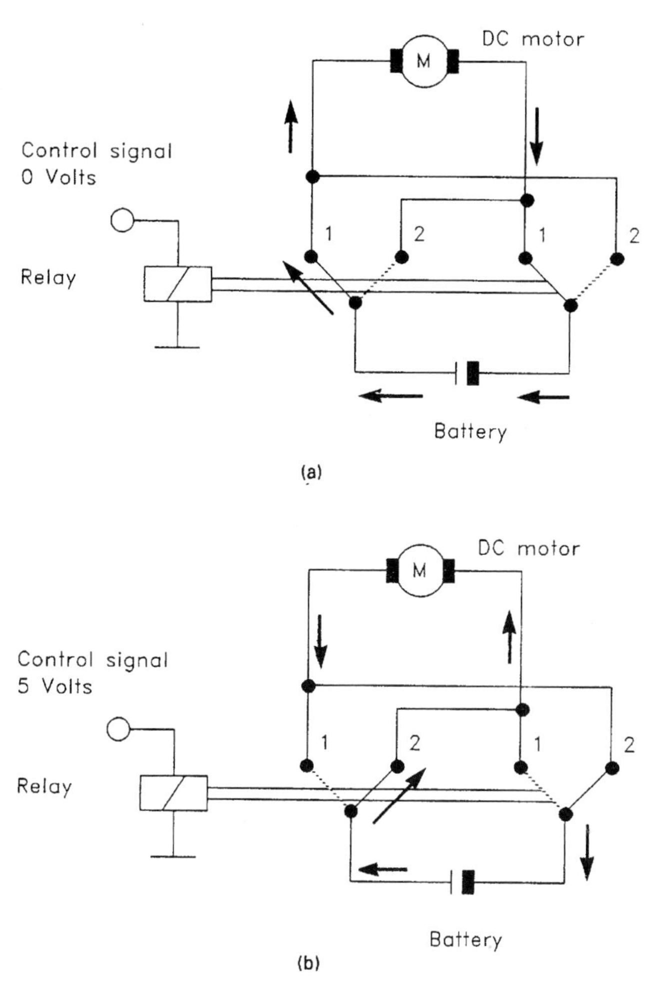

Figure 8.44 Changing direction with a solenoid relay and two changeover contacts: (*a*) Clockwise rotation, (*b*) counterclockwise rotation.

the voltage, the north and south poles move around the stator. The permanent-magnet rotor runs synchronously with the rotary field. A frequency-dependent rotary movement is produced. A change in the speed is possible only when there is a change in the frequency. Speed adjustment is more complicated than in the case of the dc motor. Therefore, synchronous motors are mainly used where no speed changes are required. However, synchronous motors develop a considerably higher speed.

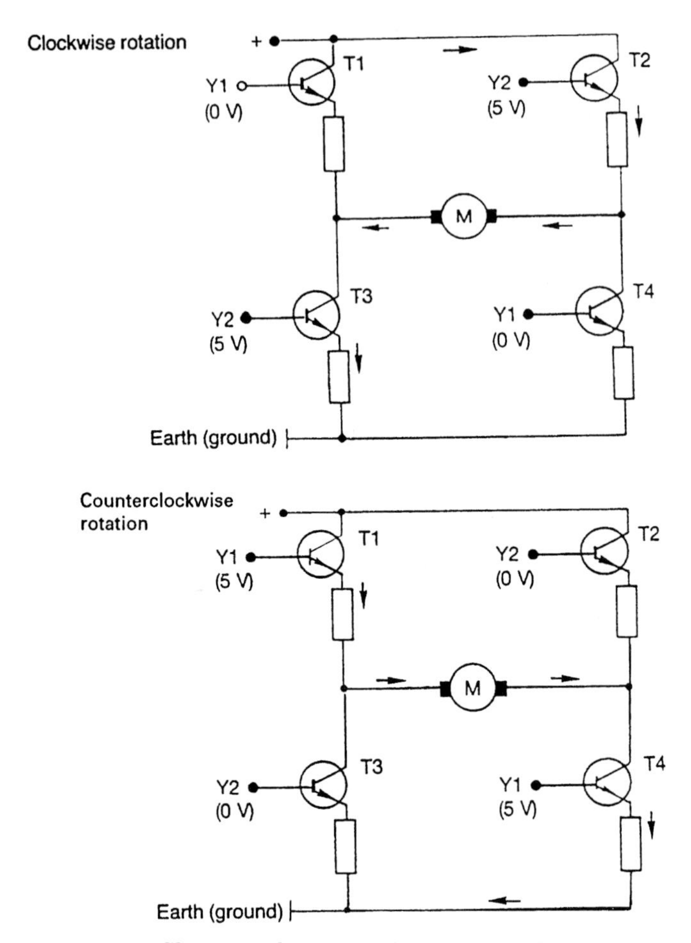

Figure 8.45 Changing direction of rotation with transistor switches.

8.12.3 Stepper motors

The stepper motor is used where precise and reproducible angles of rotation are required, as in positioning devices. The stepper motor can easily be coupled to digital systems, without the need for complicated signal processing.

8.12.4 Pneumatic motors

Pneumatic motors are frequently called *compressed-air motors.* They are suitable for all types of use, even in difficult environments, in particular where there is risk of explosion.

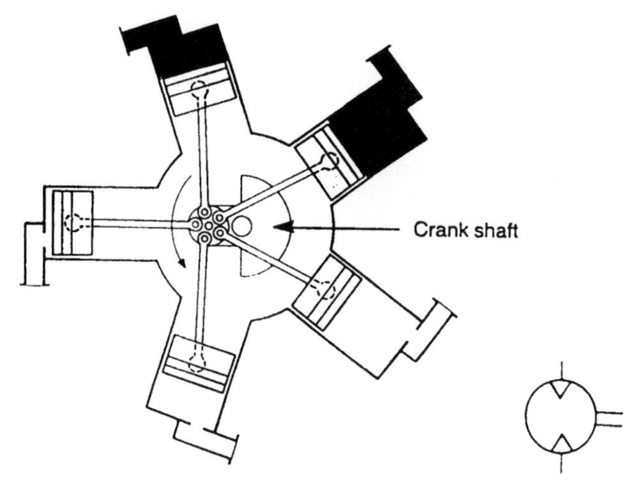

Figure 8.46 Radial pneumatic motor.

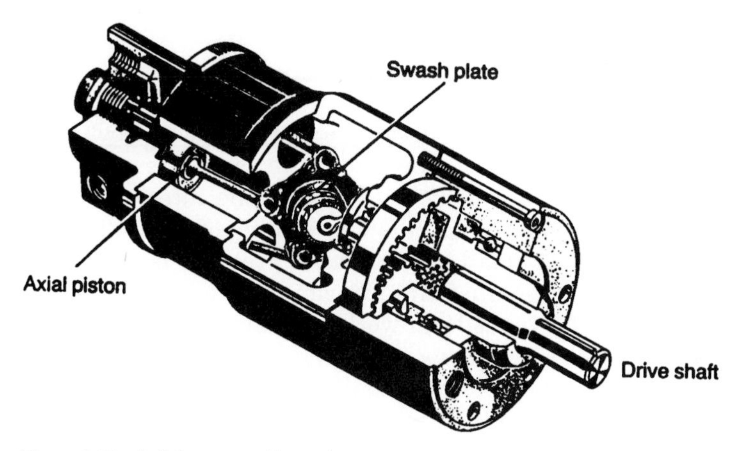

Figure 8.47 Axial pneumatic motor.

Piston motors can be subdivided into radial and axial categories. In the radial motor, pistons travel back and forth to drive a crank shaft via connecting rods. A number of cylinders are arranged in a radial pattern to ensure smooth running (Fig. 8.46). The power of the motor depends on the input pressure, the number of pistons, the piston surface area, the stroke, and piston speed.

In the axial motor, five cylinders are arranged parallel to the axis of the drive shaft. The reciprocating movements of these cylinders are converted to rotary movements through a swash plate (Fig. 8.47). The maximum speed of this type of motor can reach 5000 rpm.

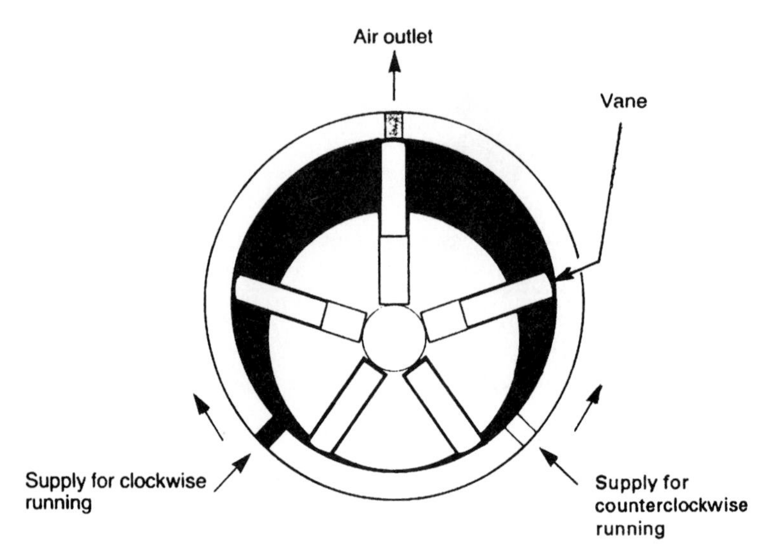

Figure 8.48 Sliding-vane motor.

8.12.5 Sliding-vane motors

The sliding-vane motor has a rotor located eccentrically in a cylindrical chamber. Slots are provided in the rotor. The vanes are guided in the slots of the rotor and forced outward toward the interior wall of the cylinder by centrifugal force. In this way sealing of the individual chambers in guaranteed. A small amount of air pressure is applied to allow the vanes to press against the interior wall even before the motor starts to move. The vanes form working chambers in which the compressed air can become effective according to the size of the surface of the vanes. The air enters the smallest chamber and expands when the chamber is enlarged (Fig. 8.48). The motor speed ranges between 3000 and 8500 rpm.

8.12.6 Hydraulic motors

Hydraulic motors are supplied with liquid under pressure, generally hydraulic oil, and transfer mechanical power to a shaft. The torque of a hydraulic motor is directly proportional to the oil pressure. The hydraulic motor rpm is directly proportional to the oil flow. The power output is

$$\text{Power} = \text{rpm} \times \text{torque}$$

Power, rpm, and torque can be interrelated by oil pressure and flow.

Hydraulic motors are subdivided into:

1. *Piston motors* with axial or radial pistons
2. *Geared motors* with internal or external gears
3. *Rotary vane motors* with liquid externally or internally supplied

The design and the operational principles of hydraulic motors are similar to those of pneumatic motors. Hydraulic motors are used in many industrial fields. They are used, for example, as drivers in vehicles and ships of all types, as rolling mills in steel works, in the presses and heavy machinery, and in injection-molding and die-casting machines.

8.13 Generation and Control of Forces in Hydraulic Motors and Actuators

The control system of a double-acting hydraulic cylinder is shown in Fig. 8.49; it uses a 4/2-way directional valve.

A power supply unit consisting of a drive motor, a pump, a pressure relief valve, and a reservoir generates the flow of oil. The pressure in a hydraulic system is not a constant value, but changes with the resistance (load resistance and internal resistance). It may increase considerably if the power component (hydraulic motor) is set in motion under load. A pressure relief valve is necessary to prevent overpressure.

A manually operated 4/2-way valve is used as the final control element for the hydraulic cylinder. During actuation, oil flows into the left-hand cylinder chamber, forcing the power piston to travel to its forward end position. In the process, the oil is forced out of the right-

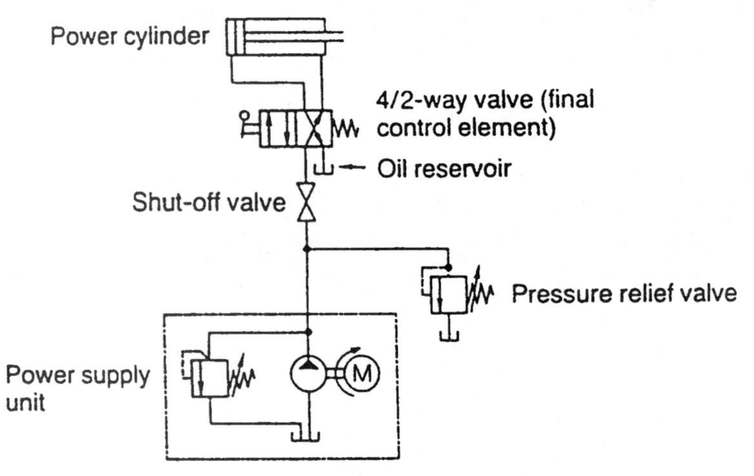

Figure 8.49 Flow controller for a double-acting hydraulic cylinder.

hand hydraulic chamber via the final control element to flow into the oil reservoir.

When the final control element has been switched over, the oil once again flows into the right-hand cylinder chamber via the final control element into the oil reservoir. Thus, the piston retracts once again.

When the cylinder retracts, the smaller annular piston area becomes effective, resulting in a higher speed for the same volumetric flow rate, i.e., the piston rod retracts more quickly.

The hydraulic cylinder can be controlled with a 4/3-way valve that recirculates in the central position as the final control element (Fig. 8.50). The function of the 4/3-way valve is illustrated in Fig. 8.51, where a pressure relief valve ensures that the pressure in the system cannot increase too rapidly. When the 4/3-way valve is in its central position, oil supplied by the hydraulic power supply unit can flow away, unrestricted, into the oil reservoir.

If the final control valve moves to the right, a path is opened for the oil to flow into the reservoir. The oil pushes the cylinder piston to the right and this displaces the oil from the right-hand chamber via the final control element to the oil reservoir. When the final control element is brought into the left-hand position, the piston retracts once again.

If, during the advance and return of the piston, the final control element is brought into the central position, the piston stops and can be moved only when there is sufficient natural oil leakage, even with the influence of external force.

The recirculating position has the advantage that, during the time that work is not being carried out by the piston, the oil flow generated by the pump is led directly away, unpressurized and without temper-

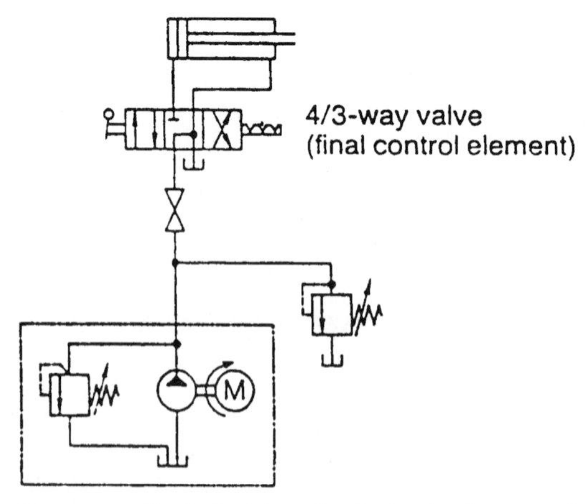

Figure 8.50 Control with recirculation in the central position of a 4/3-way valve.

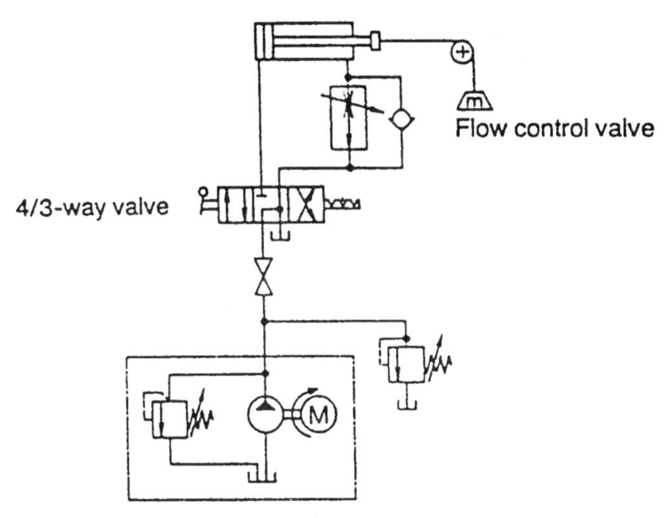

Figure 8.51 Function of 4/3-way valve.

ature rise, and not through a pressure relief valve. The arrangement is therefore energy-saving, since hydraulic oil led off through the pressure relief valve would be heated and the pump would draw against the maximum pressure (high power consumption).

The speed of the drive piston of the hydraulic cylinder can be influenced by the following factors:

1. Use of two pumps with constant output; with this method it is possible to set two speeds.
2. Use of a pump with adjustable output.
3. Use of flow control valves.

A flow control valve is an adjustable throttle valve with an integrated regulating device. With this throttle, the flow cross section is changed, causing the volumetric flow rate (volume per unit time, such as liters per minute) to be changed. The regulating device ensures that, once adjusted, a volumetric flow rate will remain constant irrespective of changing output and input pressures and the viscosity of the media. Speed control by an outlet flow control valve is illustrated in Fig. 8.52. With the outlet flow control, the oil flow led away by the consuming device can be kept constant. Once the oil flow is supplied by the power supply unit, a maximum operating pressure is set with the pressure-limiting valve. A 4/3-way valve is used as a final control element.

If the final control element is activated (moved to the right as illustrated in Fig. 8.51), then the oil flows into the left cylinder chamber and moves the piston to the right (piston rod extends). The liquid dis-

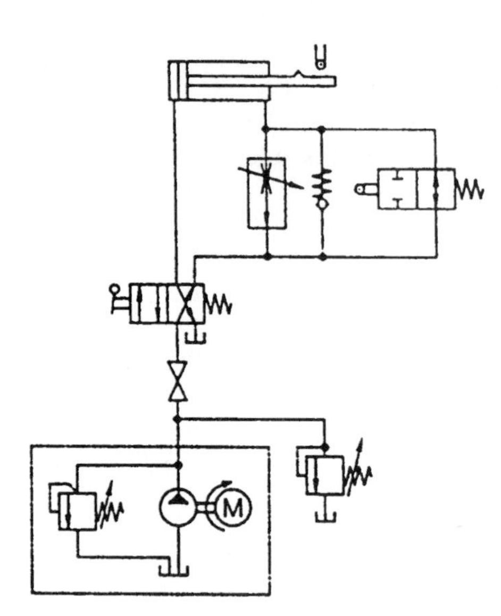

Figure 8.52 Speed control by a flow control valve.

placed from the right-hand cylinder chamber flows through the flow control valve. Also, the liquid in the final control element flows into the oil reservoir. The oil flow is kept constant by permitting the piston rod to extend at a constant speed.

If the final control element is inactive (to the far left as illustrated in Fig. 8.51), the oil flow passes unthrottled via the nonreturn valve into the right-hand cylinder chamber and moves the piston to the left. The volume of oil from the right-hand chamber flows away through the final control element into the oil reservoir.

Hydraulic speed control can be found, for example, in all machine tools where it is necessary to set and keep constant feed rates and speeds corresponding to the manufacturing program.

8.13.1 Rapid traverse control circuit

The rapid traverse feed circuit is an extension of the speed control valve circuit just discussed. It is often used with machine tools when rapid traverse and an adjustable feed are required in order to approach a work piece quickly, while the machine tool is in the operating cycle. It also can return in a rapid traverse movement (Fig. 8.52).

The oil flow is supplied by the power supply unit. The maximum oil operating pressure is set at the pressure relief valve. A 4/2-way valve serves as a final control element. In the final control element illustrated in Fig. 8.53, the oil flow is directed to the right-hand cylinder chamber of the double-acting cylinder. The oil in the left-hand chamber is displaced and permitted to flow through the final control element to the oil reservoir. When the final control element is activated, oil flows into the left-hand cylinder chamber and the piston rod extends. In this process, the liquid in the right-hand cylinder chamber flows unthrottled through the 2/2-way valve in the final control element to the reservoir container (rapid advance). If the control rail activates the roller lever on the 2/2-way valve, flow in the 2/2-way valve is blocked. The oil is directed through the flow control valve to the oil reservoir. The flow control valve regulates the flow of oil at the outlet. The flow rate of the oil and, thus, the piston speed (feed) can be set at the flow control valve. If the final control element is reversed again, the coil flows unrestricted through the nonreturn valve into the right-hand cylinder chamber.

8.13.2 Pressure-dependent sequence control

Pressure-dependent sequence control is an example of how a control system with two cylinders can be realized with purely hydraulic components (Fig. 8.53).

In the circuit diagram illustrated in Fig. 8.53, a pressure sequence valve is used as a functional module. On attaining a specified adjustable pressure, a pressure sequence valve opens to release the oil flow into a further part of the hydraulic system. Thus, in the case of increasing load resistance, the pressure buildup can be used as a switching pulse, to introduce a sequential movement, for example. A

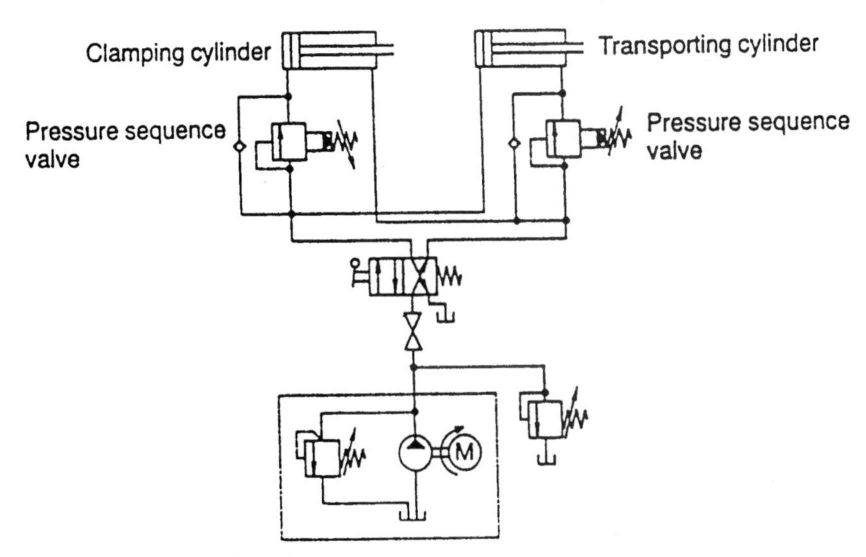

Figure 8.53 Pressure-dependent sequence control.

4/2-way valve is used as final control element. The power supply unit supplies the necessary oil flow. The operating pressure is restricted by the pressure relief valve. If the final control element is moved to the right, fluid can flow through the junction into the left-hand chamber of the transporting cylinder. The piston travels to the right and displaces the liquid from the right-hand chamber through the nonreturn valve into the oil reservoir. At the end position, pressure is built up in the left-hand chamber of the transporting cylinder as well as in the connected line of the system. The pressure sequence valve for the clamping cylinder is connected to these lines. When the pressure has reached a specific value, the pressure sequence valve opens, allowing the fluid to flow freely into the left-hand chamber of the clamping cylinder. Consequently, the piston travels to the right.

When the final control element is reversed, the piston of the clamping cylinder first retracts and displaces liquid through the nonreturn valve to the oil reservoir. Pressure builds up in the line of the transporting cylinder; when it attains a specified level, the pressure opens the connected pressure sequence valve of the transporting cylinder. Accordingly, the piston of transporting cylinder is also able to retract. With this, the cycle is completed and can be restarted (Fig. 8.54).

As in the other circuit (Fig. 8.53), one recognizes the power supply unit and the pressure relief valve for setting the maximum operating pressure. A 3/2-way valve is used as final control element. If the final control element is moved to the right, the oil flows through the flow control valve, permitting the final control element to move to the left

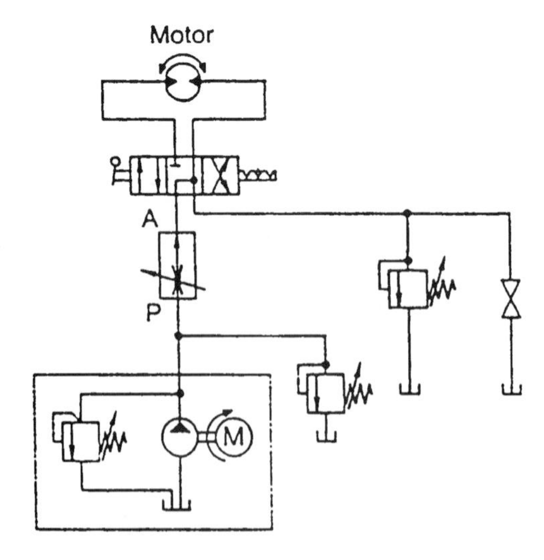

Figure 8.54 End of pressure-dependent sequence.

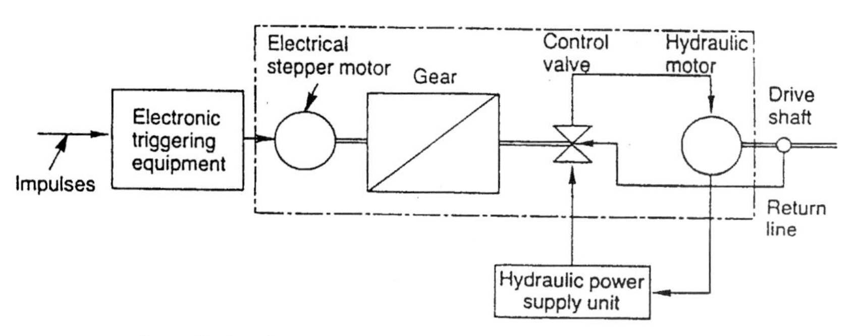

Figure 8.55 Electrohydraulic stepping motor.

side of the motor. The motor rotates in a counterclockwise direction at the speed set at the flow control valve.

In the reverse situation, if the final control element moves to the left, the motor turns in a clockwise direction.

8.14 Electrohydraulic Devices

8.14.1 Electrohydraulic stepping motor

This motor represents the integration of an electrical stepping motor and a hydraulic motor. It is used as a high-power drive in computer numerical control (CNC) machine tools (Fig. 8.55).

The electrical stepping motor moves the hydraulic control valve via a gear. When the control valve is moved, the oil flows to the hydraulic motor, which may, for example, be configured as an axial piston motor. The hydraulic motor provides the actual power.

8.14.2 Proportional valve

The proportional valve is an excellent example of how a powerful hybrid system can be produced from a combination of different technologies. With this valve it is possible to control precisely large hydromechanical forces in a matter of few milliseconds. The operational principle of a proportional valve is illustrated in Fig. 8.56.

A proportional valve consists of a control valve and a system which enables the volumetric flow rate at the valve output to be proportional to the electrical current of the input signal.

In the figure, a proportional solenoid is triggered by an electronic circuit, called the *proportional amplifier.* The proportional solenoid moves a control piston to the right or to the left through a solenoid coil in accordance with the strength of the current. For a movement to the right, the control piston of the valve blocks the left-hand return

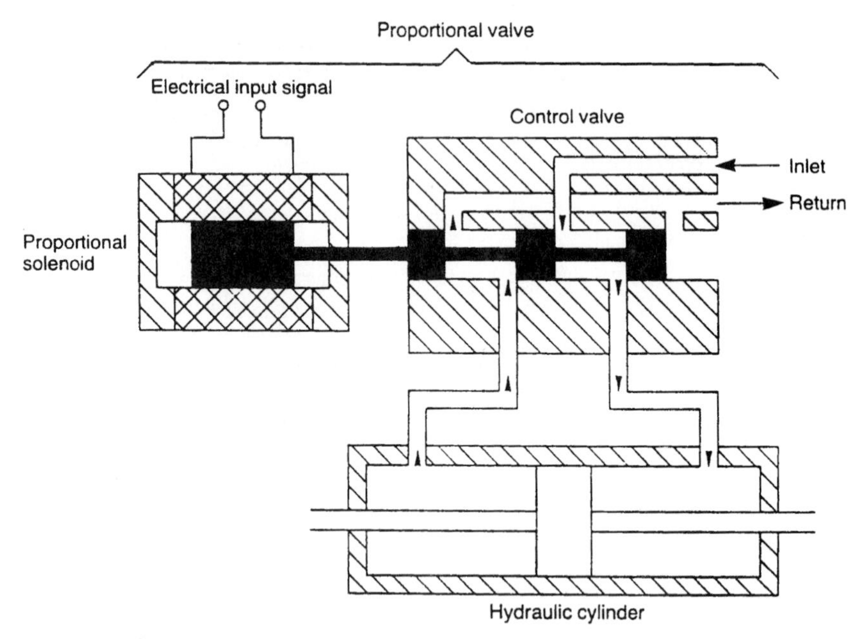

Figure 8.56 Operating principle of a proportional valve.

channel and opens the right-hand channel. This causes the flow of oil in the hydraulic power cylinder to be reversed. The oil flowing into the left-hand cylinder chamber is displaced from the right-hand chamber. The piston therefore moves to the right. When the control piston moves to the left, the power cylinder likewise moves to the left.

The controlled oil flow is proportional in flow rate and direction to the electrical control current.

8.14.3 Control of electrohydraulic systems

A circuit for electrohydraulic positioning control is illustrated in Fig. 8.57. The system forms a closed-loop configuration. A digital linear encoding device for absolute displacement determines the current position of the power piston. The value measured (actual value) and the value to be achieved (reference value) are fed to a comparator which calculates the difference between the two values.

If the reference value and actual value agree, then no movement of the power piston is needed. The system is in the correct position. If there is a difference between the reference value and the actual value (control deviation X_w is more than zero), an adjustment must be made depending on the arithmetical sign preceding the difference. The regulating amplifier generates a control signal for the proportional valve

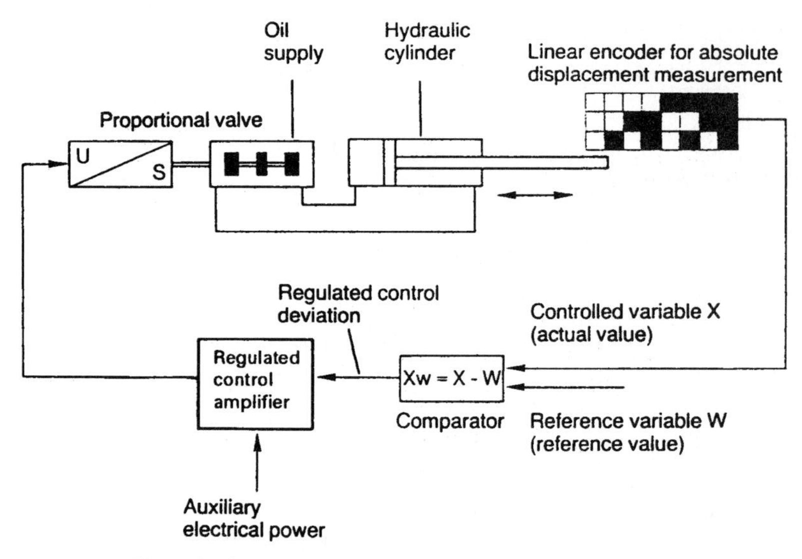

Figure 8.57 Electrohydraulic positioning control circuit.

corresponding to the divergence. The control valve piston is moved, resulting in a flow of oil through the valve and the power cylinder. Movement of the power cylinder toward the reference value gradually leads to a lessening in the divergence until it finally reaches zero. The electrical signal generated from the control amplifier directed to the proportional valve will reach zero value, allowing the oil flow and the power piston to come to a stop.

Only a modification of the reference value or a change in the actual value caused by a disturbance (e.g., changing load) leads to renewed divergence and a corresponding adjustment.

If a position is to be maintained, a continuous fluttering movement of the valve piston must take place, causing a continuous small forward and backward movement of the piston to balance out external errors (e.g., fluctuating cutting forces, changes in length because of temperature influences) and internal errors (e.g., leakage). The maximum measured error in this situation is 0.01 mm.

8.15 Signal Conversion and Transformation

Signals are often present in a form which cannot be immediately processed; they require conversion. In systems concerned with actuation, the conversion of digital signals into analog signals is often required. Such converters are referred to as digital-to-analog converters or D/A converters.

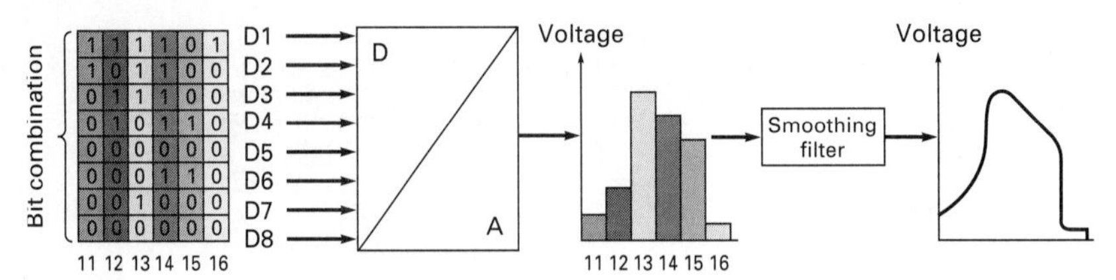

Figure 8.58 An 8-bit digital-to-analog converter.

Figure 8.58 shows an 8-bit D/A converter. Eight signal lines (D1 to D8) are connected to the input. Each signal line may transmit either a 1 (voltage present) or a 0 (voltage not present). The D/A converter reads the combination of the 8 binary signals in a very short period of time and converts them into an appropriate voltage value. This value is made available at the output. During the conversion time, the D/A converter is unable to read any further bit combinations; any modification made in the course of this time is not recognized. Hence, the output is kept at the last value converted. The temporal progression of the output voltage is therefore, stepped. The stepped progression of the voltage is converted into a continuous progression by a smoothing filter.

D/A converters exist as integrated circuits. The most important features are the word width (number of signal lines at the input) and the conversion time.

As has been illustrated, the speed of dc motors can be adjusted by using transistors as final control elements. The control signal for the transistor is an analog signal (Fig. 8.59). It is very rare for a control system to be constructed of components from a single technology.

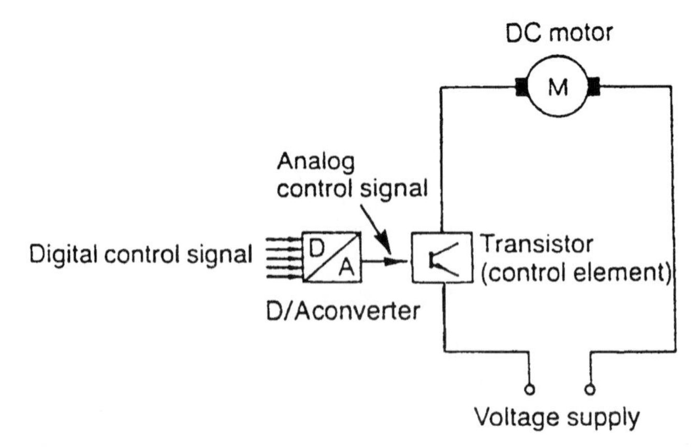

Figure 8.59 Digital-to-analog conversion for transistors in a dc motor control.

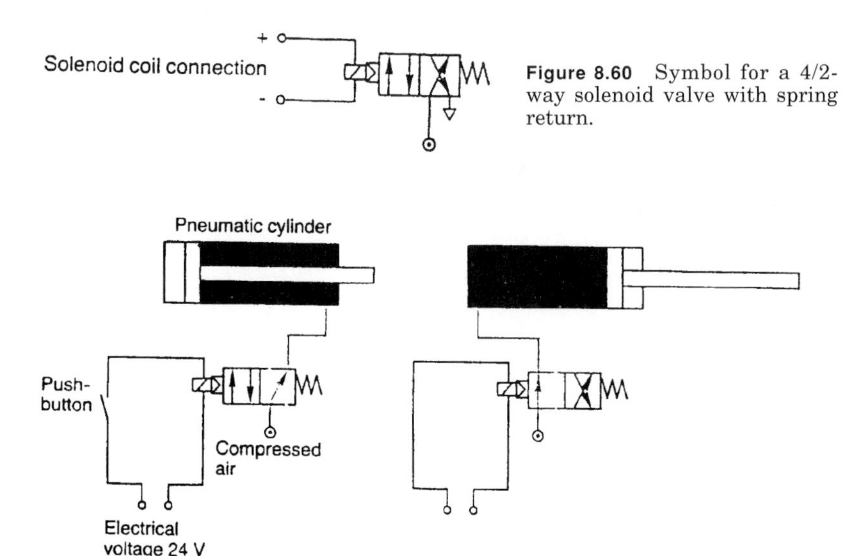

Figure 8.60 Symbol for a 4/2-way solenoid valve with spring return.

Figure 8.61 Example of a solenoid valve application.

Almost all systems are hybrid systems, e.g., combinations of pneumatic and electrical, hydraulic and electrical, etc. However, this means that at the component interfaces within the system, signal conversion should be carried out from one technology to another.

In practice these final control elements often need to react to electrical control signals. Pneumatic and hydraulic valves with solenoid activation are specially designed for these types of activation. The same principle is used here as in electric relays. A live coil acts as a solenoid and activates a mechanism within the valve for the purpose of changeover from one position to another.

Electropneumatic and electrohydraulic converters are considered solenoids. Figure 8.60 shows a symbol for the 4/2-way solenoid valve with a spring return. Figure 8.61 shows a simple example of an application for a solenoid valve, where an electrical push button triggers, for example, the movement of a pneumatic cylinder to stamp work pieces.

Chapter

9

Product Coding and Affordable Data Acquisition Systems

9.1 Bar Codes

Bar coding is a method of encoding a data identifier. While language uses sound to convey information in the form of words, bar coding uses a series of bars and spaces of varying widths to convey data identifiers to trigger database operations. Bar coding is an efficient method for quick information transfer, which accounts for its widespread use in industry.

One major business task is managing information. Bar coding is a management tool in the sense that it allows vast amounts of valid information to be gathered almost instantaneously. However, while bar coding offers this capability, management still must decide what to use the information for and how to achieve this effectively.

Bar coding has affected decision making in much the same way as the telephone. Both are technological innovations that greatly extend previous business capabilities. Just as the telephone makes instantaneous communication possible, so bar coding helps make instantaneous information gathering possible. Bar coding also has eliminated much tedious paperwork and related problems, such as revising incorrectly completed forms, deciphering illegible paperwork, and tracking forms from one location to another.

Bar coding has streamlined and revolutionized otherwise time-consuming procedures. Likewise, it has increased the accuracy of information output, without which the other gains would be neutralized.

To those industries that use it, bar coding is part of an information-gathering system that helps achieve at least some of these objectives:

1. Increased accuracy
2. Increased employee productivity
3. Increased accuracy and control of inventory
4. Decreased cost
5. Tracking of work in progress
6. Tracking of transported items
7. Increased customer satisfaction

As with all technological advances, bar coding places an ever higher premium on management decisions which can optimize use of this source for business purposes.

9.1.1 Accuracy of information gathering

Bar coding can increase the accuracy of any information-gathering task. One common example occurs very often in any supermarket. At the checkout counter, one of a customer's items is unmarked. A price check is needed. One of several possibilities will follow. As a muffled groan carries down the line of waiting customers, the clerk quickly may guess at the price, leading to an undercharge and lost revenue, or to an overcharge and customer dissatisfaction. Holding up the line for a price check itself leads to impatience all along the line.

The business that depends on price labels alone demands tedious work by operators. On the other hand, if a business has a bar-coding system in place, scanning the bar code gives automatic information on any article, such as its price.

Bar coding significantly improves the accuracy of data input through keypunching. Typically, even in the best circumstances, there is a keypunch error for every 300 entries. The tedium of repeated finger movement over the same numbered keys does take a toll. However, most bar-code types have built-in self-checking measures to ensure accuracy. As long as the bar codes conform to specifications, bar coding results in virtually error-free data input. Whatever the application, instantaneous and error-free data input is a real plus.

9.1.2 Increase of employee productivity

Bar coding increases employee productivity. Quicker data collection through a bar-code system means quicker throughput and more overall efficiency. One retailer estimated that 10 customers can be checked out in the time it used to take for 8 customers.

This gives employees more time to do tasks related to sales, displays, and customer service.

For end users, bar coding offers ease and simplicity. Reading bar code requires very little training; after a short training session and some practice, one company estimated that new hires were 80 percent as effective as experienced clerks.

By enabling new hires to become productive immediately, more experienced and more expensive staff are freer to concentrate on tasks like those mentioned above. With bar coding, increased accuracy is compounded by increased productivity.

9.1.3 Accuracy of inventory

Keeping proper inventory at all times is one of the most time-consuming aspects of running any business. The cycle of forecasting, ordering, storing, selling, tracking, and reordering inventory takes a great amount of time and energy.

Errors and miscalculation are costly. Running out of stock items can mean lost business in the short term and lost customers in the long term. On the other hand, amassing quantities of unsold items leads to attempts to remove the excess through markdowns and closeout sales. A bar-coding system allows accurate information for tracking all items throughout the ordering to reordering cycle. At the receiving dock, warehouse, and checkout stand, bar-coded information identifies goods and quantities in current stock. There is no need for a costly closing down to take inventory.

Moreover, inventory can be matched with demand. Transactions at the checkout stands identify current retail trends, what is being sold, and what is not sold. The business can determine what must be ordered, in what quantities, and when. This can be achieved from information tracked and recorded as it happens, when it happens.

By closely tailoring inventory to current customer demand, the business makes a leap in overall efficiency. Slow-moving products taking up shelf space can be eliminated; future stock orders can be aligned with customer demand.

9.1.4 Cost of bar-code installation

The cost of installing a bar-code system is returned by increased accuracy, increased productivity, and decreased expenses. Depending on the volume of throughput, initial expense, and the application itself, the system may pay for itself in a year's time or even less.

With increased productivity and optimum use of each employee, bar coding can decrease the number of staff needed to handle data entry. Data accuracy increases; the need to correct data entry mistakes almost disappears. Fewer employees per shift means less overhead. Less overhead translates into more profits for the company, lower prices, higher customer satisfaction, or a combination of these.

Accurate inventory management uses warehouse space more efficiently and decreases the need for special sales, markdowns, and closeouts. By making these efficiencies possible, a bar-coding system can decrease the cost of doing business.

9.1.5 Tracking work in progress

In addition to the general applications mentioned, bar coding specifically can track work as it occurs and track goods throughout the manufacturing process.

The ability to collect data in real time enables one to monitor the amount of time needed to complete a task. This aids in forecasting future work force needs. At any given moment, tracking work in progress reveals how many products are available for sale and how far along the remaining products are.

Tracking work-in-progress (WIP) also has implications for quality control. If components are bar-coded, repeated part failure can be identified and isolated. This facilitates appropriate corrective steps to rectify the problem. Bar coding also can increase the efficiency of two related processes, tracking transported items and tracking paperwork through a process.

9.1.6 Tracking transported items

If the shipping containers and truck's manifest are all bar-coded, a list of contents of a shipment can be transmitted electronically through a computer network to the intended receiving dock. The receiving business can route the truck to the proper unloading dock, quickly identify the shipment, identify items not ordered, and return them to the driver for proper delivery elsewhere.

As in tracking work in progress, bar codes can identify where a particular shipment is in the transportation process and where a particular truck is along its sequence of assigned stops.

9.1.7 Tracking paperwork in process

As shipments often are modified or broken down at various sights, so paperwork is modified when it is reviewed or sent through a process. Bar coding allows each paper item to be tracked and located as it passes from one area to another (e.g., files, forms, claims). Tracking paperwork in process also allows tracking employee time needed to complete a certain process.

Bar coding cuts down the amount of time needed to track these items and allows better use of employee time. Public libraries have followed this principal in switching to bar-coded systems to check out books, releasing employee time for library activities.

9.1.8 Increased customer satisfaction

Bar coding allows a business to devote more time to customer service, leading to more satisfied customers. Customer satisfaction usually translates into continued patronage.

With efficient and timely inventory control, customers can purchase those goods they really desire; the goods will be in stock at the right time. Bar coding can help ensure that cash register lines move quickly and efficiently, without unnecessary interruption. Retail clerks can spend more time helping customers.

9.2 Bar Coding Organizations

Because of the widespread acceptance of bar coding, the need arose for guidelines for specific code types and regulation of their use within particular industries. To fill this need, organizations came into existence to make recommendations, establish policy, and mandate certain conditions for bar-code use. The most prominent of these groups are

1. Uniform Code Council (UCC)
2. Automatic Identification Manufacturers (AIM)
3. Voluntary Inter-Industry Communications Standard (VICS) Committee
4. Automotive Industry Action Group (AIAG)
5. Health Industry Bar Code Council (HIBCC)
6. Logistics Applications of Automated Marking and Reading Symbols (LOGMARS)

9.3 Bar-Coding System

The major component of a bar-coding system is the bar code itself. The bar-code label is the least expensive component of the bar-code system. A bar code must be clearly readable; an unreadable bar code defeats the purpose of the bar-code system and neutralizes any advantages it may offer.

Each bar-code label has three critical aspects:

1. Symbology
2. Substrate
3. Printing technology

All three elements have a direct bearing on the bar code's reliability.

9.3.1 The bar-code symbol

Apart from primal screams and primitive utterances, human sounds take on meaning determined by a particular language. The language is the index that correlates sounds with meanings. The symbology of the bar code is its "language" type, the set of rules or standards for how data takes on meaning within its bars and spaces.

Some symbologies encode data in bars only, others in bars and spaces (collectively *elements*). In general, symbologies use bars and spaces in different widths and configurations to encode data in either fixed or variable lengths. Depending on the symbology, this data may be numeric only, alphanumeric, or potentially the entire set of ASCII characters (ASCII stands for American Standard Code for Information Interchange).

Among the most popular symbologies are these:

1. *Universal Product Code* (*UPC*). Used mostly in supermarket and retail stores, this is the most familiar code type to many. UPC is a fixed-length symbology, encoding numeric data only. Each digit is represented by two bars and two spaces, each of which can be any of four widths.
2. *Interleaved 2 of 5.* This symbology represents numeric character pairs in groups of 5 bars and the 5 interleaved spaces. In each grouping of 5 elements, 2 are wide and 3 are narrow. The location of wide elements determines what numeric value is encoded. Interleaved 2 of 5 is the standard Uniform Container Symbol (UCS) code type.
3. *Code 39, or Code 3 of 9.* This symbology uses bars and spaces to encode variable lengths of alphanumeric characters from a set of 43 characters (all uppercase letters, numerals from 0 to 9, and 7 special characters). As its name suggests, 3 of every 9 elements are wide while the other 6 are narrow. The versatility of this symbology has led to widespread use. It is the standard symbology for the U.S. Department of Defense, case coding, hospitals, and inventory control.

Code 128. This relatively new variable-length code type is compatible with Code 39. It offers higher coding density than 39 and also, as its name implies, allows encoding the full ASCII 128-character set. There is a steady increase in use of this symbology.

In many cases, industries and organizations have set clear bar-coding standards, including use of mandated symbologies. Where symbology choice is open, there are a variety of considerations. Among them are these:

- Is the bar-code data fixed or variable in length?
- Is the required data numeric or alphanumeric?
- Is the code self-checking?
- Does the symbology permit encoding the required data in the product's available bar-code space?
- Do customers require a particular bar-code symbology?
- Is the industry likely to mandate a certain symbology in the near future?

9.3.2 Substrate

After selecting the symbology, following its standards exactly, and placing bar codes where they can be read easily, the next issue is the substrate—the physical surface the bar label is printed on.

Whether it is paper, metal, plastic, or some other material, the substrate must provide enough contrast with the bar code to make it readable by scanning. A bar code reader must determine the difference between bars and spaces; this requires a contrast of at least 25 percent between bar-code elements.

Further, the substrate must resist ink spread or any other process that can disfigure the bar code elements and lead to misreads or filed reads. Ink spread fattens bars and thins spaces. A disfigured bar code cannot conform to specifications. The substrate also must not degrade in its typical environment.

If contact readers are used, the substrate and bar code must not degrade from repeated physical contact. If the bar code is on a label attached to the product, the label must adhere to the surface, and the adhesive must not degrade the substrate material or darken it so as to reduce bar-code readability.

9.3.3 Printing technology

For bar-code symbols, the important printing issues are

1. Printing tolerances
2. Density
3. Contrast
4. Expense
5. On-demand versus off-site printing

Printing tolerances are critical in ensuring that bar codes conform to symbology specifications. Bar codes not printed to specification will cause nonreads and misreads, which may block or corrupt data input, rendering the bar-code system virtually useless.

Denser bar codes make it possible to encode more data in a small space, but they require more exacting tolerances. Printing dense bar codes to specifications is beyond the capability of many printing technologies.

In general, beyond strict conformity to bar-code specifications, the printed bar code must be free of spots and voids. Since a scanning beam crosses only a small part of the bar code, a spot in the path of the beam can be misinterpreted as a bar, while a void in a bar can be misconstrued as a space. Misreadings are a like result. To eliminate any confusion regarding bar and space widths, the edges of bars must be sharply delineated: dot matrix printing can cause data confusion. For sharply defined bar edges, the interaction of printing solutions and the substrate is a constant concern.

Whether to invest in an in-house bar printing system or to pay an outside vendor for bar-code labels depends on the application and the allowable expense. To determine whether the best solution for the situation is an on-demand printer producing labels in-house, closely examine several requirements:

Quantity of bar codes

Number of different symbologies

Number of different substrates

Internal expertise about requirements

Type of data to be encoded (e.g., sequential numbers, large batches of similar or identical labels)

Printing quality

Internal or external use of bar codes

Time constraints

Price of printing equipment

9.4 Scanners

The most critical bar-code system element is the bar reader or scanner. To read a bar code, a moving-beam scanner projects light at the bar-code symbol, oscillates the light back and forth across the symbol, and measures light reflections from bars and spaces during the passes. Since bars absorb more light than the spaces, light reflected from the symbol has a pattern of levels. By analyzing this pattern of reflected light, the reading equipment can determine the pattern of bars and spaces, the code type, and then the encoded data.

Depending on criteria, there are several ways of classifying scanners. Main criteria can be

1. Light Source [light-emitting diode (LED), helium-neon gas laser, or laser diode]
2. Code presentation (hand-held versus fixed mounting)
3. Symbol presentation (contact versus noncontact)
4. Integration of decoding intelligence (in scanner or in interface controller)

9.4.1 Light source

Each light source category is divided into subcategories. LED light may be visible or infrared. LED light sources may be used individually, as in most contact wands, or may be used in an array, as with charge-coupled device (CCD) readers, whose LED array spans the entire bar code and therefore need not be moved across the bar code.

Because laser light resists dispersion, laser bar-code readers have a great depth of field. Laser light may be visible or infrared; it may emanate from a helium-neon gas laser from a more compact solid-state laser diode. While infrared laser light is perfectly adequate for most black-and-white bar-code environments, visible laser light has very few color or substrate restrictions. For many purposes, advances in visible laser-diode technology have achieved performance levels which rival those of helium-neon technology.

Use of all laser scanning products is regulated by the U.S. federal government through the Center for Devices and Radiological Health (CDRH), a division of the Food and Drug Administration (FDA). This agency classifies laser devices according to power output during operation. Basically, laser bar-code scanners use such small levels of energy that they pose no real risk to scanned goods or to the human eye.

Frequently, a scanner's power output is controlled by time-out algorithms, which limit active scanning to a brief interval. In addition, scanners frequently have an automatic shutoff feature that operates if the scan beam stops moving. These features make laser scanners as safe as ordinary light bulbs.

9.4.2 Code presentation

Since a fixed-mount scanner is stationary, the bar code must be brought into the scan field for decoding. This is an ideal arrangement for assembly-line applications. One interesting fixed-mount scanner application is operation of movie theater controls by means of bar codes appearing on the film itself. The bar codes of the film control lights, curtains and other theater operations. Generally, since the fixed-mount scanner does not have to be held, the operator's hands are free for other activities.

The hand-held scanner can be moved freely and aimed directly at the bar code; this is an advantage for bulky, hard-to-handle items. Although one hand is not free, the ability to move the scanner is a greater advantage for many applications. This is especially true if the scanner is lightweight and ergonomically well-designed.

9.4.3 Symbol presentation

Contact scanners require direct physical contact with the bar code in order to read it. Called *wands* or *light pens,* these devices are less expensive than noncontact devices and are more easily replaced.

However, frequent physical contact can degrade or ruin some barcode labels; it also can lead to dirt collection, scratching, and even breakage of the device. Tip replacement is a common necessity. The wand's fixed aperture limits it to bar codes of a certain density.

A wand is limited also by the need to be dragged at a consistent speed across a flat surface. First-read rates are decidedly lower than for noncontact scanners. Repeated attempts often are necessary.

Although significantly more expensive than wands, noncontact scanners can read bar codes of various densities over hard-to-reach curved and irregular surfaces. Typically, these scanners may have an effective scan range of up to 2 ft; special long-range versions may reach up to 5 ft. The read rates are greatly enhanced by a scan beam which moves back and forth across a bar code about 36 times per second. These features add up to a much greater flexibility in use and a much higher scanning throughput.

9.4.4 Integration of decoding intelligence

For gathering data, scanning by itself is an incomplete activity. *Scanning* simply means projecting light across the bar code and receiving the reflected light. Decoding, formatting, and transmitting the scan data compatibly to the receiving system are separate functions.

If the scanner does not contain this decode and transmission intelligence, it must attach to an interface controller, which in turn connects to the host receiving system. If the scanner contains this intelligence, it is called an *integrated scanner* and connects directly to the receiving port on the host system. When the scanner collects light reflected from the scanned bar code, it converts this light pattern into an analog signal. The signal is formed from measurements of the time it took the scan beam to cross each element, which is directly proportional to the width of each bar and space. Then digitizing circuitry refines this signal into a digital signal representing the widths of bars and spaces in the scanned bar code. The signal is sent to the interface controller.

If the bar code was printed according to specification, then the digital signal should decode accurately. The decode algorithm corrects for slight abnormalities like the halo effect, which tends to scatter light reflected from the substrate (space) but not from a bar, with the effect of shrinking spaces and flattening bars slightly. However, when they occur, printing aberrations in the scanned bar code directly affect the shape of this signal pattern.

The interface controller analyzes the digital signal to determine code type before data content. The decode algorithm compares the signal to patterns that distinguish particular code types, a process called *autodiscrimination.*

With symbology type known, the algorithm performs data analysis from the pattern of bar and space widths, according to specifications established by that symbology. The analyzed data content must be formatted for transmission to the host device. The host device may be anything from an electronic cash register to a mainframe, or portable terminal.

Since there is a wide range of possible receiving equipment, the interface function may be quite complex, involving precisely clocked signals and intercharacter delays, emulation of wand or optical character recognition signals, and preprogrammed hardware and software handshaking signaling, in addition to actual transfer of data. Compatible transfer of data depends on proper selection of the interface controller and proper programming, according to host requirements. Compatible transfer enters accurate scan data into the host system.

9.5 Information Gathering

The success of a bar-code system in gathering information depends on how well its components interact. The principal components are the bar-code symbol and the scanning device.

The symbol's quality directly affects its reliability. But a symbol's quality in turn is determined by a number of related factors: the symbol density, substrate properties, and printing quality.

To a large degree, symbol density affects symbol readability in direct relation to both printing quality and scanner spot size. The higher the symbol density, the more exacting the demands on printing quality. For any spot size, some symbol densities become more readable than others.

Spot size must create the conditions under which light reflected from the individual bars and spaces of the symbol will have clearly differentiated levels. Additional considerations for optimum spot size include wavelength, working distance, frequency band of the electronics, and choice of digitizing circuitry.

9.5.1 Symbol density

Once spot size is determined, the scanner will be able to read a certain range of symbol densities more reliably than others. This is why a particular scanner may be available in both standard and specialized versions, such as a very high density model.

Figure 9.1, illustrates how spot size and symbol element width interact. If a spot is too large when compared to bar code's narrowest element (bar or space), the spot will overlap from one element onto the

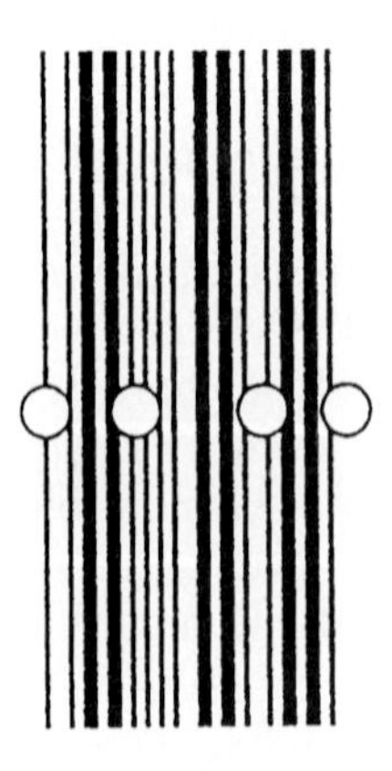

Figure 9.1 Scanner spots on a bar-code symbol.

next. This greatly weakens the differentiation between adjacent elements; light collected by the scanner's photodetector does not clearly indicate where a space or bar begins or ends. Without reliable detection of bar and space widths, decoding becomes impossible. On the other hand, if the spot is very small compared to bar and space width, it will be very sensitive to small printing imperfections. This can cause it to misinterpret a minuscule spot in a space as a bar, or a small void within a bar as a space.

9.5.2 Substrate properties

There are three interrelated substrate, or medium, characteristics:

1. The medium's surface reflectivity at a specific wavelength or range of wavelengths
2. Its light radiation pattern
3. Its degree of transparency or translucency

For successful bar-code reading, it is recommended that light reflection from a substrate be at least 70 percent. When this is not possible (e.g., with corrugated stock), successful reading depends on a greater print contrast between substrate medium and ink. Shiny surfaces and highly transparent substrate material should be avoided.

A further problem in using transparent or translucent substrate materials is paper bleed, which is caused by the scattering of incidental light rays within the medium or from the underlying surface. This phenomenon is also called *substrate scattering* or the *halo effect.* The net effect is that the bars appear larger and the spaces narrower than they were actually printed.

Because decoding is done by analyzing data on the basis of the pattern of light reflected from scanning the bar code, any distortion of the reflected light pattern can adversely affect the decoding.

Any selection of substrate material must ensure that there is sufficient print contrast between the bar code and the substrate. The print contrast signal (PCS) is a measure of the scanner's ability to capture the difference between bars and spaces according to their reflectivity:

$$\text{PCS} = \frac{R_L - R_D}{R_L \times 100}$$

where R_L = reflectivity of spaces and R_D = reflectivity of bars.

A high percentage of print contrast, combined with a lack of ink bleed, ensures a high correct read rate. Ideally, the PCS should be at least 50 percent, although many scanners can read symbols that have a PCS as low as 20 to 25 percent.

Closely related to PCS is the signal-to-noise ratio (SNR). SNR varies with the difference between the actual bars and spaces and the perceived bars and spaces. As mentioned, the reflection of light back into the scanner may distort the appearance of bars and spaces to the scanner. A poor contrast between substrate and ink, either because of colors of bar and light source or because of light absorption, can lead to noise and consequently to poor signal-to-noise ratio.

The most common form of noise occurs when the edges of bars are read. Infrequently, the middle of bars will be affected. In addition, contamination of data, once captured by the scanner, is much lower than the incidence of misreads attributable to high signal-to-noise ratio.

The relationship between SNR and the PCS is very strong. The higher the SNR, the lower the PCS is. A symbol with extensive ink bleed usually will not have a very strong print contrast signal percentage. The spaces and bars are not very distinguishable from each other, or fade into each other without sharp edges.

9.5.3 Scanner selection

Mode of operation is a prime consideration in scanner selection. For most fixed-beam scanners, either the scanner must be moved across the bar code, or the scanned bar code must be moved across the scanned field. For a moving-beam scanner in a fixed position, the bar-coded object must be brought to the scanner field; if the scanner is hand-held, the scanner's reading field can move to wherever the hand can take it. Moving-beam scanners, especially flying-spot laser scanners, automatically move the scanned beam back and forth across the bar code. This yields a major advantage in speed and data throughput. It also offers the flexibility to read bar codes through transparent surfaces, over curved and irregular surfaces, and in hard-to-reach places.

First read rate is a significant index of scanner performance. First read rate refers to the percentage of accurate decodes on the first trigger pull. The higher the percentage, the better the scanner performance. Since moving-beam scanners make multiple scans automatically, first read rate is a misnomer. There may be as many as 32 or more scans within the first second of activation. Since there are so many scans per second, a successful read will happen relatively quickly; to the user, there is no appreciable time delay between the first scan and the twenty-third scan. Overall, moving-beam scanners enjoy a much higher "first read rate" than fixed-beam scanners do.

Moving-beam scanners use a laser light source. Laser light has a single color and wavelength. It is coherent, high-energy light that can

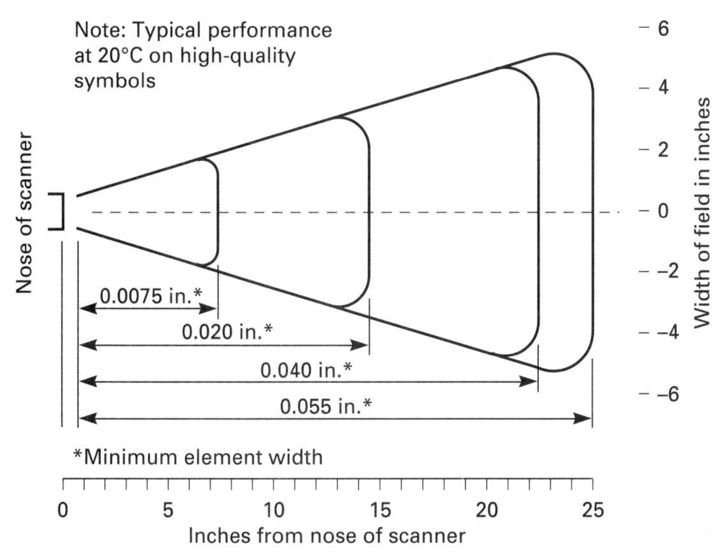

Figure 9.2 Working distance as a function of symbol density.

travel relatively long distances without significantly dispersing. By contrast, a light bulb radiates panchromatic, or "white," multidirectional light. Since laser light travels in a beam that resists dispersion, laser scanners can be used as noncontacting devices.

The distance over which a scanner can operate is called its *working range*. The longer the working range, the farther away the scanner can be held from the bar code and still read the bar code successfully. The effective working range varies with the bar-code density. The denser the code, the shorter the range (Fig. 9.2).

Related to working range is the *field of view* (FOV), defined as the width of the scanning beam's path across the bar code to be read. Up to a certain point, the FOV can be improved by pulling the scanner away from the bar code. By operating the scanner at the upper limits of the working range for a particular bar-code density, the scanned beam can spread out across the width of the bar code without loosing effective concentration. Of course, at some point the scanner reaches an outer limit where it will be unable to keep efficient focus. For a specific code density, the scanner's FOV determines the maximum bar-code length it can read.

One significant aspect of scanner design is the ability to filter out unwanted light, either ambient light that blinds the scanner or an excess of light reflecting from a substrate that floods the scanner. Either condition renders the scanner unable to differentiate between bars and spaces.

The most profound influence on scanning accuracy and SNR within the scanner component is the optimum spot size. The *spot size* refers to the diameter of the concentrated light beam as it travels across the bar code. The optimum spot is the same size as the narrowest element in the bar code. As mentioned earlier, a wide spot may overlap two narrow elements at the same time and render the scanner incapable of distinguishing between the start and stop points of an element.

Related to optimal spot size is the issue of waste size and location. The *waste* refers to the center of the depth field. This is where the highest-quality scanning takes place. Just as there is a small area in the center of a tennis racquet where optimal power and control over a tennis shot can be found, the closer to the waste the scanning takes place, the greater the scanning accuracy. The farther away the waste scanning takes place, the larger the spot becomes until, for a given symbol density, successful scanning is not possible.

9.6 Digitizing and Decoding

Without digitizing, decoding is impossible. Light reflected from the bar code is converted into an analog pulse signal. The analog signal has a variable amplitude; its pulse lengths and levels correspond to the symbol's bars and spaces. If any significant reading flaw exists, such as lack of adequate contrast, nonspecification bar-code printing, or poor printing quality, the analog signal will be useless or misleading. Flawed components at any level of a bar-coding system will affect the whole system detrimentally, if they do not cancel its effectiveness altogether.

For decoding, the continuous analog signal is digitized, converted into a signal with only two values. The zero and one values in the digital signal correspond to the scanned symbol's bars and spaces; pulse widths correspond to element widths (Fig. 9.3).

Accurate digitizing depends on accurate thresholding. There must be a cutoff value which determines whether any portion of a modulating analog pulse pattern corresponds to a bar or space. The threshold value determines what portions of a signal are recognized as bars or spaces.

9.7 Encoders

9.7.1 Glossary of encoder terms

Absolute Applied to encoders providing a unique binary word for each position.

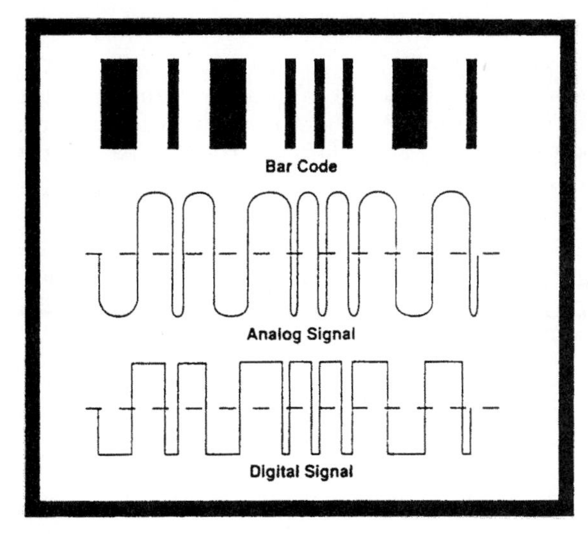

Figure 9.3 Analog and digital signals representing a bar code.

Accuracy The deviation between the actual position and the theoretical position of each bit edge; traceable to the encoding disk. Accuracy is unrelated to resolution.

Bit In incremental encoders, one quantum of data or one increment of digital code; in absolute encoders, the number of a track, which normally equates to the power of two of the final resolution (i.e., 8 bits = 2^8 = 256 positions, 12 bits = 2^{12} = 4096, etc.).

Complement The inverse of a digital signal.

CPR Cycles per revolution.

CPT Counts per turn.

Hysteresis A dead band purposely introduced in the encoder electronics. It helps to prevent ambiguities if the system happens to dither on a transition.

Incremental Applied to an encoder providing logic status 0 and 1 alternately for each successive revolution.

Index A single, separate output on an incremental encoder providing one count per revolution. It may be geared to the count channel.

Interpolation A multiplication technique for increasing encoder resolution.

Line driver An integrated circuit with differential output intended for use with a differential receiver, usually provided where long lines and high frequency are required. (Caution must be exercised with fast-switching drivers. Ringing may occur if the line is not terminated properly.)

PPR Pulses per revolution.

Quadrature Applied to two output channels out of phase by 90 electrical degrees.

Repeatability The ability to repeat bits exactly; a measure of the deviation of the actual encoder position between subsequent identical code readings. Repeatability is unrelated to resolution and is usually 4 to 10 times better than accuracy.

Resolution The number of bits or words contained in the complete code. For incremental and tachometer encoders, resolution is defined as *counts per turn*. For an absolute single turn, it is called *positions per turn*. Multiturn encoders are specified in terms of positions per turn of the input shaft and the number of turns of the internal gear. Resolution is not the same as accuracy.

State Logic level (high = 1, true; low = 0, false).

Transition Change of state occurring at a bit or word edge.

9.7.2 Universal types of encoders

Encoders are powerful tools for achieving control of automated processes economically, with high accuracy and repeatability. They can be classified into two major categories: absolute encoders and incremental encoders.

9.7.3 Absolute encoders

An absolute encoder (Fig. 9.4) provides a whole-word output with a unique code pattern representing each position. This code is derived from independent tracks on the encoder disk, corresponding into the individual photodetectors. The output from these detectors would then be high or low depending on the code disk pattern for the particular position (Fig. 9.5).

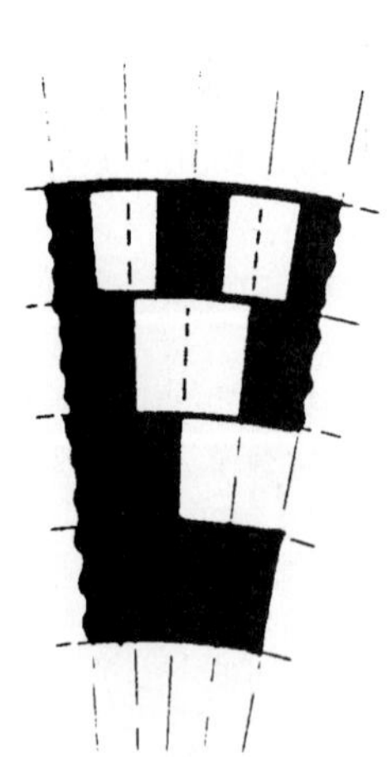

Figure 9.4 Segment of an absolute encoder.

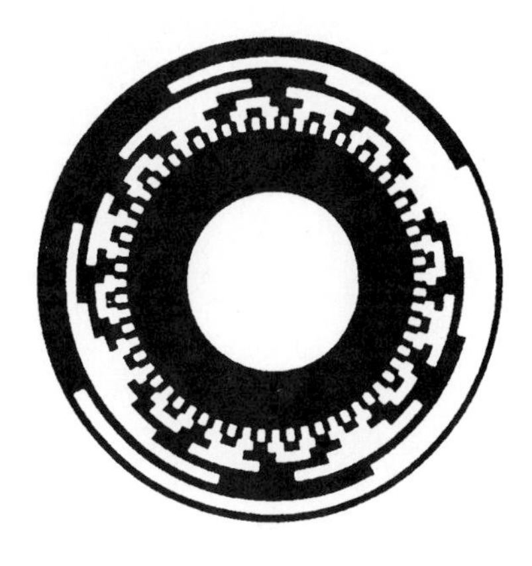

Figure 9.5 Eight-bit absolute disk.

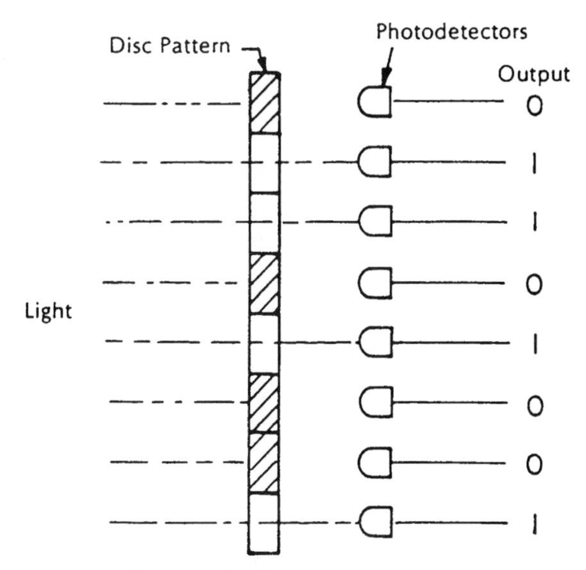

Figure 9.6 Binary code system.

Absolute encoders are used in applications where a device is inactive for long periods of time or moves at slow rates, such as flood gate control, telescopes, and cranes.

Absolute encoders are capable of using many thousands of different codes but the most common are Gray (Fig. 9.6), natural binary, and binary-coded decimal (BCD). Gray and natural binary are available up to a total of 256 counts (8 bits).

Gray code is particularly suited to optical-to-optical encoders because it is a nonambiguous code. That is, only one track changes at a time. This allows any decision during an edge transition to be limited

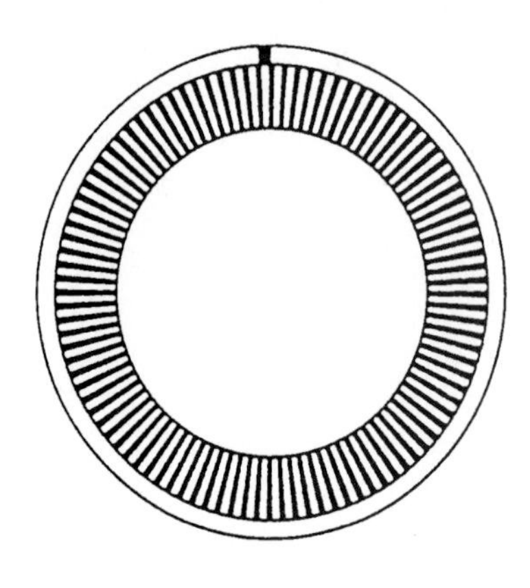

Figure 9.7 Incremental disk.

to ±1 count. Natural binary code is converted from the Gray code through digital logic. The latch option is used to lock this code to prevent ambiguities if the output changes while reading (Fig. 9.6).

9.7.4 Incremental encoders

The incremental encoder (Fig. 9.7) produces a series of square waves. The number of square waves can be made to correspond to the mechanical increment required. For example, to divide a shaft revolution into 1000 parts, an encoder could be selected to supply 1000 square-wave cycles per revolution. By using a counter to count those cycles one could tell how far the shaft has rotated: 100 counts would equal 36°, 150 counts, 54°, etc. The number of cycles per revolution is limited by the physical line spacing and the quality of light transmission. The incremental resolution can be divided into 2540 cycles per turn directly placed on the encoder disk. Higher resolutions are available through various multiplication techniques.

Generally, incremental encoders provide more resolution at a lower cost than their absolute encoder cousins. There are also fewer interface problems because they have fewer output lines. Typically, an incremental encoder would have four lines: two quadrature signals and two power lines.

A 13-bit absolute encoder would require 13 output wires plus 2 power lines. To use complementary signals for noise immunity would require 28 conductor cables.

There is often confusion about the different types of incremental encoders. There are two main incremental encoders, the tachometer encoder and the quadrature encoder.

Tachometer encoder. A tachometer encoder (Fig. 9.8) is sometimes called a *single-track incremental encoder* because it has only one out-

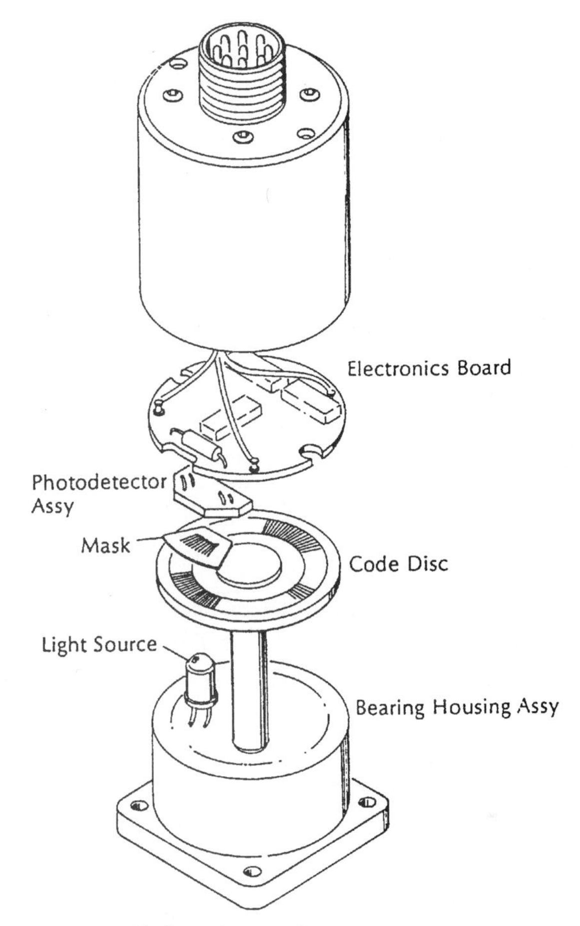

Figure 9.8 Tachometer encoder.

put and cannot detect direction. The output is usually a square wave (Fig. 9.9).

Velocity information is available by looking at the time interval between pulses or at the number of pulses within a time period. When using the interval between the pulses, the encoder should provide good edge-to-edge accuracy. Any inaccuracy would cause the servo system to constantly correct "errors" caused by disk pattern irregularities.

Quadrature encoder. Most incremental systems use two output channels in quadrature for position sensing (Fig. 9.9). This allows both counting the transitions and viewing the state of the opposite chan-

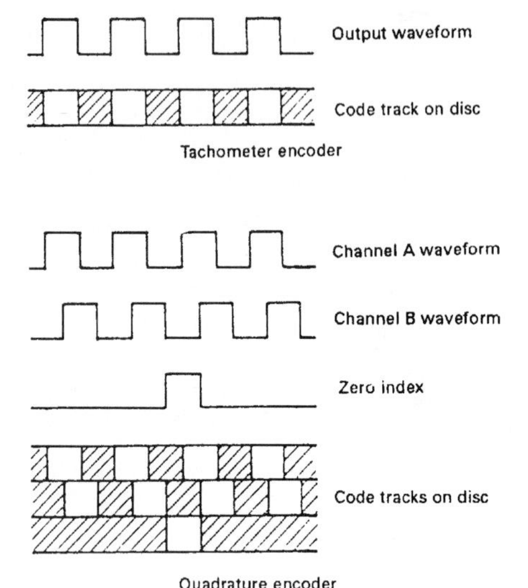

Figure 9.9 Encoder outputs.

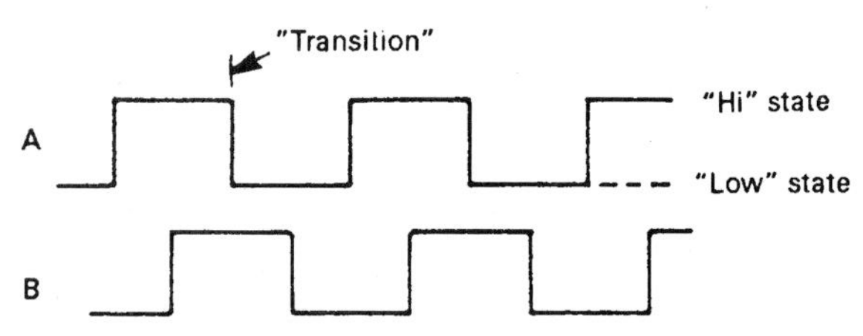

Figure 9.10 Comparison of A and B outputs of quadrature encoder.

nel during these transitions. Using this information one can determine if A leads B, and thus derive direction information (Fig. 9.10).

It is important to have this directional information available because of the vibration inherent in almost any system. An error in count will occur should an encoder with a single channel (tachometer-type) stop on a transitional edge. As vibrations force the unit back and forth across this edge, the counter will count the transitions even though the system is virtually stopped. But by utilizing quadrature detection on a two-channel encoder and viewing the transition in its relationship to the state of the opposite channel, one can generate reliable directional information.

9.7.5 Directional sensing pulse multiplication

Once the quadrature signal is decoded, pulses can be generated of fixed duration at selected edge transitions within a cycle. These pulses can be fed via clockwise and counterclockwise output lines to an up-down counter or programmable counter input port.

Many counter and personal computer manufacturers include a quadrature detection circuit as part of the electronics. This allows the use of a two-channel quadrature input without further conditioning.

With quadrature detection one can derive 1×, 2×, or 4× the basic code disk resolution. For example, 10,000 pulses per turn can be generated from a 2500-cycle two-channel encoder. With a quality disk and properly phased encoder, the 4× signal will be accurate to better than one-half count (Fig. 9.11).

9.7.6 Interpolation

The output accuracy characteristics of interpolating encoders are somewhat different than those of standard incremental encoders.

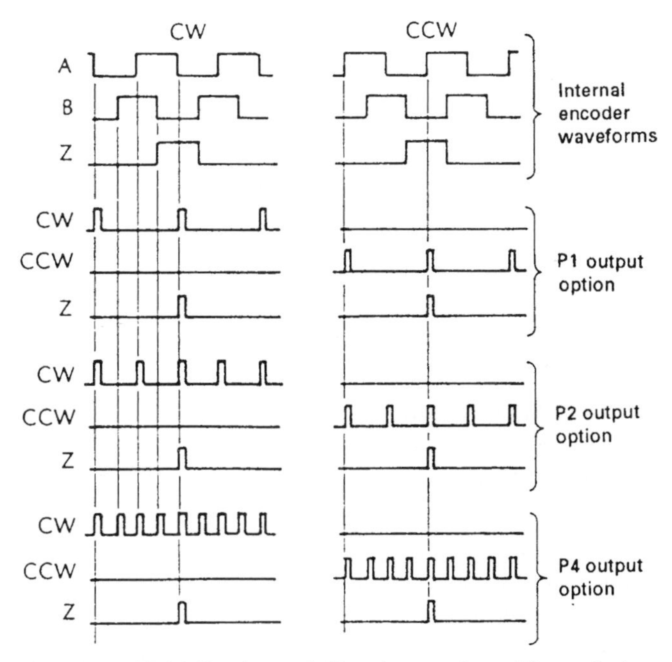

Figure 9.11 Multiplication and direction sensing with quadrature pulses.

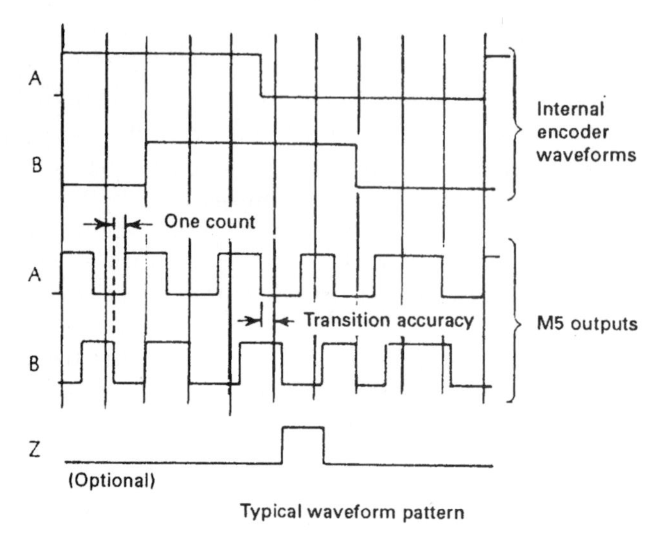

Figure 9.12 Typical interpolation waveforms.

Electronic multiplication provides a higher degree of angular resolution while trading off some pattern regularity. While these encoders are not suited for velocity servos, they are ideal for position readouts and position servos.

The major specifications of interest for this type encoder are transition accuracy and frequency response. In position readout situations the system usually does not know where it is within a particular count, so tight transition accuracy specifications tend to be wasted. What is desired in these applications is the ability to subdivide a count into smaller divisions to improve system resolution. Interpolation provides this ability without any sacrifice in mechanical integrity or internal electrical signal strength, thereby retaining safety factors associated with lower resolution specifications (Fig. 9.12).

9.8 Selecting Encoders for Innovative Affordable Automation

Industrial encoders are available for use over a wide range of incremental conditions. The large variety of available designs allows the user to customize an encoder to the desired requirements. This also allows the specifying engineer to select only the options needed without incurring unnecessary additional costs.

There are a number of factors that must be considered to ensure reliable, consistent encoder operations in industrial applications. In particular, the encoder must have a high degree of mechanical and electrical stability. In order to achieve this stability for affordable automation the encoder must have a solid foundation. The encoder disk, shaft, and bearings must be of the highest quality to ensure the ultimate accuracy of the device.

The encoder disk interrupts the light as the encoder shaft is rotated, and it is the code pattern etched on the disk which is primarily responsible for the accuracy of the electrical signal generated by the encoder. Should the disk pattern be inaccurate, the resulting signal will reflect that inaccuracy. Many manufacturers have developed sophisticated divided-circle machines capable of accuracies in the sub–arc-second range. Originally intended for the military and aerospace industries, this quality is now automatically incorporated into industrial products.

The shaft and bearings maintain accurate rotation of the disk and help to eliminate such errors as wobble and eccentricity, which would be translated into position errors. Eccentricity, for example, can cause inaccuracies in the encoder output that will not be apparent to the user during electrical testing, but will cause false position information. In order to eliminate eccentricity errors some manufacturers have developed an electronic centering fixture capable of centering accuracies up to 40 millionths of an inch.

9.8.1 User abuse

Even with the appropriate package, shaft, bearings, and disk, the user must exercise care to avoid undue shock and abuse. In particular, the glass code disk can be damaged if the encoder is dropped or a pulley is hammered on the shaft. The typical specification for an industrial encoder allows a 50*g* shock for 11 ms and vibration of 20*g* from 2 to 2000 Hz.

9.8.2 Bearing loads

In applications utilizing gears or drive shafts, excessive radial (side) loading on the shaft can shorten bearing life. Therefore, encoders should be specified in accordance with the anticipated side loading. Typical loads for industrial encoders are 5, 40, 80, and 100 lb.

Ultraheavy-duty encoders are available to withstand heavier loads as well as shock up to 200*g*.

9.8.3 Protective housing

To adequately protect the optical and electronic components from exposure to the environment, encoder case thickness should be consistent with the severity of expected abuse. In applications where the housing may be struck by tools or debris, a heavier housing or protective shroud should be considered.

9.8.4 Finish

Aluminum encoder housings with chemical coating (Iridite or Alodine) finish are sufficient for most applications. However, if the encoder is intended for operation in a corrosive environment, a hard anodized finish with a dichromate seal should be considered.

9.8.5 Electrical shielding

Most industrial environments will be electrically noisy. The internal electronics are protected by the conductive metal housing; however, the output electronics and the output cable need to be selected according to expected conditions.

9.8.6 Temperature and other specifications

The temperature specification of the selected encoder must be consistent with the application; 0 to 70°C is the standard operating temperature for industrial encoders. Extended temperature ratings from −40 to +80°C are available.

Some applications may require special certification—for example, that an encoder is explosionproof. Basically, testing for this certification determines that if certain flammable gases infiltrate the encoder housing and are ignited by the internal electronics, the resulting flame or explosion is not able to escape from the housing and ignite the surrounding atmosphere. Specifically designed encoders are available that meet the appropriate specification (NEMA Class 1, Group D, Division 1).

9.8.7 Environmental sealing

If the application requires operating in a liquid or dusty environment, the encoder must be selected accordingly. Adequate sealing is imperative to ensure against contamination from liquids or dust, particularly through the shaft/bearing assembly. Contaminants that infiltrate the shaft bearing can rapidly degrade encoder performance; they may also work their way to the encoder interior where they can disrupt the optical components or damage the circuit board. A sealed bearing is sufficient protection against dust, but a shaft seal must be used in applications where liquids are present. If liquid contamination is expected, the user should request a leak test.

9.9 Examples of Encoders for Affordable Automation

Example 1: Encoding the motion of a lifting mechanism The encoder is to be integrated with a programmable logic controller (PLC) having a high-speed counter module requiring 5-V line drivers and complementary inputs.

In order to encode the motion of a lifting mechanism using wire of 0.008-in. diameter and a pulley of 4-in. circumference, the following parameters may be applied to either of the configurations in Figs. 9.13 and 9.14:

Speed	20 rpm
Resolution	0.001 in.
Counts per inch travel	1/0.001 = 1000 counts/in.
Effective radius R_{eff}	Circumference ÷ $2\pi = 4/2\pi = 0.63662$ in.
Actual radius R_{act}	$R_{eff} - \frac{1}{2}$ wire diameter $= 0.63662 - 0.004 = 0.63262$ in.

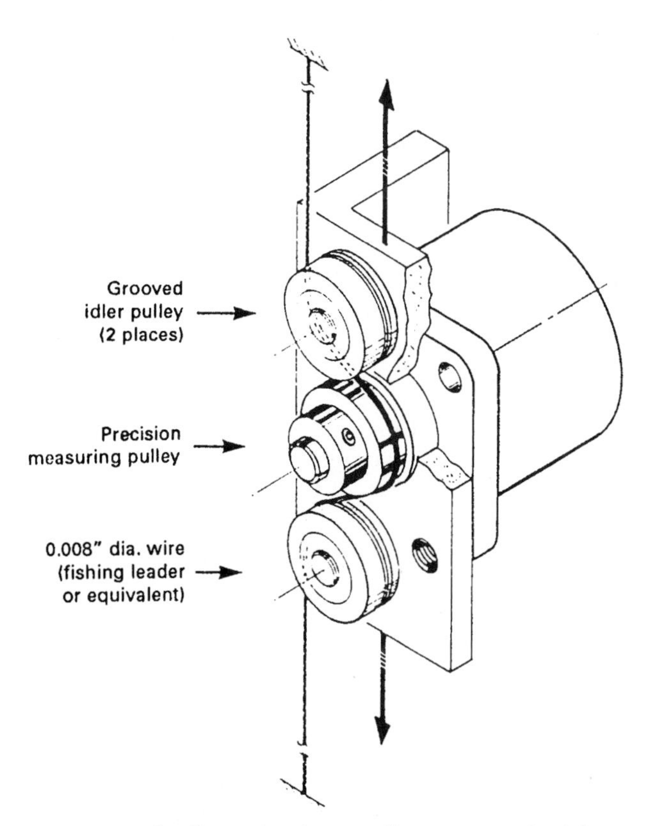

Figure 9.13 Configuration for encoding motion of a lifting mechanism.

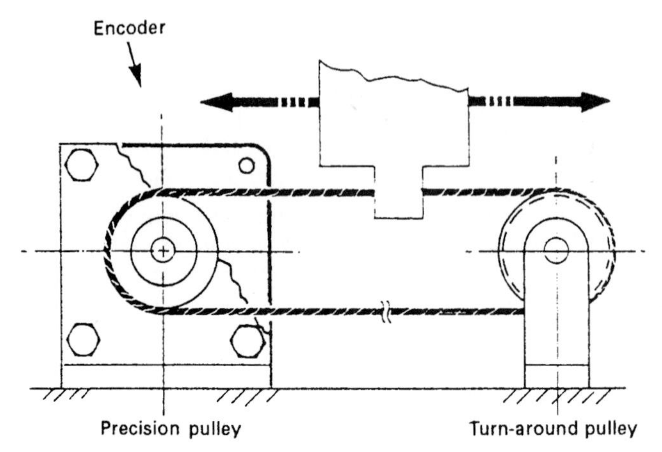

Figure 9.14 Alternative configuration for encoding motion of a lifting mechanism.

Thus the pulley diameter = $2 \times 0.63262 = 1.2652$ in.

For manufacturing plant implementation, it is recommended that a 100-ft electrical cable be connected to the programmable logical controller.

Encoder specifications Use a heavy-duty encoder with a sealed shaft. The number of cycles per turn can be determined as follows:

$$4 \text{ in./turn} \times 1000 \text{ counts/in.} = 4000 \text{ counts/turn}$$

Since the counter has 4× multiplication, a 1000 cycle/turn encoder will suffice.

Example 2: Display of numerically controlled lathe The display is to be integrated with a 6-decade quadrature counter and 5-V power supply. In order to encode the motion of the lead screw of the numerical-control (N/C) lathe mechanism (Fig. 9.15), the following parameters may be applied:

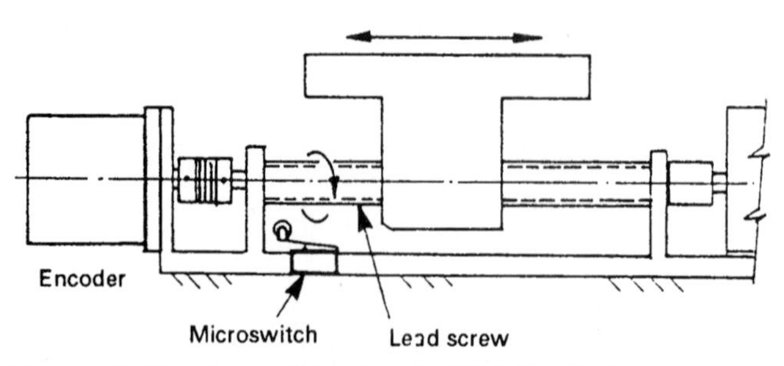

Figure 9.15 Encoder configuration for NC lathe display.

Speed of lead screw	500 rpm
Pitch	¼ in.
Travel	20 in.
Resolution	0.0005 in.
Power supply	5 V
Cable length	20 ft
Environment	Oil mist

Encoder specifications Use a heavy-duty encoder with a sealed shaft. The number of cycles per turn can be selected as follows:

$$\text{Counts/turn} = \text{pitch/resolution} = 0.025/0.0005 = 500 \text{ counts/turn}$$

Use a 250 counts/turn encoder with built-in multiplication to obtain 500 counts/turn.

Example 3: Encoding a linear actuator The encoder is to be integrated with a 6-decade pulse counter and 5-V power supply. In order to encode the motion of the rack and pinion of the linear actuator (Fig. 9.16) the following parameters may be applied:

Number of teeth	40
Pitch	1/20
Stroke	20 in.
Resolution	0.0002 in.
Environment	Clean
Cable length	18 in.

Encoder specifications Use a light-duty encoder. No sealing is required. The number of counts per turn is

$$\text{Counts/turn} = \text{(in/turn)/resolution} = \frac{2}{0.0002} = 10{,}000$$

Use a 2500 counts/turn encoder with 4× built-in multiplication to obtain 10,000 counts/turn.

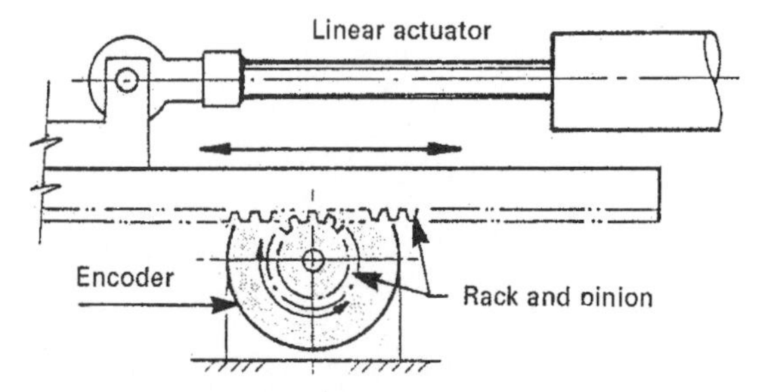

Figure 9.16 Encoder configuration for a linear actuator.

Chapter

10

Economic and Social Concerns

This chapter is particularly written to those who may possess the talents of innovation—and need the economic means to implement their dreams. It is written to all my students, my peers, and the entrepreneurs who wish to grasp the simple yet practical understanding to build their new manufacturing business with sound financial strength—attempting to make America a better America.

This chapter is dedicated to my students, the hope of tomorrow. Their eyes hold the sparks of inspiration and see the challenges to our country and rise to them. They can take vision and creativity, mixed with the discipline of science and thereby, to form the future.

May they take the fire of their enthusiasm and turn their ideas into reality. As I have said to my postgraduate engineering students, "Engineering in the absence of economics is a degenerate science." My sincere desire is that students foster their creativity, nurture their confidence, and take the risk to establish their enterprise.

10.1 Financial Planning and Control of Manufacturing Operation

The small manufacturing firm, just as surely as the larger manufacturer, must have yardsticks for measuring the results of every activity of its business operation. For this purpose, management should take the following steps:

1. Define its objectives precisely
2. Prepare a written statement of the steps to be taken to accomplish these objectives

3. Refer continually to its statements of objectives to see that its operation is proceeding according to schedule

For some manufacturers, the budget may be simply an informal sales forecast, production schedule, and profit forecast. If only one person is responsible for seeing that activities proceed in accordance with plans, the budget becomes a simple basis for action by a top executive. It helps an executive make on-the-spot decisions that are in line with an overall plan. It charts the course a top manager should take in carrying many different functions.

If the business grows to the point where selling, producing, and perhaps financing responsibilities are delegated to different persons, a more specific and formal plan becomes important. The budget must still direct the decisions of the owner-manager and key executive of the firm, but in addition it must coordinate the actions of the different members of the group.

10.2 Developing a Plan

A manufacturer's budgeting procedure must be simple. It must be focused on a limited number of functions that are important to the special circumstances. All individuals who are to have any responsibility for seeing that it works should take part in planning it. The overall budget for most manufacturers should include the following parts:

1. Sales budget
2. Production budget
3. Selling and administrative budget
4. Capital budget
5. Financial budget

The first four of these budgets taken together provide a basis for projecting profits, since the sales budget provides an estimate of revenues and the other three of costs. The financial budget provides a plan for financing the first four plans. Each of the five budgets should be made in advance for periods of 6 months or a year.

10.2.1 Sales budget

The accuracy of any budget or plan for profitable activity depends first on the accuracy of the sales forecast. In many small plants, the volume of sales may be affected substantially by the addition of a single new customer or the loss of a single old one. The sales forecast in these cases has to include a good deal of guesswork.

TABLE 10.1 Sales Forecast

	PRODUCT 1		PRODUCT 2		
Sales territory	Product units	Dollar sales	Product units	Dollar sales	Total sales
A	1,850	$14,800	2,200	$48,400	$ 63,200
B	575	4,600	500	11,000	15,600
C	1,400	11,200	1,000	22,000	33,200
Total	3,825	$30,600	3,700	$81,400	$112,000

Nevertheless, a forecast should be made. It will usually be based mostly on past sales figures, but sales tendencies and economic conditions that are likely to increase or decrease sales volume in the coming period should be studied.

The sales forecast or budget that might be suitable for a typical small plant is shown in Table 10.1. Forecasted sales are stated in both product units and dollars and are broken down by sales territories.

10.2.2 The production budget

Once the sales forecast has been made, the production budget can be prepared. Enough units of products 1 and 2 must be produced to supply the estimated sales volume and maintain reasonable stock levels. In this example, it has been decided that about a 2 months' supply should be kept in the stock. If the inventories at the beginning of the period are below that level, production will have to exceed the sales forecast. Table 10.2 shows how the quantity to be produced can be figured.

On the basis of the number of production units required, budgets or standards for material, labor, and overhead costs are prepared. These production costs should be figured in detail so that cash requirements, material purchasing schedule, and labor requirements can be set up. Detailed plans are then made for monthly production levels.

TABLE 10.2 Required Production Based on Sales Forecast

	Product 1	Product 2
Sales forecast	3825	3700
Desired inventories (1/6 of annual sales)	637	617
Product units required	4462	4317
Less: beginning inventories	200	400
Total production required	4262	3917

10.2.3 The selling and administrative budget

The sales volume of a small plant usually does not provide a large enough margin over production cost to overcome high selling and administrative costs. These nonmanufacturing overhead items must be watched constantly to see that commitments for fixed costs are kept low.

This can be done best by using a selling and administrative budget. All cost items expected in these areas should be listed and classified according to their fixed and variable tendencies. Account classifications in the accounting system should correspond to the classifications used in the budget.

At the end of each month, the fixed, variable, and total selling and administrative costs budgeted for the expected level of operation should be adjusted for the level actually reached during that month. Actual costs should be matched against these adjusted figures and any sizable differences investigated. If possible, action should be taken to prevent repetition of any unfavorable variances.

10.2.4 The capital budget

Plans for acquisition of new buildings, machinery, or equipment should be developed annually. A list of amounts to be spent at specified times during the year will provide information for use in a firm's financial budget as well as cost and depreciation figures to be shown on its projected income statement and balance sheet.

10.2.5 The financial budget

Any business owner must be certain that funds are available when needed for plant or equipment replacements or additions. In addition, enough working capital must be available to take care of current needs. This requires a financial plan or budget.

Funds for major replacement or addition can ordinarily be planned for on the basis of each individual expenditure. Management must know in each case whether surplus working capital will be available or new long-term financing will be required. If new financing is the answer, a source of funds must be found. Plans for a loan or an additional investment by owners should be made well before the time when the funds will be needed.

To have enough working capital at all times without a large oversupply at any time requires detailed planning.

Planning the receipts and expenditures for each month must take into consideration expected levels of activity, expected turnover of accounts receivable and accounts payable, seasonal tendencies, and any other tendencies or circumstances that might affect the situation at any time during the budget period. Relating the inflow plus beginning balance to the outflow for each month shows whether a need for outside funds is likely to arise during the budgeted period.

10.2.6 Income statement and balance sheet projections

From information provided by the four basic budget summaries—sales, production, selling and administrative, and financial—estimated financial statements can be drawn up. These projected statements bring the details together. They serve to check the expected results of operations and the estimated financial position of the business at the end of the budget period.

10.3 Planning for Profit

Projection for income statement and balance sheet figures will indicate profit, return on investment (ROI), and return on assets (ROA) management can expect. For example, these statements in highly abbreviated form may be as follows:

Net profit before interest and taxes			$430,000
Interest cost			80,000
Net profit after interest			$350,000
Income taxes (40%)			140,000
Net profit			$210,000
Assets	$3,000,000	Debt	$1,000,000
		Capital or stock investment	$2,000,000
	$3,000,000		$3,000,000

The return on investment, i.e., the profit earned in relation to the value of the capital required to produce the profit, is expected to be

$$\frac{\$210{,}000}{\$2{,}000{,}000} = 10.5\%$$

Further, the return on assets is expected to be

$$\frac{\$258{,}000}{3{,}000{,}000} = 8.6\%$$

Note: In this latter calculation, the return is calculated after taxes and before deduction of interest cost. The $258,000 is determined either by adding the after-tax interest cost of 60% of $80,000 or $48,000 to the net profit $210,000 or by subtracting the 40% tax on the net profit before interest or taxes of $430,000 from this amount. The 8.6% return on asset is increased to 10.5% return on investment as the result of the leverage provided by borrowed funds at an interest of less than 8.6%.

These ROI and ROA figures may or may not be considered adequate. If they are thought to be inadequate, various parts of the budget should be reviewed to identify possible areas of improvement. In any case, the profit goal should be high enough to provide a reasonable return and at the same time be realistically attainable.

10.3.1 Controlling operations

Budgeting and profit planning are necessary for control of operations, but the forecasts and budgets do not of themselves control operating events. A budget is useless without comparisons with actual operating results, and the comparisons are useless unless the manager takes action on deviations from the budget.

Suppose, for example, that the sales of product 1 in territory A are budgeted at 155 units per month (1850 divided by 12), and actual sales for January and February are 120 and 110 units. (Assume that sales of product 1 in territories B and C are in line with the budget.) On investigating, the manager finds that sales are being lost to a new competitor in territory A.

The actions the manager might take include these:

1. Put more money into the selling effort and try to win back the old customer.
2. Lower the price of product 1 to obtain more sales.
3. Recognize that the sales are lost, and lower the production budget accordingly.

Each of these actions will have an impact on the budget-planning operations of the company. An increase in the selling effort will affect the selling and administrative budget and the financial budget—more funds will be needed. Lowering the selling price if the unit sales level originally forecast is just regained, will result in lower cash receipts than in the original financial budget. A decision to reduce production to avoid piling up inventories will affect purchase orders, labor costs, variable overhead costs, income taxes, and other budgeted items. The actions have been signaled by a variation from the sales forecast, but a change in plans affects all the budgets and forecasts.

Any sizable variance from budgeted figures, favorable or unfavorable, should be looked into. Management should find the cause so that when it takes action it will be reasonably sure that it is acting wisely. For example, a favorable variance in direct labor costs might be due to a special incentive program. In such a case, management should put off any plans to use the surplus funds until it is sure that the reduction in labor costs will be permanent.

Unfavorable variances from the budget should not be studied in isolation from related budget items. For example, suppose the actual selling expense is much higher than the amount budgeted. It may turn out that the extra expense produced a large volume of new orders. In that case, the added cost may be entirely justifiable. Hasty action to reduce the expense without investigation might result in loss of sales and a net reduction in profits.

Great care must be given to seeking out the causes of variances, and the consequences of possible control action should be analyzed carefully in order to avoid unintended and unwanted results.

10.3.2 Cost and profit analysis

In direct cost system costs are classified as variable (product costs) and fixed (period costs). With these classifications, management can determine which of its products contributes the most and which the least to covering fixed costs and providing a profit. Direct costing provides information for decisions to expand, reduce, or continue production of given product lines.

In a standard cost system, product costs for materials, labor, and overhead are budgeted or planned before the actual production takes place. The preestablished costs then provide a point of reference for analyzing the costs of actual production. By analyzing the variances from these standards, the manufacturer can identify the sources of excessive cost and investigate the causes. This kind of deficient process—the identification of a problem area and the discovery of its cause—can point the way to actions that will reduce costs and improve profit.

10.4 Information Assimilation and Decision Making

A business manager should plan ahead for all the activities of the business. That is, the manager should develop a set of budgets for sales, production, selling and administrative expense, and financial requirements. As indicated in the standard cost system, the actual results of various activities can be compared to the budgeted or planned results. Analysis of the variances found will then help the manager to make reasonable and thoughtful decisions.

These analytical methods—direct cost, standard cost variance analysis, and budgeting-planning control—are ways of putting cost-accounting information to good use. The manufacturer can make decisions on the basis of information provided by these analytical tools. There are, however, several other important ways of analyzing costs and profits. Two of these—break-even analysis and incremental analysis—are illustrated below.

10.4.1 Break-even analysis

Sales forecasts are uncertain at best. For this reason, it is important for management to know approximately what cost changes can be expected to go along with volume changes. It must know what levels of production and sales volume are necessary for profitable operation. It must know how much effort or cost is justified in order to keep volume at a high level. Break-even analysis provides this sort of information.

Break-even analysis is so-called because the focal point of the analysis is the break-even point—the level of sales volume at which revenues and costs are just equal. At this level, there is neither profit nor loss.

10.4.2 Break-even chart and formula

A break-even chart is illustrated in Fig. 10.1. A break-even chart can be prepared from budget figures and a knowledge of capacity levels.

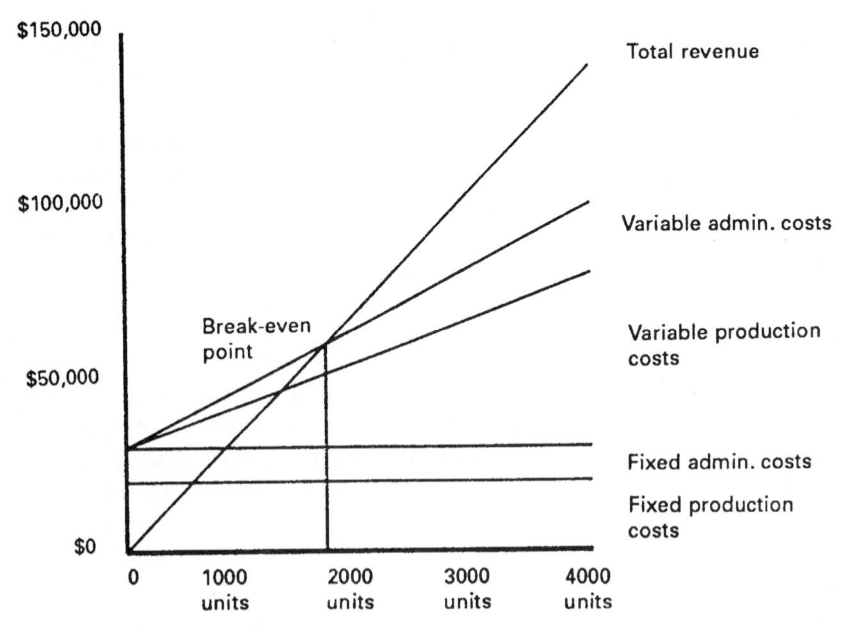

Figure 10.1 Break-even chart.

In Fig. 10.1, it is assumed that fixed costs for production are $20,000 and for selling and administrative activities, $10,000. Variable production costs are $12.50 per unit, or a total of $50,000 at the maximum a capacity of 4000 units. Variable selling and administrative costs are $5.00 per unit, or $20,000 at the 4000-unit level. This gives minimum costs of $30,000 (the total fixed costs) at zero production and maximum total costs of $100,000 at 4000 units. The selling price is $35 per unit.

A total cost line is drawn from the $30,000 cost level for no production to the $100,000 level for maximum production and sales. A total revenue line is drawn from the zero line for no revenues to $140,000 for the maximum sales of 4000 units. The point where these two lines cross is the break-even point—the level of operation at which there will be neither profit nor loss.

The area enclosed by the two lines below the break-even point represents loss and the enclosed area above the break-even point represents profit.

Figure 10.1 shows that the volume at which the business can be expected to break even is slightly below one-half the maximum point, or between 1500 and 2000 product units. The exact level can be calculated as follows:

$$\frac{\text{Fixed costs}}{\text{Selling price} - \text{variable cost per unit}} = \frac{\$30{,}000}{\$17.50} = 1714.29 \text{ units}$$

Verification:

Gross revenue (1714.29 units at $35)		$60,000
Variable costs (1714.29 units at $17.5)		30,000
Contribution margin	$30,000	
Fixed costs	30,000	
Net profit or loss	$ 0	

10.4.3 Utilization of break-even analysis

A break-even chart can give approximate answers to many questions. For example, if the variable production costs are expected to increase 10 percent in the coming year without any change in the selling price of the product, the total cost line will be drawn steeper, and a new, higher break-even point will result. This will show what increase in sales volume is needed to offset the increased costs.

Or suppose a change in selling price of the finished product is being considered. The total revenue line will now be steeper for an increase or less steep for a decrease in selling price. The chart will then show the effect on profit if the sales volume remains the same or if, for example, it drops—as might occur if the price is increased.

Suppose the manufacturing firm whose figures are used in Fig. 10.1 is producing and selling 2000 units, thus making a profit of $5000. Now suppose it wants to give its employees a general wage increase. As planned, the increase will have the effect of adding $1 to the variable costs of each unit (thereby decreasing the contribution by the same amount) and $3000 to fixed costs. The owner wants answers to these questions: How many units must the owner produce and sell at $35 each to realize the same profit—$5,000? Can the owner produce this volume without investing any more in plant and equipment?

The required units are calculated by using the following formula. (The calculation can be verified as described in the preceding section.)

$$\frac{\text{Fixed costs} + \text{decreased profit}}{\text{Selling price} - \text{variable costs per unit}} = \frac{\$33{,}000 + \$5000}{\$16{,}000} = 2303 \text{ units}$$

Thus, the manufacturing firm will have to produce and sell 2303 product units instead of 2000 in order to realize the same $5000 profit. Since the plant capacity is 4000 units, management can increase production to 2303 units without any further investment in plant and equipment.

Break-even charts give quick approximate answers. They should not take the place of detailed calculations of the results of anticipated changes, but they do encourage careful consideration of the effects of any increases in either fixed or variable costs. They can also be a constant reminder of the importance of maintaining a high sales volume.

10.4.4 Incremental analysis

Incremental costs and revenues are costs and revenues that change with increases or decreases in the production level. Fixed costs as well as variable costs may change under certain conditions. Incremental analysis is an examination of the changes in the costs and revenues related to some proposed course of action, some decision that will alter the production level or change production activity. Two kinds of incremental analyses are explained here.

The large new production order. Sometimes it happens that a manufacturer has an opportunity to land a new customer and a very large order for goods. Usually, such a large order cannot be handled without adding new production employees and supervisors and increasing overhead costs. The new order thus creates a decision-making problem in which basic information differs somewhat from the current cost data. A study of the incremental costs and revenues—those that arise from the new contract—provides information on which the manufacturer can base the decisions to accept or reject the new order.

Suppose that a manufacturer currently produces 10,000 product units a year. Fixed costs are $20,000 and variable costs are $5 per unit—$2 for materials, $1 for labor, $1 for variable production overhead, and $1 for variable selling expenses. The unit selling price is $10. A new customer wishes to buy 10,000 units a year at $9 per unit. A new night shift would be required to double the plan's production as required by the new customer's order.

Costs at the original level of production (10,000 units) are not necessarily accurate for the new level (20,000 units). Assume that the costs of producing the 10,000 additional units required by the new order are estimated as follows:

Variable costs per unit	
Materials	$2.00
Labor (including night shift premium)	2.00
Variable production overhead	1.25
Variable selling costs	0
Total variable costs	$5.25
Incremental fixed costs	
Additional supervisors	$10,000
Power, light, heat	5,000
Office expenses	3,000
Total incremental fixed costs	$18,000

Note: For accuracy, it is important to use current or projected prices of materials, labor, and overhead in this type of incremental analysis.

Average prices or allocated costs derived from historical cost accounting records (for example material costs based on the first-in, first-out method) may not produce current figures which can be relied on in making a decision.

The financial results of accepting the new contract would therefore be as shown below:

	Original production	New order	Total
Gross revenues	$100,000	$90,000	$190,000
Variable costs	50,000	52,000	102,000
Contribution margin	50,000	37,000	87,000

Fixed costs	20,000	18,000	38,000
Profit before taxes	$30,000	$19,500	$49,500
Return on sales	30.0%	21.7%	26.1%

The incremental analysis shows that the new contract will be profitable, but the rate of return will not be as high as for the original level of production. The computation would have given different results if the original variable costs of $5 per unit and the fixed cost of $20,000 had been used. Incremental analysis makes it possible for a manufacturer to examine the probable results of making major changes in the production level and in the arrangement and scheduling of production activity.

10.4.5 Make-or-buy analysis

Most manufacturing firms buy some parts from outside suppliers for use in assembling the finished products. Often the manufacturing firm could make the part in its own plant instead of buying it from an outside source. The question is whether the cost of making the product itself would be less than the cost of purchasing it from an outside supplier. An analysis of these two alternatives is known as a *make-or-buy analysis.*

Take, for example, a manufacturing firm that now purchases 10,000 small electrical motors a year at $4.80 each after deducting quantity discounts. The question it faces is this: can it make the motor itself for $4.80 or less?

Assume that the costs of producing 10,000 motors a year in its own plant would be as follows:

Variable costs	
Materials per unit	$2.00
Overhead costs per unit for the new activity (power, suppliers, etc.)	0.75
Total variable cost per unit	$2.75
Total variable cost per 10,000 units	$27,500
Fixed costs	
Salaries of 3 technicians (capable of producing 15,000 units a year)	$21,000
Depreciation on special machinery ($40,000 spread over a 5-year life)	8,000
Total fixed costs for the new activity	$29,000
Total costs	
Total cost of producing 10,000 motors in plant	$56,500
Total cost per unit	$ 5.65

Thus, with a production level of 10,000 units a year, the manufacturing firm would be better off buying the motor from a supplier at $4.80 each. However, with the 3 technicians and the $40,000 in special equipment, it has the capacity to produce 15,000 motors a year. At this level, its total cost would be $70,250 and its unit cost $4.68. It would save at least $0.12 per unit, or $8750 over the 5-year useful life of the special equipment. So the decision to make or buy the motor would depend on the volume needed.

On the other hand, the owner might decide that the $1750 a year would not be enough to offset the extra burden that policing the motor activity would place on key people. This assumes, of course, that the present supplier delivers quality motors on time.

Another analysis is also involved in this decision, however—the determination of whether the $8750 saving is enough to justify investing $40,000 in special equipment, or whether the money could be invested to better advantage elsewhere. This decision is known as a *capital budgeting decision.*

10.4.6 The importance of the basic data

The techniques of cost and profit analysis discussed above are intended as guides. In such a case, the validity of the method depends largely on the validity of the basic data—the fixed and variable cost classifications, cost estimates, and volume estimates. Errors in any one of these factors will cause mistakes in the computations and may result in erroneous decisions.

Every manufacturing firm should continually review the way its costs are classified and accumulated, and it should be very careful in making cost estimates. Its general accounting system and the cost accounting system should be reviewed periodically to ensure that basic cost data are collected and classified according to an established accounting plan.

10.5 Communication

In a business organization, gathering and using information about the operation of business may be a fairly simple matter. A purchasing agent, production manager, or sales manager who has only a few employees under his/her direction can observe day-to-day activities and exercise direct supervision and control. When a business becomes larger, however—when more employees and several levels of supervision are necessary—the processes of obtaining business information and directing the business become more complex.

The larger business requires a system for collecting information about operations and distributing it to managers in the company. Such a system includes four processes:

1. Accumulating
2. Analyzing
3. Reporting
4. Communicating

So far, discussion has emphasized techniques for accumulating, analyzing, and reporting cost accounting data. Attention must be given to methods of communicating the information to those in the firm who need it in order to accomplish their tasks.

Communication involves both a sender and a receiver of information. A manager sets up an accounting system to be the sender. The manager assures that a system produces certain reports and analyses of the basic accounting data.

If the information is to be useful, the needs of those who will receive it must be known and considered. Thus, communication depends on a two-way exchange between the senders and the receivers. Figure 10.2 illustrates a communication network in a business organization; the arrows show the two-way flows of information between senders and receivers.

Communication also requires that the information sent and received be useful. To be useful, information must meet the following standards:

1. It must be suited to the needs of the receivers in form and content
2. It must be free from clerical errors, incorrect classification, and unrealistic estimates
3. It must be presented in clear language and not buried under a mass of unnecessary details
4. It must be received soon after the actual events, in time for suitable action to be taken.

10.5.1 The need for timeliness

There is bound to be some time lag from the time an activity is completed to the time a report of that activity can be made available. It takes time for the clerical staff to sort and total materials requisitions and time tickets for posting to job-cost sheets or for use in cost-system calculations. It takes time to prepare data for electronic data possessing (EDP), deliver it to a service bureau, and obtain the printed output.

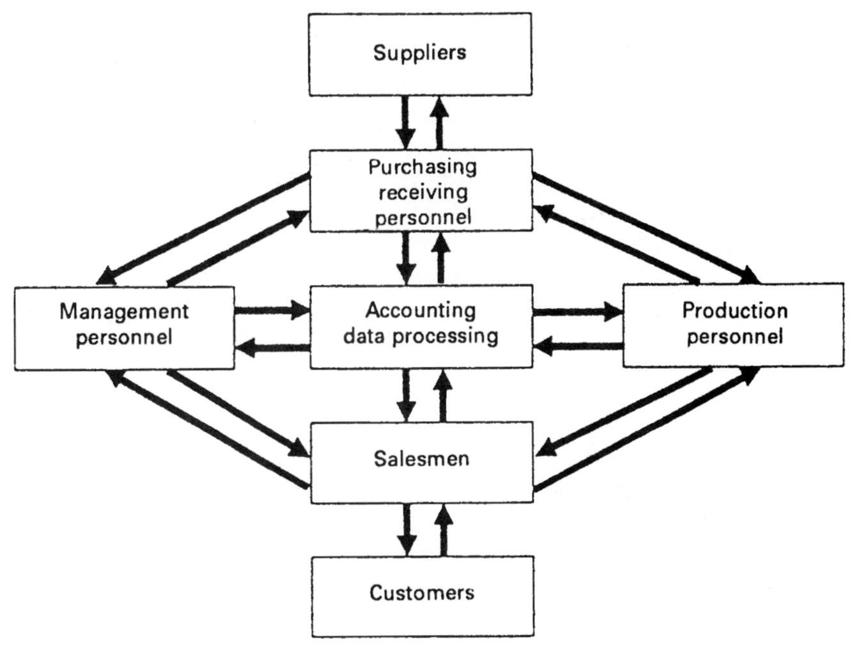

Figure 10.2 Communication network.

If a report is delayed very long, it may be useless by the time it finally reaches the one who needs it.

Each manufacturing firm must decide in which areas of operation quick reports are needed. For example, if a machining department is overloaded with work, this should be known promptly so that workers can be shifted to prevent a production bottleneck. Or perhaps work can be routed to another department where machine time is available. For this kind of scheduling, daily or even hourly performance may be needed. Activity reports must be communicated quickly where the situation is likely to change often.

Other situations may require reports on different schedules. Since labor efficiency usually changes slowly, reports on labor hours and labor costs may be prepared weekly or monthly. Material price variance reports may be prepared at the end of a periodic buying season or after the second week in a month when vendors' invoices are received. Materials usage and efficiency reports may be prepared daily, weekly, or monthly according to the characteristics of the manufacturing process and the needs of the production manager.

In general, the amount of clerical time, effort, and expense put into a communication system is related to the importance of the activities being reported. If the manufacturing process uses a large amount of raw materials and parts, automated equipment, and a minimum of labor time, the accounting system should produce frequent reports on material prices and usage and on equipment maintenance and repair costs. Labor costs can be reported less often because they are not a large part of total production cost.

On the other hand, if labor costs do make up a large portion of the total costs, the communication system should emphasize labor efficiency and cost. Material cost can be reported less frequently. The manufacturing firm must decide how much it can afford to spend on its communication system and how valuable the information is to its managers.

10.5.2 Reporting strategy

It takes time to accumulate, calculate, and record cost data; therefore, cost information is generally communicated in monthly, quarterly, and annual reports. The reports usually cannot be delivered in time to be useful on an hourly, daily, or weekly basis. Consequently, the more frequently issued reports contain only physical data—material units, labor units, machine hours, number of employees, machine downtime, work-in-process backlog, material spoilage, etc.

These reports in terms of physical units (without dollar figures) are useful for managers, who can take direct action to affect physical usage of materials, labor hours, and some kinds of overhead spending. Cost reports in dollar terms are more useful for managers who can take action that affects prices, wage rates, overhead spending, etc.

The form in which information is communicated can also be important. It varies according to the circumstances and the one who is to receive it. Sometimes an oral report is enough to get the necessary action started, but more often reports are written or printed. Reports issued frequently may be handwritten or printed on standard forms in order to get them out quickly. Other reports may be typed or printed out by electronic data processing equipment (EDP).

In some cases, displays in the form of large graphs or drawings are used. For instance, displays showing employee output, employee safety performance, and other information of interest for workers can be shown in plant areas. These displays, when accompanied by the name and picture of an employee, help to keep the employees informed and interested in their job performance.

All but the smallest of manufacturing businesses are usually organized so that different managerial duties are assigned to different persons. Along with each manager's authority goes a responsibility for managing a segment of business efficiently. Figure 10.3 shows a simple organization chart for a typical manufacturing plant.

President	
Controllable costs	
Selling expenses	x
Accounting and office expenses	x
Production costs	$73,000
	$101,000

Production manager			
Controllable costs	**Prod A**	**Prod B**	**Total**
Supervisory level	$22,000	x	x
Controllable at manager level:			
Salaries	x	x	x
Depreciation	x	x	x
Overhead	x	x	x
	$35,000	$38,000	$73,000

Production Supervisor A			
Controllable costs	**Dept A-1**	**Dept A-2**	**Total**
Department level	$2,500	x	x
Controllable at supervisory level			
Material	x	x	x
Labor	x	x	x
Overhead	x	x	x
	$10,000	$12,000	$22,000

Foreman A-1	
Controllable costs	
Material	x
Labor	x
Overhead	x
	$2,500

Figure 10.3 Simple Organization Chart.

In responsibility reporting, information is reported on the basis of areas and responsibility. For example, reports contain information useful to the foremen for performing their job will be directed to departmental foremen.

This means that the responsibility accounting and reporting system will be only as detailed as the company's organization plan. In a plant with the organization shown in Fig. 10.3, reports could be prepared for the foremen, the supervisors, the purchasing agent, the production manager, the sales manager, the salespeople, and the president. A company that has other organization units must work these into the plan if those units are to be served by the responsibility-oriented reports.

A responsibility-oriented accounting system should generate reports that provide information on which each manager can base the decisions required by the decision-making authority. The system should also provide reports showing the results of these decisions.

10.6 Responsibility centers

The responsibility centers may be cost centers, revenue (sales) centers, or profit centers. In Fig. 10.3, each producing department and each of the two service departments (purchasing and accounting) can be considered a cost center if the costs of operations are analyzed along those lines. Each product can be considered a profit center, with the responsibility held jointly by the sales manager (for revenues) and the product supervisors (for product costs). The sales office can be considered a revenue center that produces the gross sales revenues of the company.

10.6.1 Controllable and uncontrollable costs

Costs and revenues are reported to managers according to their controllable and uncontrollable characteristics. A cost item is controllable by a manager if the action directly affects the amount of cost incurred.

Assume, for example, that the product managers have sole authority to purchase new or replacement equipment, thus incurring depreciation costs. The depreciation cost is then controllable at the manager's level but uncontrollable at the department foremen's level. Consequently, the department foremen cannot be held responsible for depreciation charges, and that cost is not reported as part of their responsibility. The principle of controllable-uncontrollable classification should be treated consistently at all management levels.

An illustration of responsibility cost reporting and the pyramiding construction of reports is shown in Fig. 10.3.

10.6.2 Construction of the reports

Each report delivered to a manager should be constructed according to the usefulness to that manager of information provided by the accounting system. The actual costs can be compared with budgeted figures, standard-cost variances can be shown, unit-cost calculations can be made. Any presentation of the information that will make for sound decision making is suitable.

The classification of costs as controllable or uncontrollable can be difficult. One of the most troublesome problems is obtaining the agreement and cooperation of the foremen, supervisors, and managers who will be assigned certain responsibilities. It is important to make certain that the reports will be used for making decisions and directing activities. To this end, the supervisory employees involved should take part in planning the reporting system. Preparation of format and content of responsibility reports can serve to involve all the managers in the communication-information system, and the whole process can serve to bring the managers together into a smooth-working team.

10.6.3 Analyzing cost and profit data

The responsibility-center accounting scheme is just one way to organize cost and profit data for analyzing the results of operations. For a firm that produces and sells more than one product in more than one geographical area to customers that have a variety of characteristics, there are many other possible types of analysis. Several are listed below:

1. *Product lines.* Analysis by product lines can show which products are most profitable. It can point up the need for cost-saving programs or identify products for maximum sales promotion.
2. *Special orders.* The results of the production of special orders should always be analyzed to determine whether producing to custom-design specifications is really profitable. Such analyses could lead to special pricing formulas, better cost-estimating procedures, or more efficient use of workers and equipment.
3. *Sales analysis.* The costs of selling products can be analyzed according to various customer characteristics, taking into account the cost of various selling efforts, as follows:

a. *Customer groups.* Sales costs and profits can be classified by customer groups. This is especially valuable when a few customers account for a large part of the sales.
b. *Industry.* Analysis of sales, costs, and profits of sales to certain industries may be useful in planning efforts to sell to growing industries.
c. *Geographical areas.* The cost of shipping goods to faraway customers may consume a large part of the gross or contribution margin. Analysis by area can help the sales manager assign salespeople where they can obtain the most profitable results.
d. *Order size.* Quantity discounts must be watched. Also, classification of sales, costs, and profits by order size may bring out opportunities for new selling categories.
e. *Distribution channels.* Classification of sales by distribution channels—wholesalers, retailers, and retail customers—may open up profit opportunities and areas for further investigation.
f. *Combinations.* Combinations of the analyses outlined in items *a, b, c, d,* and *e* may give useful information. For example, sales, costs, and profits classified by both order size and geographical area might lead to the development of a new selling strategy.

These costs and profitability analyses could absorb a great deal of a clerical time and perhaps cost more than they would be worth. Management of each firm must decide for itself which ones are worthwhile for its operation. Electronic data processing can play an important part. With proper programming, coding, and processing of the data, a manufacturer can have a vast amount of information and have it more promptly than by manual processing. The task is then that of using the information to plan production and sales strategies.

10.6.4 Communication of business financial status

The primary service of accumulating and analyzing cost accounting information is to have a sound basis for management decisions. But the information is also important for financial reports and presentation to outsiders.

The financial reports most often issued to outsiders are the balance sheet and the income statement. These statements of financial conditions and the results of operations are of interest to creditors (bankers, suppliers) and to investors and prospective investors.

If the manufacturer takes government contracts, government agencies will ask to audit the cost of production for various reasons—renegotiation of prices, establishing cost-plus prices, etc.

Because of the close interaction between internal cost accounting and external financial reports, careful attention should be given to the accounting systems and the support cost accounting records.

10.7 Mathematical Methods for Planning and Control

The cost records described in the preceding sections provide the data for more advanced mathematical methods of analyzing and planning business operations. This section outlines briefly some that can be used by many manufacturers:

1. Dealing with uncertainty (probability)
2. Capital budgeting analysis
3. Inventory analysis (economic order quantity)
4. Linear programming
5. Project management (PERT and CPM)
6. Queueing
7. Simulation

10.7.1 Dealing with uncertainty

Business management always involves uncertainty. The manager is never entirely sure what will happen in the future. This uncertainty is of special concern in preparing budgets, establishing standard costs, and analyzing budget variances, as well as in other decision situations.

Uncertainty means that the manager must try to predict or evaluate any value within a reasonable range of estimated values.

The most important concept in dealing with uncertainty is probability. There are two kinds of probability: objective and subjective.

Objective probability is a measure of the relative frequency of occurrence of some past event. It can be illustrated by the example in Table 10.3.

TABLE 10.3 Objective Probability and Expected Values

(1) Hours per unit	(2) Units observed	(3) Expected probability	(4) Expected hours	(5) Expected cost at $5
1	100	0.1	0.1	$ 0.50
2	200	0.2	0.4	2.00
$2\frac{1}{2}$	400	0.4	1.0	5.00
3	300	0.3	0.9	4.50
Total	1000	1.00	2.4	$12.00

Assume that a manager observes production, counts the number of the units produced, and tabulates the count according to the number of direct-labor hours used for each unit. This has been done in columns 1 and 2 of Table 10.3. Column 3 indicates the objective probability for each time classification—that is, the odds that each unit will require the number of direct-labor hours. It is computed by dividing the number of observations for each classification by the total number of observations. For example, of the 1000 units whose production was observed, 400 required 2½ direct-labor hours per unit. Thus, 2½ direct-labor hours were needed for 4 out of 10 or 0.40 of the units (400 divided by 1000).

The objective probabilities in column 3 are used to accumulate the expected direct-labor hours in column 4 (column 3 times column 1). The expected cost figures in column 5 are found by multiplying the labor rate of $5.00 an hour by the expected hours in column 4. The totals in column 4 and 5 are expected values—the average direct-labor hours per unit and direct-labor cost per unit that can be expected on the basis of past production experience.

Subjective probability is not based on observation of past events. It is a manager's estimate of the likelihood the certain events will occur in the future.

Assume that a sales budget is being prepared. The manager estimates that chances are even (0.5) that sales will be the same as last year—1000 units. The probability of selling 1100 units is estimated to be 0.25 and the probability of selling 1200 units 0.15. There is a 1 in 10 chance, the manager estimates, of selling 1300 units.

The expected sales, based on these estimates of the future, are calculated in Table 10.4. The resulting figure for expected sales—1085—can be used in the sales budget.

Notice that use of probabilities enables the manager to reduce a range of values to a single value. The manager can still retain for later reference the original data used to make the estimate. The use of probability measures, together with the data a manager must accu-

TABLE 10.4 Expected Sales

(1) Sales (in units)	(2) Subjective probability	(3) Expected sales (2)×(1)
1000	0.5	500
1100	0.25	275
1200	0.15	180
1300	0.10	130
	1.00	1085

mulate in using them, helps to deal with the uncertainty that is characteristic of every business operation.

10.7.2 Capital budget

Capital is invested in a fixed asset only if the asset is expected to bring in enough profit to recover the cost of the equipment and provide a reasonable return on the investment.

The purpose of capital budgeting analysis is to provide a sound basis for deciding whether the asset under consideration will in fact do this if it is purchased.

These elements must be considered in managing a capital budgeting decision:

1. The cost of the fixed asset
2. The expected net cash flow provided by use of the asset
3. The opportunity of investing funds in capital equipment
4. Present value

Cost of the fixed asset. This cost is usually the purchase price, transportation, and installation cost of the asset. However, buying the asset may mean that other investments will have to be made—additional inventories, for example. If so, these too should be included as part of the total investment outlay.

Expected net cash flow. The net cash flow may be either (1) the cash cost savings or (2) the increased sales revenue minus the increased cash costs due to using the new asset. No deduction for depreciation is made in calculating cash flow.

Opportunity cost. This is the return that the business could realize by investing the money in the best alternative investment. For example, if the best alternative is to place the money in a savings account that pays 5 percent annually, the opportunity cost is 5 percent on the amount of the investment.

Present value. Scientific capital budgeting is based on the concept of present value. For example, $1.00 put in a savings account at 5 percent interest will be worth $1.05 at the end of one year, $1.1025 at the end of the second year, and $1.1576 at the end of the third year.[1]

---

[1]First year: $\$1.00 + 0.05(\$1.00) = 1.05 \times \$1.00 = \$1.05$
Second year: $1.05(1.05 \times \$1.00) = 1.05^2 \times \$1.00 = \$1.1025$
Third year: $1.05(1.05^2 \times \$1.00) = 1.05^3 \times \$1.00 = \$1.1576$

TABLE 10.5 Alternative Application of Present-Value Concept to Capital Expenditure Decisions

(1) End of year	(2) Amount to be received	(3) Discount factor	(4) Present value
1	$2.00	1.05	$1.90
2	2.00	(1.05)^2	1.81
3	3.00	(1.05)^3	2.59
Total	$7.00		$6.30

Suppose an opportunity arises to invest some money with the expectation of receiving $1.1576 (including the amount invested and the interest or other income) at the end of the third year. Investment must not be more than $1.00, because the present value of $1.1576 to be received 3 years hence is $1.00 (based on the best alternative investment at 5 percent).

This present value is found by discounting the expected future value by the opportunity cost rate. This is accomplished by reversing the process for finding a future value, as follows:

$$\$1.1576 \div 1.05^3 = \$1.00$$

or, in general terms,

$$\text{Present value} = \frac{\text{value } n \text{ years hence}}{1 + \text{discount rate}}$$

Table 10.5 shows another way of applying the present-value concept to capital expenditure decisions. Consider investment of $6.50 in an asset that would yield the amounts shown in the table. The amounts received from the asset investment will total $7.00 (including salvage value of the asset), whereas the investment, if placed in a savings account at 5 percent, at the end of 5 years will be worth $6.50 × 1.05^3, or $7.52. Therefore, the $6.50 should not be invested in the asset.

The present value analysis in column 4 of Table 10.5 shows that the present value of the asset investment is $6.30, which is 20 cents less than the amount that would be invested. In other words, the value of the investment opportunity is minus 20 cents. If the amount of the investment were $6.29 or less and brought in the same amount, the present value of the investment opportunity would be positive. The general rule is that investment opportunities with a positive present value should be taken.

TABLE 10.6 Revenue and Cost Characteristics.

	Year 1	Year 2	Year 3
Sales revenue	$10,000	$11,000	$12,000
Variable cost	6,000	7,000	8,000
Fixed costs*	2,000	2,000	2,000
Cash flow from production	$ 2,000	$ 2,000	$ 2,000
Sales of equipment	0	0	1,000
Net cash flow	$ 2,000	$ 2,000	$ 3,000

*Not including depreciation charges.

Example Assume that an opportunity arises to buy a piece of equipment for $5000. With this equipment, products can be manufactured with the revenue and cost characteristics described in Table 10.6.

The present value of these cash flows, as computed previously in Table 10.5, is $6300. The value of the investment opportunity is therefore $6300 minus $5000, or $1300. This positive investment opportunity value shows that the equipment investment will have a higher rate of return than the 5 percent available on the next best investment (assumed to be a savings account paying 5 percent). The depreciation is not considered a cash flow item and thus should not be included as a cash cost. The example illustrated does not include depreciation charges. In addition, these figures do not include income tax payments (which are a cash flow item).

However, the cash flow figures in Table 10.7 do take taxes into account. Once taxes are included, depreciation should be considered as well because depreciation is tax-deductible.

The present value is $6092, rather than the $6300 calculated from the earlier figure which did not include taxes. The value of the investment opportunity,

TABLE 10.7 Cash Flow Figures.

	Year 1	Year 2	Year 3
Sales revenue	$10,000	$11,000	$12,000
Variable production costs	$ 6,000	$ 7,000	$ 8,000
Fixed production costs (including depreciation)	3,833	3,833	3,834
Net profit before taxes	167	167	166
Income taxes @ 48%	80	80	80
Net profit after taxes	87	87	86
Cash flow from sales of equipment (add back depreciation)	$ 1,833	$ 1,833	$ 1,834
Net cash flow	$ 1,920	$ 1,920	$ 2,920

after income taxes, is $1092, which is still higher than what could be earned from a savings account carrying a 5 percent interest rate.

The present value method of analyzing a capital investment of opportunity makes it possible to take into account the timing of the expected cash returns and to compare them with those of other investment opportunities. Note that the cash returns are expected values. The probability methods discussed earlier are used in calculating them.

10.7.3 Inventory analysis

Maintaining adequate inventories of raw materials and parts can tie up capital for a manufacturer. Careful management of inventory levels offers possibilities for large cost savings and the release of funds from inventory investment. Two major kinds of costs are associated with maintaining inventories: holding costs and ordering costs.

Holding costs are costs associated with inventory investment. They include insurance, taxes, rent on warehouse space, and the opportunity cost of alternative uses of the funds invested in inventory.

Ordering costs are the costs of processing purchase requisitions and vendors' invoices and of receiving the goods in the warehouse (receiving department salaries, etc.).

The *economic order quantity* (*EOQ*) is the quantity of goods that should be ordered at one time to ensure the lowest total inventory costs (holding cost plus ordering cost). There is a tradeoff between holding costs and ordering costs. When the order quantity is small and the inventory is kept low, the holding costs are also low. The ordering costs, however, are high, because orders must be placed often and more clerical and receiving department time is needed. When the order quantities are large and inventory is kept high, the holding costs are high (more insurance, taxes, rent), but the ordering cost is low—fewer requisitions have to be processed, and receiving operations are less frequent. This cost tradeoff is shown graphically in Fig. 10.4.

The basic data for the inventory problem are the holding costs, the ordering costs, and the expected demand for raw materials and purchased parts. Other factors that may enter into the purchasing decision are the lead time (the length of time between purchase order and receipt of the materials or parts) and the availability of quantity discounts.

10.7.4 Linear programming

Linear programming is a mathematical method for making the best possible allocation of limited resources (labor hours, machine hours, materials, etc.). It can help management decide how to use its production facilities most profitably. Suppose, for instance, that the firm pro-

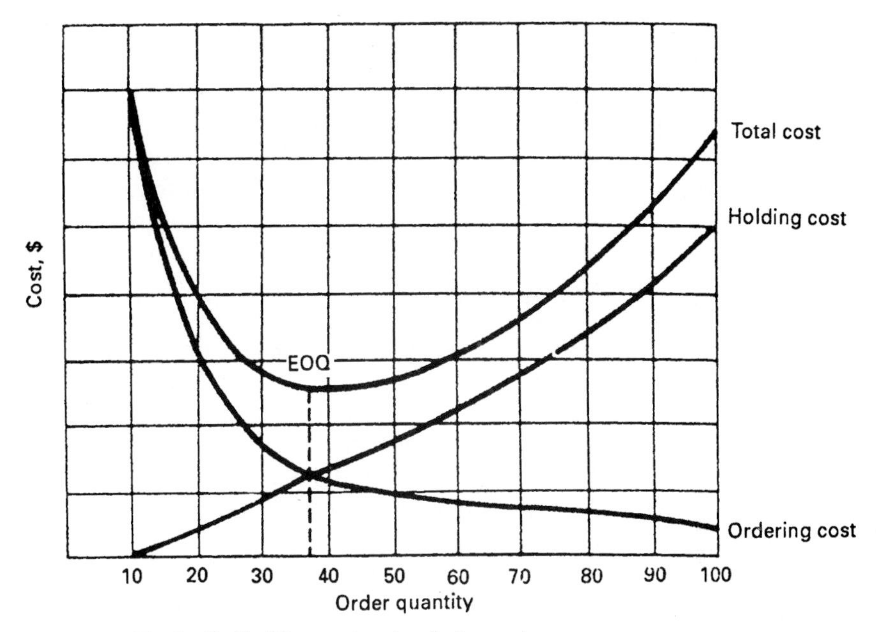

Figure 10.4 Tradeoff of holding cost and ordering cost.

duces more than one product, each with a different contribution rate. Management needs to know what combination of quantities produced, given the limitations of its facilities, will bring the highest profit.

Three concepts must be considered in seeking a solution to the problem:

1. The profit function
2. The constraints of the problem
3. The production characteristics

The profit function. The profit function is a mathematical expression used to show how the profit will vary when different quantities of product 1 and product 2 are produced. It requires knowledge of the contribution of each product toward overhead and profit.

If the symbols x_1 and x_2 stand for the number for each product to be produced, and if product 1 makes a profit contribution of \$2 per unit and product 2 makes a profit contribution of \$5 per unit, the profit function is $2x_1 + 5x_2$. The problem is to find the values of x_1 *and* x_2 that will yield the highest total value (profit), given the constraints of the problem and the production characteristics.

The constraints. Sometimes the profit-maximizing solution is obvious—simply produce as many product units as possible with the resources available. However, the manufacturing process may have characteristics that prevent the use of resources in this way.

The constraints of this problem are the limitations imposed by the scarcity of production resources. Suppose that the manufacturer uses lathes, drilling machines, and polishing machines in the production process. The lathes can be used for a maximum of 400 hours a month, the drilling machine for 300 hours, the polishing machine for 500 hours. The machines cannot be used more than the hours indicated, but they need not be used to the limit of this capacity in order to maximize the profit function.

The constraints are applicable only as long as the production capacity remains constant. The manufacturer could purchase new machines or put on additional work shifts and expand the machine time available. But both of these actions will probably change the statement of the problem. The profit contribution of the product might be changed (by higher night-shift direct-labor costs, for example). Also, a capital budgeting analysis might be required. The linear programming problem is applicable for a time period during which constraints are fixed.

Production characteristics. The term *production characteristics* refers to the machine times used in producing the product. Suppose that product 1 requires 1 hour on a lathe, no time on a drilling machine, and 1 hour on a polishing machine. Product 2 requires no lathe time, 1 hour on a drilling machine, and 1 hour on a polishing machine. As stated above, the lathe can be used for a maximum of 400 hours a month, the drilling machine for 300 hours, and the polishing machines for 500 hours. The production characteristics would then be expressed as follows:

$$x_1 \leq 400$$

which means that no more than 400 lathe-hours are available for product 1 and 1 hour is required for each unit. (If 2 hours per unit were required, the expression would be $2x_1 \leq 400$.)

$$x_2 \leq 300$$

which means that no more than 300 drill hours are available for product 2, and 1 hour per unit is required.

$$x_1 + x_2 \leq 500$$

which means that no more than 500 polishing hours are available for both products, with 1 hour per unit required in both cases.

The problem for the inequalities given above is as follows:

$$\text{Maximize:} \quad 2x_1 + 5x_2$$

$$\text{Subject to:} \quad x_1 \leq 400$$

$$x_2 \leq 300$$

$$x_1 + x_2 \leq 500$$

There are many solutions that would satisfy all three constraints; there is only one maximizing solution. The possible solutions are shown graphically in Fig. 10.5. The axes represent the number of product units, and the solid lines represent the production characteristics. Any point in the shaded area will satisfy the constraints, but the maximizing solution will be at the edge of this area—point *A*, *B*, *C*, or *D*.

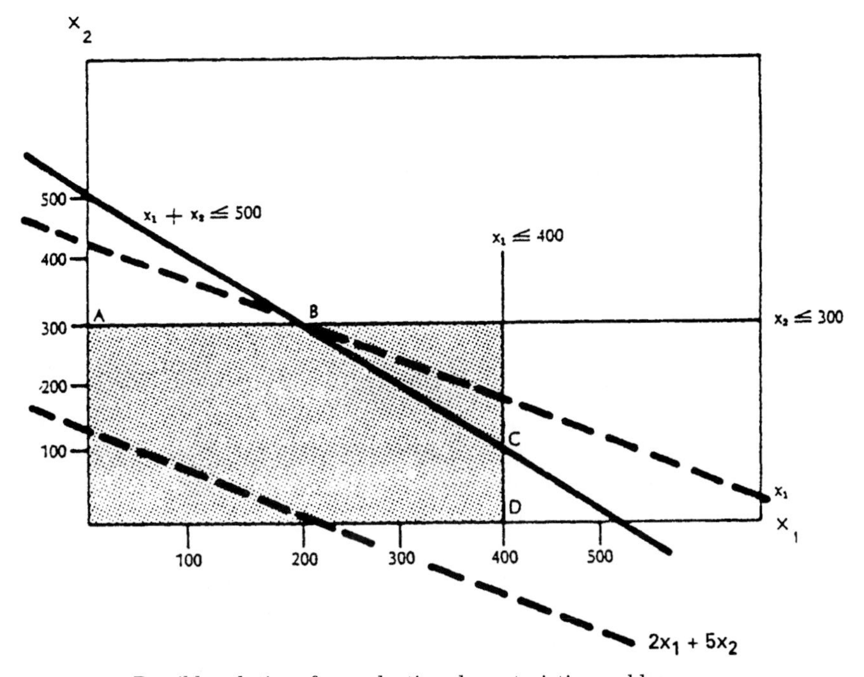

Figure 10.5 Possible solutions for production characteristics problem.

Which of these points is the maximizing solution can be found by drawing a line that represents the profit function $2x_1 + 5x_2$ (the lower dashed line in Fig. 10.5). When this line is shifted upward, always parallel to its original position, it finally reaches the point where it passes through the shaded solution space at only one point—point *B*. This point, where $x_1 = 200$ and $x_2 = 300$, is the solution that gives the highest possible total contribution within the limited resource constraints. The solution gives the following results:

Total contribution	\$1900 (200 × \$2 + 300 × \$5)
Lathe time used	200 hours (200 × 1 hour)
Drill time used	300 hours (300 × 1 hour)
Polishing time used	500 hours (200 × 1 hour + 300 × 1 hour)

There are 200 idle lathe hours, but these cannot be used under existing conditions without reducing the output of product 2. If more lathe time is used for product 1, then more polishing time must also be used; and the polishing time must be taken away from product 2, thus reducing product 2 output. Any reduction in product 2 output presents a loss of \$5 per unit, whereas producing additional units of product 1 will add only \$2 per unit. The net loss from such a substitution is therefore \$3 per unit. On the other hand, the output of product 2 cannot be increased, since there is no more drill time available.

The graphic solution described above is a rough approach to a fairly simple problem. More complex problems can be programmed into a computerized automated system.

10.7.5 Project management

Manufacturing and construction firms may undertake one-time-only jobs, such as the construction of a building or the design and manufacture of a special piece of a large assembly system. In such projects, at any one time much of the company's time, resources, and prospects may be tied to a single large job. And whether the job will result in profit or loss for the company may depend on its being completed on time.

If the completion data is to be met, the various segments of the job must be coordinated. Subunits must be scheduled, and they must be finished according to the schedule date.

Project management techniques have been developed for planning the subunits that make up such a job. One of the most important of these methods is the Project Evaluation and Review Technique (PERT). Another one, very similar to PERT, is known as the Critical Path Method (CPM).

PERT. PERT is based on what is called a network plan. A flow chart—the network—is used to illustrate how the individual part of a project (the activities and the starting and completion events) depend on one another and which task must be finished before others can be started. It is particularly concerned with dovetailing individual parts of a project into the schedule of the project as a whole.

PERT analysis makes the following estimates from data about the activities and events that make up the project:

1. Earliest expected time for completion of each activity, or task.
2. The latest allowable time for each completion event.
3. The critical path of the network; that is, the path through a flow chart that has the least total amount of slack. (*Slack* is the difference between the latest allowable time and the expected time of completion.)
4. The probabilities that events will occur on a schedule.

A simplified PERT network is shown in Fig. 10.6. The same information is shown as a bar chart in Fig. 10.7. (The bar chart brings out the pattern of activities and slack time clearly, but would be unwieldy for use in a more complex network).

The circled numbers in Fig. 10.7 represent events—the start or completion of an activity—for example, excavation for a building or erection of the steelwork. The lines represent work activities extending over time and leading from one event to another. Other symbols used in the figure and discussion have the following meanings:

t_e = expected time
T_E = sum of the expected times up to a given event

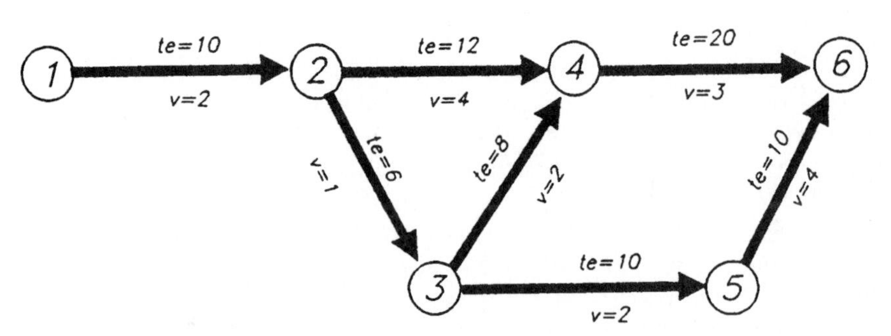

Figure 10.6 Simplified PERT network.

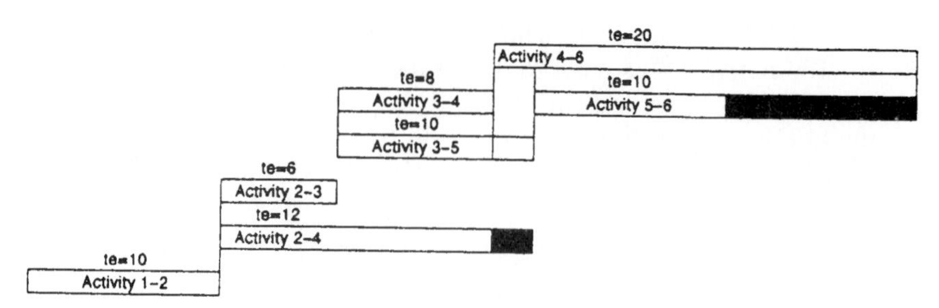

Figure 10.7 PERT bar chart.

T_L = latest allowable time (cumulative)
v = variance

The values for t_e and v in the figures are assumed.

If 1 represents the beginning of excavation for a building, 2 completion of excavation, and 3 completion of pouring a foundation, then the line from 1 to 2 represents the activity of excavation and the line from 2 to 3 the activity of pouring a foundation. The activity of steelwork erection leading to completion of event 4 cannot begin until after both preceding completions 2 and 3 have occurred. Other intermediate completion events must take place before any entire project is finally completed, event 6.

Time estimates must be made for each activity. Since actual activity time is uncertain, probability calculations are used in computing the expected time t_e and the variance v associated with the expected time.

The problem of project management is to determine the expected time for completion of the entire project and to identify the activities that could cause bottlenecks. PERT enables the management to accomplish this and thus schedule work activities for the earliest possible completion.

For example, in Fig. 10.6, work leading to completion event 2 is expected to be completed 10 weeks after the start. From this point, two network paths lead to event 4—the path 2–4, with an expected completion time of 12 weeks, and the path 2–3–4, with a completion time of 14 weeks. The total expected completion time for event 4 is the larger of the total completion times on the two network paths that lead to the event. The path 1–2–4 has a total expected time of T_E = 22, and the path 1–2–3–4 has total expected time of T_E = 24. Thus, the T_E for event 4 is 24. The variance associated with T_E = 24 is V = 5, the sum of the variances along the 1–2–3–4 path.

Thus, the manager can expect completion event of 2 in 10 weeks, event 4 in 24 weeks, and event 6 in 44 weeks after work is started. One other time variable must be taken into account, however: the latest time T_L that can be allowed for an event without disturbing the T_E of the final event in the network (event 6).

If event 6 must be completed in 44 weeks, then T_L for event 6 is 44. Since T_E for event 6 is also 44, the lag time allowable is zero ($T_L - T_E$). In order to complete event 6 in 44 weeks, event 4 must be completed in 24 weeks. Thus, T_L for event 4 is 24 and the slack time allowable is also zero. The T_L and slack time can be calculated for each event by working backward from event 6.

The critical path through the network—the path in which there is the least amount of slack—is identified by noting the events where allowable slack time is zero. In Fig. 10.6, the critical path lies through events 1–2–4–6. Any construction delay in these events will increase the total completion time for the project. Event 5, however, is not on the critical path. The tasks on the path segment 3–5–6 could be delayed for as much as 8 weeks in all without affecting the final completion time for event 6. The activities represented by the path segment 2–4 could be delayed 2 weeks without delaying the final completion of the project. Resources could, if necessary, be shifted from these activities to other tasks to ensure completion of the entire project on schedule.

Note that T_E and T_L are expected values. There are statistical variances associated with each of these measures. The expected time estimates and variance estimates are used in calculating the probabilities that events will occur on schedule.

CPM. The Critical Path Method of project management is closely related to PERT. Both PERT and CPM involve determining expected times of completion for individual events and for the entire project. PERT goes further to include variance analysis; CPM does not. CPM, on the other hand, goes beyond PERT in another direction. It uses cost data to assess the financial effects of setting up crash programs in the network's critical path segments to ensure completion on the schedule.

PERT/cost. The PERT techniques can be augmented with cost data to facilitate financial planning for large projects. The cost data are budgeted according to work activity and completion-event classifications. Actual cost should be accumulated according to the same classifications so that they can be compared with the budget figures.

Queueing analysis. A queueing, or waiting-line, situation exists when there is a service center of some kind that receives customers, pro-

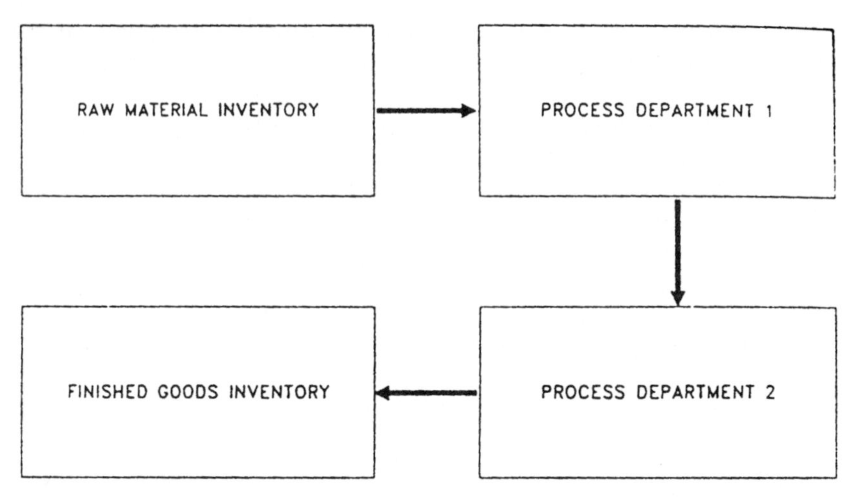

Figure 10.8 Manufacturing queueing situation.

cesses their work, and sends them on their way. Three types of waiting lines often operate in business. One is seen in a doctor's office where arriving patients form one line and are served through only one station—the doctor. Another is seen in a barber shop where people form one line and are served through several stations—any of the barbers. A third type of queueing is in a supermarket where customers form many lines and are served through many stations—the checkout counter.

A waiting line can occur in a manufacturing plant when product units—"customers"—arrive at a processing department or work area for completion and transfer to the next department. A manufacturing queueing situation is illustrated in Fig. 10.8.

The arrival of raw materials at processing department 1 can be controlled by the timing of materials requisitions, so it is unlikely that a waiting line of materials units will form at department 1. However, the processing or service rate in department 1 determines the timing of the release of the semifinished units to department 2.

As long as department 2 can accept the units for processing, no waiting line will form. Units will be processed as they arrive at the work area. However, if department 2 is busy, a bottleneck will occur, and semifinished units will form a backlog. The time spent in line carries a cost to the manufacturer similar to the holding cost of inventory. (Note that the backlog of semifinished products is a part of the work-in-process inventory.) The queue, and therefore, the waiting cost, can be reduced by speeding up the processing or service rate in department 2.

The processing or servicing of additional units, however, adds to the cost of manufacturing process. A move to accelerate processing in department 2 and shorten the queue may add costs beyond normal production costs for materials, labor, and overhead. This would be the case, for instance, if it were necessary to add a night shift or new equipment. Thus, there is a tradeoff between waiting costs and some processing costs similar to the EOQ tradeoff between holding and ordering costs.

Queueing analysis takes into consideration the following factors:

1. Arrival rate
2. Service rate
3. Queue lengths
4. Utilization of the service facility
5. Total service time for product units
6. The cost of waiting
7. The cost of servicing

These variables are combined in such a way as to provide data on the efficiency of the present service and the economic feasibility of adding service units.

Simulation. Simulation is a method used to represent a real situation in an analytical framework. Verbal descriptions, diagrams, flowcharts, and systems of equations are all forms of simulation. A budget is the result of simulating the financial results of operations, based on assumptions about volume, cost relationships, and planned production activities.

Simulation allows management to try out various actions without risking its resources in the real marketplace. With simulation, it can calculate events based on alternative assumptions on paper and then attempt to select the best course of action for real application.

In the terminology of modern management science, simulation refers to a computerized model of a company or some segment of it. The model includes sets of mathematical equations that quantify volume-cost-profit relationships. The computer can accept charges in the relationships and calculate the effects of the change on operations. Suppose, for example, that management is contemplating substituting a new raw material that will reduce materials requirements but increase labor usage. Management can program the changes into a computerized simulation and learn what the overall effect will be on manufacturing operations.

Simulation is a powerful management tool, but it is complicated by the fact that the operating characteristics have to be specified in great detail. Nevertheless, it can be used to great advantage to test and examine alternative courses of action without having to put them into actual practice. The volume-cost-profit relationships and physical production characteristics described in this chapter can provide much of the data required for simulation models.

10.8 Where Do Sensor and Control Systems Take Us?

Currently, most industries employ equipment with limited intelligence, but in the near future, the level of adaptive sensors and control systems and of computer technology will rapidly improve this equipment. On the other hand, as computer-integrated manufacturing strategies are introduced on a wide scale and as technology moves forward, the size of the work force involved in engineering will be reduced dramatically, while the pattern of working seems destined to change completely (Fig. 10.9).

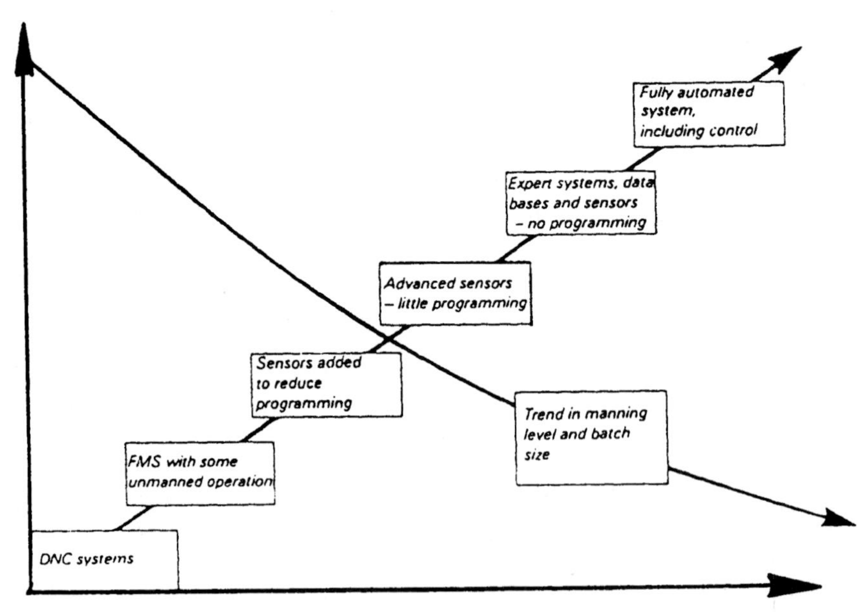

Figure 10.9 The role of affordable sensors and control systems in a totally automated factory.

Already sensors and control systems play a significant role in manufacturing. For example, tool wear and breakage is monitored by sensors that measure the torque or power needed at the spindle to control tool wear and breakage; Macotech[2] has taken that concept a step further to adjust the feed rate and torque to obtain the optimum cutting rate whatever the conditions. As a result, cutting times have in many cases been reduced by 50 percent, while the need to program the tool in detail is eliminated; the machine no longer needs to be told the material, cutting speed, and feed rate.

In one case, the concept of employing sensors and control systems in manufacturing has been applied to deep drilling, where a constant load is applied. The feed rate changes automatically to maintain the constant load on the cutting tool, but if, as a result of the buildup of swarf, the feed rate is reduced to preset level, the drill is withdrawn to clear the swarf, and then started again. Not only is the speed optimized through sensors and control systems, but as a result, drill wear is more consistent, and tools do not break.

These are really elementary adaptive sensors and control systems, but are already here. The aim in increasing the level of intelligence of a machine is to reduce the need for attention and programming. For example, with a vision system, the controller for a robot or machine will be able to recognize the workpiece, determine its precise position, and instruct the machine to carry out the necessary operations. Thus, it will not be necessary to load a workpiece precisely on a pallet, nor install limit switches and other sensors to identify the workpiece and check its position. Since location need not be precise, robot loading will be easier.

Force sensors will optimize the cutting speed whatever the material, and in the case of the robot, will ensure that a component is installed with the correct amount of force. The use of such sensors and control systems will allow planning of manufacturing operations to be automated, with such programming as is needed accomplished off-line.

A lathe/milling machine would be equipped with adaptive sensors and control systems on the spindle, to adjust the speed for turning, and on the tool to adjust the speed for drilling, tapping, and milling. Vision sensors would be used to identify the work piece for the robot. Lathe/milling machines can already be programmed off-line, and with vision sensors, the robot could be programmed off-line with the sen-

[2]Macotech Machine Tool Co., Inc., Seattle, Wash., U.S.A.

sors adjusting the position and size of the gripper, or changing the gripper to suit the workpiece. Also, the controller could assess whether the new set of chuck jaws were needed, and if so, instruct the robot to change them. Thus, the lathe/milling machine would be able to machine a wide variety of components completely unmanned.

Manufacturers can expect the other machines to be improved through sensors and control systems as well. For example, several machining centers have an extra attachment to allow machining of the fifth face, whereas some Mandelli[3] machines can machine the fifth face without the need for an attachment. This is a worthwhile development, since it reduces downtime. With the introduction of modular universal machining centers for turning, milling, boring, and grinding operations, manufacturers will encourage more moves in this direction. In some cases, handling equipment to permit the fifth face to be machined without the need for a full resetting operation is also likely to come into use, as are laser welding[4] and laser sensor measurement systems on the machine tool itself.

But the major development will continue to come in sensors and control systems, computers, and software. In the next stage, the computer at each flexible manufacturing system might have access to an expert system, which is in effect a huge database. Included in the database will the data concerning all the components to be produced—input from the computer-aided design (CAD) system—such as machining methods for different speeds, surface finishes, and tolerances. In fact, the database will constitute a complete encyclopedia of machining. This will be updated continually and directly from the CAD system and from a remote database on machining (Fig. 10.10).

Once sensors and control systems are perfected, it will be possible to perform machining without programming. When a workpiece is fed into the system, a vision sensor will identify it in detail. Then the data will be fed back to the database and compared with data for the workpieces that are to be machined, and full dimensional data on the workpiece will be accessed. The vision sensor will then scan the workpiece, and its dimensions will be compared with those of the finished part in the part library.

The data will then be fed back to the controller which will instruct the machine how to start machining, by reference to the database and not to a specific program. When necessary, the workpiece will be transferred to another plant for automatic heat treatment or grind-

---

[3]Mandelli Machine Tool, Italy.

[4]S. Soloman, *Modern Welding Technology,* TAB Book Publishing Company, Blue Ridge, Pa., 1982.

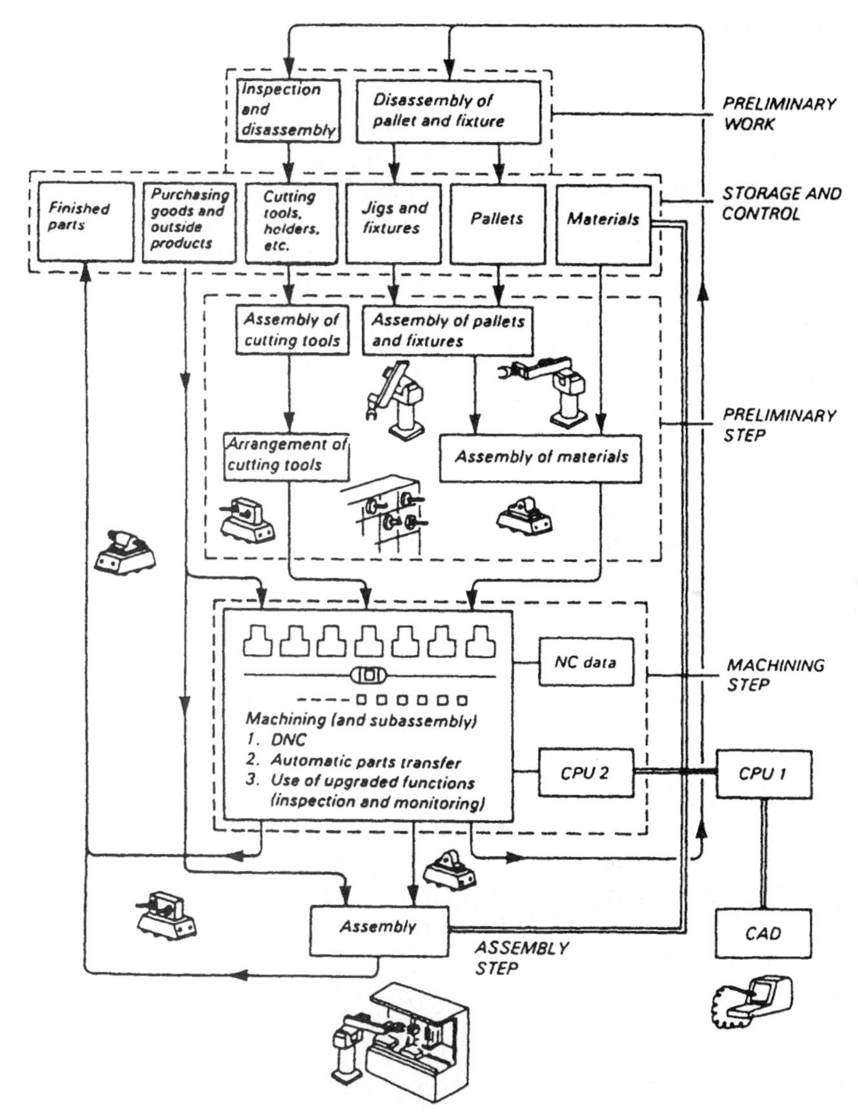

Figure 10.10 Updating a machining expert system.

ing, and later to assembly. In this way, machining without programming and without manual intervention will be practical. Instead, the database will contain a vast number of programs, the best one being selected automatically for the shape of the component to be produced.

Of course, these advanced systems in controls and sensors will not be needed to produce all components, nor will they be installed in all flexible manufacturing systems. Lower-level machines would still continue in use for more routine medium-volume parts.

In assembly, similar sensor and control systems will come into use. In these systems, there will be more stages, because the right group of components must be presented in suitable positions to be assembled. One handling robot, on a trolley, might be used to supply components to the assembly cell, where some robots and standard units, such as presses and CNC nut runners, would carry out assembly.

With the combination of these expert systems and sensor and control systems, flexible manufacturing systems will be able to machine or assemble in batches of one with the efficiency expected in high-volume production runs. Thus, the real aims of flexible manufacturing systems—high productivity with batches of one, produced within a very short time after the order is placed—will be achieved.

Another major development will be in the computer systems from purchasing through delivery. All the systems will be connected together, and will perhaps be incorporated in one computer (which will have voice recognition and voice response ability), so that, for most jobs, the keyboard will be unnecessary. Indeed, by the end of this decade computers with conventional languages and vocal human/machine interfaces should be available. Any one will be able to make the most complex transactions and programs using normal conversation, so the problem of computer literacy will disappear. It will also have access to databases of products, materials, stock, costing, and other relevant information.

In the case of production control and order processing, this means that once an order is accepted, the data will be fed to the CAD system, where the designer will produce a new design, or modify or merely call up an existing design. Planning will be automatic; from the data generated, the schedule will be produced, components and materials ordered, and on arrival they will be fed through the system to be packed and delivered with hardly any manual intervention. Needless to say, the materials will not wait around in storage, but will come straight from the delivery truck into the system to start being processed.

Overall, therefore, we are at the beginning of an era which will lead to the end of the factory as we know it. Some machines will be able to carry out simple routine maintenance themselves through advanced sensory and control systems. Very complex robots with a number of sensors will be available to carry out repairs, so that in theory, operators will be needed only to work with the software—in design and in updating the databases and expert systems. Both the machine and robot will be extremely complex, so that the operators who are responsible for their installation, let alone maintenance staff, will probably not be able to understand the complete system—that is already the situation with large-scale computers.

Some engineers expect to continue to rely on people to carry out some jobs. They argue that there is little point in taking everyone out of the plants just for the sake of doing so, and the devices that can do these jobs are likely to be very expensive. In addition, even with all these elaborate sensors, there is possibility of some catastrophic failure, and it will be a very long time before managers are prepared to go home at night and leave a plant completely unattended. In any case, working conditions will be very good, and working hours will be short, so there is every reason to employ some people in the plants.

However, other experts say that there is no point in thinking in terms of staffing the plant at all, except to provide external supervision. They say that because the machines will be quite complex, maintenance fitters will not know how to rectify faults. Then, because the sensors and control systems will be so advanced, the machine will in any case be better able to trace faults and put them right, so rectification by humans may not be helpful at all. Clearly, that stage is still a long way away, but managers need to be ready for it.

But just how many people will be needed in manufacturing by the year 2000? Some experts estimate that work force in manufacturing will be cut by 60 to 75 percent in the next 10 years, while the number of people actually employed on the shop floor is likely to drop by about 90 percent. The effects of a 70 percent reduction in the work force in the major industrialized nations is shown in the table below.

Country	Industrial work force, current	Industrial work force after reduction by 70%
United States	30,000,000	9,000,000
Japan	19,500,000	5,800,000
West Germany	11,500,000	3,400,000
Italy	7,700,000	2,300,000
France	7,500,000	2,250,000
United Kingdom	5,300,000	1,600,000
Sweden	1,350,000	400,000

Although these figures appear alarming, they reflect similar trends that have taken place in agriculture, process industries, automotive industries, and computer industries. However, those changes took place far more slowly, and the rate of change is bound to cause considerable problems to politicians and industrialists alike. With the prospect of 20,000,000 jobs being lost in the United States, 13,000,000 in Japan, 5,000,000 each in France and Italy, and 3,500,000 in the United Kingdom in around 10 years, people are bound to be alarmed.

Of course, the transition will take some time, but it is likely that the introduction of computer-integrated manufacturing strategies with the current level of technology of sensor and control systems will produce this reduction in the work force. By the year 2000 or so, a substantial reduction in workers can be expected as computerization goes further ahead.

In this period, though, many new products and systems will go into production, and, in theory, with the reduction in prices that the increased productivity will bring, demand worldwide should increase. Equally, it is clear that the development and maintenance of all this software will provide a large number of skilled jobs. In fact, there is likely to be a shortage of skills in many areas, which will have the effect of retarding development of new systems.

Clearly, the nations that train the most people in the relevant skills will have a head start. In any case, the changes in the number of employees are based on the assumption that each nation maintains its current share of the market for manufactured goods. In practice this is unlikely, since some countries are more competitive than others, while newly industrialized nations will also take some of that business.

Although it is imperative that people be educated in the right skills, that is not enough; the whole way in which people are educated and think of work needs to be changed. The introduction of computer-integrated manufacturing strategies and computerized sensors and control systems in manufacturing, transport, and perhaps some other services, will lead to the need for people to work fewer hours. It is likely that a 3-day week will appear quite soon, both to spread work around and also to ensure that people are prepared to work the unsocial hours that are occasionally needed.

But what will people do when they are not working? People talk glibly of extra leisure, but that requires a quantum jump in the standard of living, and we certainly need something to give us that. The other worry is that as we move from active to supervisory roles, we shall tend to lose skills in dexterity and become less active. The actual education level needs to be raised, so that people can cope with the jobs available in this new society, while it also needs to be expanded so that people are better placed to face life with more time on their hands—but perhaps with the need to do more for themselves in maintaining their homes and their possessions.

Perhaps many people will have more than one job, one in a company for 3 days a week, and another outside for 2 days; of course, others will not want much at all, and they will probably be able to fulfill just that. None of these futuristic notions will be possible unless much value is being added in manufacturing and business.

We are entering an exciting era of significant opportunities, opened up by sensor and microelectronic control systems and computer-integrated manufacturing strategies. But the competition will be intense, and only those that take big chances now will reap the benefits. As the psalmist once said, "They that sow in tears shall reap in joy. He that goeth forth and weepeth, bearing precious seed, shall doubtless come again rejoicing, bringing his sheaves with him."[5]

---

[5]Psalms 126, Verses 5–6, *The Bible.*

Index

ABOUT THE AUTHOR

Sabrie Soloman is President and CEO of American $Sensor_X$, Inc., as well as Professor of Advanced Manufacturing Technology at Columbia University, where he lectures on Affordable Automation, Sensors and Control Systems in Manufacturing, Flexible Manufacturing Systems, Design for Manufacturability, and Computer-Integrated Manufacturing. He saved in excess of $7 million within a few months of operation at American $Sensor_X$, Inc. by applying his innovative techniques for affordable automation. He was the first to introduce and implement unmanned flexible synchronous/asynchronous manufacturing systems in the microelectronics industry. Dr. Soloman holds numerous patents, and is also the author of *Introduction to Electromechanical Engineering, Modern Welding Technology*, and *Sensors and Control Systems in Manufacturing*. He is a fellow of the Royal Society of Mechanical Engineers and L'Ordre des Ingenieurs du Quebec (Canada). He received several awards from the American Society for Mechanical Engineers (ASME), the Society of Manufacturing Engineers (SME), and the American Management Association. Dr. Soloman is considered an international authority on advanced manufacturing technology and automation in the microelectronics and automotive industries. He has been and continues to be instrumental in developing several industrial modernization programs in European and African countries. He is the first to incorporate advanced vision technology to a wide array of robot manipulators. Dr. Soloman was selected by Vice President J. Danforth Quayle to deliver the closing address, "Innovative Automatic Aided Systems and Sensors for Universal Design" at the Universal Design Conference, New York, May 14, 1992.